INTERNATIONAL SERIES OF MONOGRAPHS IN
NATURAL PHILOSOPHY

GENERAL EDITOR: D. TER HAAR

VOLUME 76

LIGHT SCATTERING IN PLANETARY ATMOSPHERES

LIGHT SCATTERING
IN
PLANETARY ATMOSPHERES

BY

V. V. SOBOLEV

Translated by

WILLIAM M. IRVINE

with the collaboration of
Michael Gendel
and
Adair P. Lane

PERGAMON PRESS

OXFORD · NEW YORK · TORONTO · SYDNEY
BRAUNSCHWEIG

Pergamon Press Ltd., Headington Hill Hall, Oxford
Pergamon Press Inc., Maxwell House, Fairview Park, Elmsford,
New York 10523
Pergamon of Canada Ltd., 207 Queen's Quay West, Toronto 1
Pergamon Press (Aust.) Pty. Ltd., 19a Boundary Street,
Ruschutters Bay, N.S.W. 2011, Australia
Pergamon Press GmbH, Burgplatz 1, Braunschweig 3300,
West Germany

First edition 1975

Library of Congress Cataloging in Publicaton Data

Sobolev, Viktor Viktorovich.

Light scattering in planetary atmospheres.

(International series of monographs in natural philosophy, v. 76).
Translation of Rasseyanie sveta v atmosferakh planet
Includes bibliographies.
1. Planets—Atmospheres. 2. Light—Scattering. 3. Radiative transfer. I. Title.
QB603.A85S6213 1975   551.5′66   74–13852
ISBN 0–08–017934–7.

Printed in Hungary

CONTENTS

LIST OF TABLES

FOREWORD

IN RECENT times the launching of space rockets has led to a significant growth of interest in planetary investigations. Whereas astrophysicists once studied primarily stars, nebulae, and galaxies, at present a considerable part of their effort is directed toward the determination of the nature of the planets. The application of a variety of astrophysical methods has already yielded valuable results in this area.

Progress in the study of the planets depends not only upon the collection of new observational data, but also upon development of the theory used to interpret the data. It is from solar radiation scattered in the planetary atmospheres that we derive most of our information about the planets. Consequently, the theory of multiple scattering of light, otherwise known as the theory of radiative transfer, is necessary for the interpretation of planetary observations.

This theory is important for the physics of planets for another reason. The process of transfer of solar radiation in the planetary atmosphere determines to a significant extent the physical state of the atmosphere. In order to calculate various quantities characterizing this state, it is thus necessary to use the theory of radiative transfer.

Theoretical astrophysicists have been developing radiative transfer theory for a long time. However, they have been primarily concerned with stellar atmospheres, within which the scattering of light is isotropic. In the atmospheres of the planets, light scattering by an elementary volume is anisotropic. This fact severely complicates the theory. Nevertheless, in recent years the theory of radiative transfer for anisotropic scattering has made considerable progress and has been increasingly used in the study of planetary atmospheres. The present monograph has been written for the purpose of summarizing the results of work in this area.

The monograph is concerned mainly with the theory of radiative transfer for anisotropic scattering. The first eight chapters deal with the general problem of multiple scattering of light in an atmosphere consisting of plane-parallel layers illuminated by parallel radiation. In the following two chapters, the theory is applied to the determination of the physical characteristics of planetary atmospheres. The last chapter discusses the theory of radiative transfer in spherical atmospheres, which is necessary for the interpretation of observations made from spacecraft.

The emphasis in the monograph on the theory rather than its application is easily understood; the theory is designed not only for the interpretation of existing observational data, but also for that to be gathered in the future. One must also bear in mind that the theory of radiative transfer is utilized in related sciences, such as meteorology and oceanology, and also in certain branches of physics and chemistry.

This monograph is the second book by this author dealing with the theory of radiative transfer. The first book, entitled *Radiative Transfer in the Atmospheres of Stars and Planets* (Gostekhizdat, 1956),[†] is devoted to a wide range of questions (radiative transfer with frequency redistribution, nonstationary scattering of radiation, etc.). Since then, the theory has developed so fast that at present the examination of each of these questions may be the subject for a separate monograph. The solution of one such problem is examined in the present book. The previous book, for the sake of brevity, will henceforth be referred to as *TRT* (with reference to the chapter and page).

The author wishes to express his thanks to his colleagues in the Department of Astrophysics at Leningrad University for their assistance in the preparation of the manuscript and for their critical remarks.

V. V. SOBOLEV

[†] Published in English as *A Treatise on Radiative Transfer* (D. Van Nostrand, 1963).

TRANSLATOR'S PREFACE

THE present monograph is the first complete text devoted solely to the study of radiative transfer in planetary atmospheres. It has clearly been written at an appropriate time and fills an important need. In fact, as the author points out, the theory also has direct applications in fields as varied as oceanology and neutron physics.

The order of presentation is described most fully by the author in his Concluding Remarks. The fundamental physical problem considered is the multiple scattering of radiation within an atmosphere. The discussion is thus applicable to the region of the spectrum from the near ultraviolet through the near infrared, for which atmospheric thermal radiation is either negligible or assumed known *a priori* and all scattering processes are assumed to be coherent in frequency. The treatment is a macroscopic one, so that the microscopic processes of excitation and emission by atoms and molecules are described only in terms of the average properties of an elementary volume. The author concentrates his attention on analytical procedures, approximate and asymptotic as well as exact. The reader interested in the special problems associated with the transport of thermal radiation in the terrestrial atmosphere may wish to consult the textbook by R. M. Goody: *Atmospheric Radiation* (Oxford, 1964).

In the translation we have attempted to remain faithful to the style of the original, while making the book easily accessible to as wide as possible an English speaking audience. The result will hopefully be useful to graduate students and to scientists in fields other than astronomy, as well as to the professional astrophysicist. We have taken some liberties with paragraphing, where English and Russian practice differ. Of greater importance, we have added new or explanatory material at several points, most obviously in the Addendum to Chapter 10, but also to a small degree in Chapter 1 and at a few other places. Some recent work by the author, published since the Russian edition of this monograph, has been included in the Appendix.

Only two points of terminology require comment. Following the classic example of Chandrasekhar we have used *phase function* for the Russian *scattering indicatrix* (sometimes translated as scattering diagram). The choice of *illumination* (see page 16) rather than such alternative English terms as *irradiance* or simply *incident flux* was less clear-cut, but follows the precedent set in the translation of Professor Sobolev's earlier monograph on radiative transfer.

We are very grateful to Professor Sobolev for providing the initial impetus for this translation, and to him and his colleagues for their labors in correcting a first draft of the manuscript and for proof-reading the final equations. The completion of the English manuscript would have been impossible without the dedicated hard work of Ms. Kathleen Carr and Ms. Maryellen Maesano. One of us (W. M. I) wishes to gratefully acknowledge support from the U.S. National Aeronautics and Space Administration and the gracious hospitality of Professor O. E. H. Rydbeck, Director of the Onsala Space Observatory, Chalmers University of Technology, Sweden, where this work was completed.

University of Massachusetts, Amherst   W. M. IRVINE
                                                                         M. GENDEL
                                                                         A. P. LANE

LIST OF SYMBOLS

Symbol	Description	Introduced on page
a	particle radius	3
a	surface albedo	74
$A(\zeta)$	atmospheric (plane) albedo	18
A_g	geometric albedo	178
A_s	spherical albedo	18
$\bar{A}_s$	spherical albedo in presence of planetary surface	83
$b(\eta)$	relative source function deep within a semi-infinite atmosphere	25
B	source function	9
$\bar{B}$	source function in presence of reflecting surface	75
B^*	source function for inhomogeneous atmosphere illuminated from below	67
B^0	source function averaged over azimuth	11, 25
B^m	source function for mth azimuthal component of radiation field	11
$c(\zeta)$		25
E	illumination	16
E_k	exponential integral	14
g	parameter characterizing forward elongation of Henyey–Greenstein phase function	5
H	radiation flux	16
$H^m(\eta)$	H-function for mth azimuthal component of radiation field	21, 94
H_*	scale height	215
$i(\eta)$	relative intensity of radiation deep within a semi-infinite atmosphere	25
I	intensity of radiation	6
$\bar{I}$	intensity of radiation in presence of reflecting surface	75
$\bar{I}$	mean intensity of radiation	161
I^0	intensity of radiation averaged over azimuth	11, 25
I^m	intensity of mth azimuthal component of radiation field	11
I_R	intensity of radiation diffusely reflected by planetary surface	75

Chapter 1

BASIC EQUATIONS

THE problem of multiple scattering of light in a planetary atmosphere requires the simultaneous solution of two equations. One of these is the equation of radiative transfer, which determines the change in the intensity of radiation along a ray if the coefficients of absorption and emission are given. However, in planetary atmospheres the emission coefficient is not known *a priori*. This coefficient depends upon the intensity of radiation falling on an elementary volume and upon the law of light scattering for this volume. The integral relation expressing the emission coefficient in terms of the intensity of radiation is commonly known as the equation of radiative equilibrium. These two equations form the basic system with whose solution we shall concern ourselves.

The present chapter initially considers the elementary act of light scattering; the equations of radiative transfer and radiative equilibrium are derived subsequently. The basic problem of the theory of light scattering in planetary atmospheres is formulated with the aid of these equations. We then consider the integral equation for the "source function", which relates the emission coefficient to the absorption coefficient. If the source function is known, all of the characteristics of the radiation field in the atmosphere may be easily determined. The chapter ends with the presentation of several methods for the solution of the basic problem.

1.1. The Scattering of Light by an Elementary Volume

We assume in what follows that the law of radiation scattering by an elementary volume of the atmosphere is known. Let us introduce several quantities with the aid of which that law may be expressed. Consider radiation falling normally on the boundary of a plane layer of infinitesmal thickness ds. From the energy incident on the layer, a certain fraction $\alpha \, ds$ will be removed from the original beam. The quantity α is called the *absorption coefficient*. In general, the quantity $\alpha \, ds$ consists of two parts: the first part gives the fraction of energy scattered by the layer, while the second gives the fraction of energy that has experienced true absorption, i.e. has been transformed into other forms of energy. We shall denote these parts by $\sigma \, ds$ and $\varkappa \, ds$, respectively, and we shall call σ the *scattering coefficient* and $\varkappa$ the *coefficient of true absorption*. It is clear that

$$\alpha = \sigma + \varkappa. \tag{1.1}$$

We note that the quantity α is sometimes referred to as the extinction coefficient, and the quantity $\varkappa$ as the absorption coefficient. However, we shall use the previously mentioned terminology.

Instead of the quantities σ and $\varkappa$, it is possible to use the absorption coefficient α and the ratio of the scattering coefficient to the absorption coefficient:

$$\lambda = \frac{\sigma}{\alpha}. \tag{1.2}$$

The quantity λ represents the probability that a photon which interacts within an element of volume will be scattered rather than truly absorbed. It is usually denoted $\tilde{\omega}_0$ in English language texts, and is called the *single scattering albedo* or the particle albedo. When $\lambda = 1$, we refer to the case of *pure scattering* or conservative scattering.

The intensity of radiation scattered by an elementary volume in general depends upon direction relative to that of the incident beam. We signify by $x(\gamma)\, d\omega/4\pi$ the probability that radiation is scattered into a solid angle $d\omega$ about a direction forming an angle γ with the direction of the incident radiation. The quantity $x(\gamma)$ is called the *phase function* [*indikatrisa rasseyania*]. Clearly,

$$\int x(\gamma)\, \frac{d\omega}{4\pi} = 1, \tag{1.3}$$

where the integration is carried out over all directions. Since with the present choice of coordinates $d\omega = 2\pi \sin \gamma \, d\gamma$, then instead of (1.3) we have

$$\tfrac{1}{2} \int_0^\pi x(\gamma) \sin \gamma \, d\gamma = 1. \tag{1.4}$$

The quantities α, λ, and $x(\gamma)$ determine the light scattering law for an elementary volume. Each of these quantities depends on the frequency of the radiation. In the future, however, we shall not indicate this dependence (except for particular examples), but it should always be understood.

It is known that a planetary atmosphere consists of molecules and aerosols (water drops, dust, etc.). We present a brief examination of their capability for scattering radiation.

Rayleigh was the first to develop the theory of molecular light scattering [1]. He found that the scattering coefficient for molecules is inversely proportional to the fourth power of the wavelength. Consequently, molecules scatter light much more strongly in the violet portion of the spectrum than in the red (this explains the blue color of the sky). The phase function in this case has the form

$$x(\gamma) = \tfrac{3}{4}(1+\cos^2 \gamma). \tag{1.5}$$

True absorption is ordinarily absent for light scattering by molecules (i.e. $\lambda = 1$). Only for frequencies characteristic of molecular bands does true absorption play a role (see Chapter 9).

The theory of light scattering by spherical particles with sizes comparable to the wavelength of the radiation was developed by Gustav Mie [2]. In this case the scattering law depends on the refractive index m of the particle material and on the value of the parameter

$$z = \frac{2\pi a}{\tilde{\lambda}},$$ (1.6)

where a is the particle radius and $\tilde{\lambda}$ is the wavelength.[†] The refractive index is in general complex. For metallic particles the refractive index has a nonzero imaginary part, while for dielectrics the index is real. Thus, dielectric particles give rise to pure scattering of radiation ($\lambda = 1$), while metallic particles produce both scattering and true absorption ($\lambda < 1$). The theoretical basis for light scattering by small particles was set forth in great detail by K. S. Shifrin [3] and H. van de Hulst [4]. In recent years, many computations using the Mie theory have been carried out with the aid of electronic computers for various values of m and z.

Since there are particles of various sizes in an elementary volume, the phase function found through the Mie formulas must be averaged over the particle sizes. The phase function obtained in this manner is called *polydisperse*. In these computations it is often postulated that the number of particles with values of the parameter z in the interval from z to $z+dz$ is equal to

$$\frac{(n+1)^{n+1}}{n!} \left(\frac{z}{\bar{z}}\right)^{n} e^{-(n+1)\,z/\bar{z}} \frac{dz}{\bar{z}},$$ (1.7)

where n is an integer and $\bar{z}$ is a mean value of the parameter z related to the mean radius of the particles $\bar{a}$ by equation (1.6). Tables of phase functions for polydisperse media are found in the book by Deirmendjian [5].

Table 1.1 presents examples of polydisperse phase functions for particles with refractive index $m = 1.38$ and for a particle size distribution given by equation (1.7) with $n = 2$ (calculations by V. M. Loskutov). From the table we see that the forward elongation of the phase function increases with an increase in the mean radius of the particles. For planetary clouds, the parameter $\bar{z}$ is of order 10.

In works dealing with the theory of multiple scattering, the usual method is to expand the phase function in Legendre polynomials. In other words, the quantity $x(\gamma)$ is represented as a sum

$$x(\gamma) = \sum_{i=0}^{n} x_{i} P_{i}(\cos \gamma)$$ (1.8)

with a finite or infinite number of terms. The formula

$$x_{i} = \frac{2i+1}{2} \int_{0}^{\pi} x(\gamma)\, P_{i}(\cos \gamma) \sin \gamma \, d\gamma$$ (1.9)

serves for the determination of the x_i coefficients. It is apparent that x_0 always equals one. It is also easy to show that $|x_i| < 2i+1$ for any $i \geq 1$.

[†] We shall denote the wavelength by $\tilde{\lambda}$ in order to distinguish it from the quantity λ determined by equation (1.2).

TABLE 1.1. PHASE FUNCTION $x(\gamma)$
($m = 1.38$ and $n = 2$)

γ	$\bar{z}$						
	2	4	6	8	12	16	20
0	18.6	37.9	56.0	122	266	464	715
10	15.0	20.5	20.1	17.8	13.4	11.0	9.75
20	8.59	7.16	5.39	4.60	4.18	4.08	4.02
30	4.13	3.08	2.66	2.43	2.29	2.26	2.26
40	1.94	1.53	1.48	1.40	1.31	1.28	1.27
50	0.96	0.82	0.84	0.81	0.75	0.72	0.70
60	0.52	0.47	0.50	0.49	0.44	0.41	0.39
70	0.31	0.29	0.31	0.30	0.26	0.24	0.22
80	0.20	0.19	0.21	0.20	0.17	0.15	0.13
90	0.14	0.14	0.15	0.14	0.12	0.10	0.09
100	0.11	0.11	0.12	0.11	0.09	0.07	0.06
110	0.09	0.10	0.10	0.09	0.07	0.06	0.06
120	0.08	0.10	0.10	0.09	0.07	0.05	0.05
130	0.09	0.11	0.12	0.10	0.08	0.06	0.05
140	0.10	0.14	0.17	0.17	0.15	0.13	0.12
150	0.11	0.20	0.26	0.29	0.30	0.31	0.32
160	0.12	0.28	0.38	0.38	0.33	0.29	0.26
170	0.13	0.19	0.36	0.49	0.52	0.46	0.40
180	0.14	0.21	0.44	0.65	0.79	0.82	0.84

The coefficient x_1 of the series (1.8),

$$x_1 = \tfrac{3}{2} \int_0^\pi x(\gamma) \cos \gamma \sin \gamma \, d\gamma, \tag{1.10}$$

is of special interest. As was mentioned in the Foreword, a principal source of complexity in planetary scattering problems is the anisotropy of the phase function. The presence of aerosol particles produces a maximum of the scattered radiation in the forward direction (the direction of the incident radiation). This forward peak appears in a polar diagram as an *elongation* of the phase function, which may be characterized by the factor x_1; the greater the forward elongation of $x(\gamma)$, the greater is the value of x_1 (but always $x_1 < 3$).

Phase functions represented by a small number of terms in equation (1.8) are often used in the qualitative analysis of a radiative transfer problem. Isotropic scattering is the simplest case:

$$x(\gamma) = 1. \tag{1.11}$$

The two-term phase function is also used:

$$x(\gamma) = 1 + x_1 \cos \gamma, \tag{1.12}$$

as is the three-term one

$$x(\gamma) = 1 + x_1 \cos \gamma + x_2 P_2(\cos \gamma). \tag{1.13}$$

The Rayleigh function (1.5) represents a special case of (1.13) where $x_1 = 0$ and $x_2 = \tfrac{1}{2}$.

It should be understood, however, that the elongated phase functions which often appear in practice cannot be adequately represented by equations such as (1.11)–(1.13), but require use of expression (1.8) with $n \gg 1$. In order to encompass all possible phase functions in theoretical studies in a simple manner, formulas for the phase function are sometimes used which give different degrees of elongation by the change of only one parameter. The following are examples of such formulas:

1. Binomial phase function

$$x(\gamma) = \frac{n+1}{2^n}(1+\cos \gamma)^n, \tag{1.14}$$

where n is an integer.

2. Ellipsoidal phase function

$$x(\gamma) = \frac{2b}{\ln \dfrac{1+b}{1-b}} \frac{1}{1-b \cos \gamma}, \tag{1.15}$$

where b is a parameter.

3. Henyey–Greenstein phase function

$$x(\gamma) = \frac{1-g^2}{(1+g^2-2g \cos \gamma)^{3/2}} \tag{1.16}$$

where g is a parameter $(-1 \le g \le 1)$.

Formula (1.16) was introduced by the above-named authors in their study of diffuse light in the Galaxy. It is convenient because it may be very simply expanded in Legendre polynomials. Thus, we have

$$x(\gamma) = \sum_{i=0}^{\infty} (2i+1) g^i P_i (\cos \gamma). \tag{1.17}$$

In addition, formula (1.16) approximates real phase functions rather well. In order to achieve a better approximation, it is possible to use a combination of two expressions of type (1.16) with different values of the g parameter (one positive and the other negative). We shall employ the Henyey–Greenstein function in the future in the solution of various problems.

1.2. The Equation of Radiative Transfer

If the scattering law for an elementary volume of the medium is known and the energy sources in it are given, then it is possible to find the diffuse radiation field in the medium. Let us obtain the necessary equations for this purpose.

The intensity of radiation is the basic quantity characterizing the radiation field. At a given place in space, the intensity in a particular direction is defined in the following manner. Let us take an elementary area which is perpendicular to the direction we have chosen. If the area is $d\sigma$ and the radiation falls in the frequency interval from ν to $\nu+d\nu$ in the solid angle $d\omega$ in the time dt, then the amount of radiant energy $d\mathscr{E}$ falling on the area from the given direction will be proportional to $d\sigma \, dv \, d\omega \, dt$, i.e. it will be

$$d\mathcal{E} = I \, d\sigma \, dv \, d\omega \, dt. \tag{1.18}$$

The proportionality coefficient I occurring in this formula is called the *intensity* of radiation. Generally speaking, the quantity I depends on the coordinates of the given point, on direction, and on the frequency v.

If the radiation intensity is known, then other quantities characterizing the radiation field may be determined: the density of radiant energy, the flux of radiant energy, the surface illumination, etc. The corresponding formulas will be presented later. An important characteristic of the radiation intensity is the fact that in empty space it does not change along a ray with distance from the source. We shall now obtain an equation showing how it changes in regions where absorption and emission of radiant energy occur.

We must next introduce the *emission coefficient* ε. Emission here refers to any process which adds to the intensity in the direction under consideration, and thus includes both scattering into the beam from other directions, as well as thermal or other emission processes within the volume. The quantity $\varepsilon \, dV \, d\omega \, dt \, dv$ defines the energy emitted by the volume element $dV = d\sigma \, ds$ within the solid angle $d\omega$ in the frequency interval from v to $v + dv$ in the time period dt.

Let us consider an elementary cylinder, the axis of which is directed along a given ray. Let the area of the cylinder's base be $d\sigma$ and the height be ds (the height is small in comparison with the linear dimensions of the base). Consider the radiation entering and leaving the cylinder within a solid angle $d\omega$ in the frequency interval from v to $v + dv$ in the time period dt. We shall denote the radiation intensity where the ray enters the cylinder by I and the intensity leaving the cylinder by $I + dI$. Then the difference between the energy leaving and the energy entering the cylinder will be equal to

$$dI \, d\sigma \, d\omega \, dv \, dt.$$

This difference is a result of energy losses (absorption) and gains (emission) within the cylinder. Since the absorbed energy is equal to

$$\alpha \, ds \, I \, d\sigma \, d\omega \, dv \, dt$$

and the emitted energy is equal to

$$\varepsilon \, d\sigma \, ds \, d\omega \, dv \, dt,$$

we then have

$$dI \, d\sigma \, d\omega \, dv \, dt = -\alpha \, ds \, I \, d\sigma \, d\omega \, dv \, dt + \varepsilon \, d\sigma \, ds \, d\omega \, dv \, dt.$$

It follows that

$$\frac{dI}{ds} = -\alpha I + \varepsilon. \tag{1.19}$$

This is the equation we are seeking which determines the changes in the intensity of radiation as it passes through an absorbing and emitting medium. It is known as the *equation of radiative transfer*.

It must be emphasized that equation (1.19) refers to a specific frequency. For simplicity, however, we shall not indicate that dependence, but we shall simply write I (e.g.), instead of I_ν.

If the emission coefficient ε and absorption coefficient α are given, the radiation intensity I is easily determined by integration of equation (1.19). However, in the case of a medium which scatters radiation, the quantity ε depends on the intensity falling on the elementary volume from all directions. Since precisely such media will interest us in the future, we must add to equation (1.19) an expression for the quantity ε in terms of the radiation intensity.

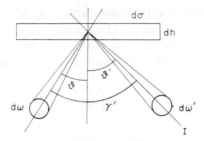

FIG. 1.1. Interaction of a beam of light with an element of volume $dV = d\sigma\, dh$.

Let us take an elementary volume with a cross section $d\sigma$ and a height dh (Fig. 1.1). Let an intensity I fall on this volume within the solid angle $d\omega'$ in the direction forming an angle ϑ' with the normal to the base. It is apparent that the energy falling on the volume per unit frequency interval per unit time is equal to $I d\omega'\, d\sigma \cos \vartheta'$. Since the path travelled by radiation in the volume is just $dh \sec \vartheta'$, a fraction $\alpha\, dh \sec \vartheta'$ of the energy falling on the volume is absorbed by it (i.e., is removed from the incident beam), so that the energy absorbed equals

$$\alpha\, dh\, d\sigma\, I\, d\omega'.$$

The energy scattered by the volume within the solid angle $d\omega$ in a given direction is found by multiplying the absorbed energy by the quantity $\lambda x(\gamma')\, d\omega/4\pi$, where γ' is the angle between the directions of the incident and scattered radiation. As a result, we obtain the following expression for the scattered energy

$$\lambda x(\gamma')\, \frac{d\omega}{4\pi}\, \alpha\, dh\, d\sigma\, I\, d\omega'.$$

Since radiation falls on the elementary volume from all sides, this expression must be integrated over all directions of the incident radiation. It is evident that the result of this integration must be equated to the quantity $\varepsilon\, dh\, d\sigma\, d\omega$, which gives the total emitted energy per unit frequency per unit time. Thus, we find

$$\varepsilon = \lambda \alpha \int I x(\gamma')\, \frac{d\omega'}{4\pi}\,. \tag{1.20}$$

Equation (1.20) gives the desired expression for the quantity ε in terms of the intensity of radiation I for a medium in which both absorption and scattering of radiant energy occur.

If there are internal energy sources in the medium producing true emission, then instead of equation (1.20) we must write

$$\varepsilon = \lambda\alpha \int Ix(\gamma') \frac{d\omega'}{4\pi} + \varepsilon_0. \tag{1.21}$$

This relation may be called the *equation of radiative equilibrium*. (In stellar atmospheres this terminology is used for equation (1.21) integrated over frequency, but this integration is not necessary in the present context since there is no interaction between the radiation fields at different frequencies.) The quantity ε_0 is called the coefficient of true emission.

In this manner, the problem of determining the diffuse radiation field is reduced to finding the quantities I and ε from equations (1.19) and (1.21). For this solution, the internal and external sources of radiation must be given, and also the optical properties of the medium, i.e. the quantities α, λ, and $x(\gamma)$. Substituting (1.21) into (1.19) we obtain

$$\frac{dI}{ds} = -\alpha I + \lambda\alpha \int Ix(\gamma') \frac{d\omega'}{4\pi} + \varepsilon_0. \tag{1.22}$$

Equation (1.22) is the integro-differential *equation of radiative transfer* in a medium that emits, absorbs, and scatters radiant energy.

1.3. The Basic Problem

Moving directly to the discussion of radiative transfer in planetary atmospheres, let us make certain assumptions (some of which we shall later abandon). We begin by assuming that the atmospheric thickness is much smaller than the radius of the planet. This will allow us to consider an atmosphere consisting of plane-parallel layers instead of a spherical atmosphere. We shall take the absorption coefficient α to be a function of altitude only. The phase function $x(\gamma)$ and the single scattering albedo λ will be considered to be constant in the atmosphere. Let us also assume that the atmosphere is illuminated only by parallel solar radiation and that internal energy sources are absent. At this time we shall not consider light reflected by the surface of the planet.

The problem of multiple scattering of light for the given assumptions will be called the basic problem. Later we shall show that a number of other problems, such as the inclusion of internal energy sources and the reflection of light by the surface of the planet, may be reduced to this basic problem.

Let us also agree on the conventional symbols. In the usual manner, instead of the geometric altitude h above the planetary surface, we introduce the *optical depth* τ defined by the formula

$$\tau = \int_h^\infty \alpha(h)\, dh. \tag{1.23}$$

In addition, we introduce the atmospheric *optical thickness*

$$\tau_0 = \int_0^\infty \alpha(h)\, dh. \tag{1.24}$$

In order to specify direction at every point of the atmosphere we shall use a spherical co-ordinate system with polar axis oriented toward increasing optical depth. We shall denote the polar angle by ϑ and the azimuthal angle by φ (Fig. 1.2).

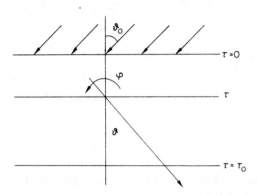

FIG. 1.2. Definition of coordinates within a plane-parallel layer.

In place of the emission coefficient ε, we introduce the function B, which is defined by the equation

$$B = \frac{\varepsilon}{\alpha}. \tag{1.25}$$

The function B is generally known as the *source function*. The radiative transfer equation (1.19) in this terminology takes the following form for an atmosphere consisting of plane-parallel layers:

$$\cos \vartheta \, \frac{dI}{d\tau} = -I + B. \tag{1.26}$$

Up to this point, we have used the symbol I to represent the total intensity of radiation, including the direct (unscattered) solar radiation. Because the latter involves a delta-function of direction for incident parallel light, it is convenient to redefine I to signify only the *diffuse* (singly or multiply scattered) intensity. We shall take the angle of incidence of the direct solar radiation equal to ϑ_0 and the corresponding flux through a surface perpendicular to the incident beam equal to πS at the top of the atmosphere. With the aid of (1.25), we may then rewrite equation (1.20) as

$$B = \lambda \int Ix(\gamma') \frac{d\omega'}{4\pi} + B_1, \tag{1.27}$$

where the source term B_1 results from scattering of the direct solar radiation. From equations (1.21) and (1.25) we see that

$$B_1 = \frac{\lambda}{4} Sx(\gamma) e^{-\tau \sec \vartheta_0}, \tag{1.28}$$

where γ is the angle between the directions of the incident parallel solar radiation and the scattered radiation.

The unknown functions I and B entering equations (1.26) and (1.27) depend on τ, ϑ, ϑ_0 and φ. These equations may therefore be rewritten in the following detailed manner:

$$\cos \vartheta \, \frac{dI(\tau, \vartheta, \vartheta_0, \varphi)}{d\tau} = -I(\tau, \vartheta, \vartheta_0, \varphi) + B(\tau, \vartheta, \vartheta_0, \varphi), \tag{1.29}$$

$$B(\tau, \vartheta, \vartheta_0, \varphi) = \frac{\lambda}{4\pi} \int_0^{2\pi} d\varphi' \int_0^{\pi} I(\tau, \vartheta', \vartheta_0, \varphi') \, x(\gamma') \sin \vartheta' \, d\vartheta' + \frac{\lambda}{4} Sx(\gamma)e^{-\tau \sec \vartheta_0}, \tag{1.30}$$

where

$$\cos \gamma' = \cos \vartheta \cos \vartheta' + \sin \vartheta \sin \vartheta' \cos (\varphi - \varphi'), \tag{1.31}$$

$$\cos \gamma = \cos\vartheta \cos \vartheta_0 + \sin \vartheta \sin \vartheta_0 \cos \varphi. \tag{1.32}$$

For simplicity, the azimuth of the direct solar radiation has been set equal to zero.

It is necessary to add boundary conditions to the above equations. Since the direct solar radiation is not included in I, and diffuse radiation does not fall on the atmosphere either from above or below, we have

$$\left. \begin{aligned} I(0, \vartheta, \vartheta_0, \varphi_0) &= 0 \quad \text{for} \quad \vartheta < \frac{\pi}{2}, \\[2mm] I(\tau_0, \vartheta, \vartheta_0, \varphi) &= 0 \quad \text{for} \quad \vartheta > \frac{\pi}{2}. \end{aligned} \right\} \tag{1.33}$$

In this manner, the basic problem formulated earlier is reduced to finding the functions $I(\tau, \vartheta, \vartheta_0, \varphi)$ and $B(\tau, \vartheta, \vartheta_0, \varphi)$ from equations (1.29) and (1.30), under the boundary conditions (1.33).

With the aim of simplifying our notation we shall often use the following designations:

$$\cos \vartheta = \eta, \quad \cos \vartheta_0 = \zeta. \tag{1.34}$$

Likewise it should be understood that by $I(\tau, \eta, \zeta, \varphi)$ we mean the diffuse radiation intensity at the optical depth τ moving in the direction specified by the polar angle arccos η and the azimuth φ, for parallel solar radiation incident at an angle arccos ζ and azimuth $\varphi_0 = 0$, and by $B(\tau, \eta, \zeta, \varphi)$ we mean the corresponding source function.

Thus, equations (1.29) and (1.30) may be written in the following manner:

$$\eta \, \frac{dI(\tau, \eta, \zeta, \varphi)}{d\tau} = -I(\tau, \eta, \zeta, \varphi) + B(\tau, \eta, \zeta, \varphi), \tag{1.35}$$

$$B(\tau, \eta, \zeta, \varphi) = \frac{\lambda}{4\pi} \int_0^{2\pi} d\varphi' \int_{-1}^{1} I(\tau, \eta', \zeta, \varphi') \, x(\gamma') \, d\eta' + \frac{\lambda}{4} Sx(\gamma)e^{-\tau/\zeta}, \tag{1.36}$$

where

$$\cos \gamma' = \eta\eta' + \sqrt{(1-\eta^2)(1-\eta'^2)} \cos (\varphi - \varphi'), \tag{1.37}$$

and

$$\cos \gamma = \eta\zeta + \sqrt{(1-\eta^2)(1-\zeta^2)} \cos \varphi. \tag{1.38}$$

In solving equations (1.35) and (1.36) it must be remembered that $I(0, \eta, \zeta, \varphi) = 0$ for $\eta > 0$ and $I(\tau_0, \eta, \zeta, \varphi) = 0$ for $\eta < 0$.

As has already been mentioned, the phase function in light scattering theory is frequently expanded in Legendre polynomials; i.e. it is represented by equation (1.8). We may then separate off the azimuthal dependence of I and B by expanding these quantities in a Fourier cosine series. The system of equations (1.35) and (1.36) then separates into distinct pairs of equations that determine the coefficients of these expansions. This procedure reduces by one the number of independent variables in the functions being sought (from τ, η, φ, to τ, η; ζ is a parameter).

In order to obtain the above-mentioned separation, we shall take advantage of the addition theorem for Legendre polynomials, which states that if $\cos \gamma'$ is given by formula (1.37), then

$$P_i (\cos \gamma') = P_i(\eta)P_i(\eta')+2 \sum_{m=1}^{i} \frac{(i-m)!}{(i+m)!} P_i^m(\eta)P_i^m(\eta') \cos m(\varphi-\varphi'), \qquad (1.39)$$

where the $P_i^m(\eta)$ are the associated Legendre polynomials. By substituting (1.39) into (1.8) we have

$$x(\gamma') = p^0(\eta, \eta')+2 \sum_{m=1}^{n} p^m(\eta, \eta') \cos m(\varphi-\varphi'), \qquad (1.40)$$

where

$$p^m(\eta, \eta') = \sum_{=m}^{n} c_i^m P_i^m(\eta)P_i^m(\eta') \qquad (1.41)$$

and

$$c_i^m = x_i \frac{(i-m)!}{(i+m)!} \qquad (1.42)$$

for $m = 0, 1, 2, \ldots, n$.

If $x(\gamma)$ is given by formula (1.40), then from equations (1.35) and (1.36) it follows that the functions $I(\tau, \eta, \zeta, \varphi)$ and $B(\tau, \eta, \zeta, \varphi)$ may be represented in the following form:

$$B(\tau, \eta, \zeta, \varphi) = B^0(\tau, \eta, \zeta)+2 \sum_{m=1}^{n} B^m(\tau, \eta, \zeta) \cos m\varphi, \qquad (1.43)$$

$$I(\tau, \eta, \zeta, \varphi) = I^0(\tau, \eta, \zeta)+2 \sum_{m=1}^{n} I^m(\tau, \eta, \zeta) \cos m\varphi. \qquad (1.44)$$

Using the orthogonality of the functions $\cos m\varphi$, we find that the expansion coefficients $B^m(\tau, \eta, \zeta)$ and $I^m(\tau, \eta, \zeta)$ are determined by the equations

$$\eta \frac{dI^m(\tau, \eta, \zeta)}{d\tau} = -I^m(\tau, \eta, \zeta)+B^m(\tau, \eta, \zeta), \qquad (1.45)$$

$$B^m(\tau, \eta, \zeta) = \frac{\lambda}{2} \int_{-1}^{1} p^m(\eta, \eta') I^m(\tau, \eta,' \zeta) \, d\eta' +\frac{\lambda}{4} Sp^m(\eta, \zeta)e^{-\tau/\zeta} \qquad (1.46)$$

with the boundary conditions

$$\begin{rcases} I^m(0, \eta, \zeta) = 0 \quad \text{for} \quad \eta > 0, \\ I^m(\tau_0, \eta, \zeta) = 0 \quad \text{for} \quad \eta < 0. \end{rcases} \qquad (1.47)$$

Therefore, when the phase function $x(\gamma)$ is represented by formula (1.8), we can replace the equation system (1.35) and (1.36) by the pairs of equations (1.45) and (1.46) for $m = 0, 1, 2, \ldots, n$.

The number of equation pairs obtained is equal to the number of terms in the expansion of the phase function in Legendre polynomials. Thus, with increasing elongation of the phase function, i.e. with increasing number of terms in the given expansion, the change from equations (1.35) and (1.36) to equations (1.45) and (1.46) becomes less convenient. However, for a number of problems it is not necessary to know all the terms in expansions (1.43) and (1.44) but only the first few. As we shall see later, the terms I^0 and B^0 are especially significant.

1.4. Integral Equations for the Source Function

From equations (1.35) and (1.36) it is possible to obtain one integral equation that determines the source function $B(\tau, \eta, \zeta, \varphi)$. In order to do this we must solve equation (1.35) for $I(\tau, \eta, \zeta, \varphi)$ and substitute the solution into equation (1.36).

From equation (1.35) and our assumption that there is no diffuse radiation incident on the atmosphere, we obtain

$$I(\tau, \eta, \zeta, \varphi) = \int_0^\tau B(\tau', \eta, \zeta, \varphi) e^{-(\tau-\tau')/\eta} \frac{d\tau'}{\eta} \qquad (\eta > 0), \qquad (1.48)$$

$$I(\tau, \eta, \zeta, \varphi) = - \int_\tau^{\tau_0} B(\tau', \eta, \zeta, \varphi) e^{-(\tau-\tau')/\eta} \frac{d\tau'}{\eta} \qquad (\eta < 0). \qquad (1.49)$$

Substitution of expressions (1.48) and (1.49) into equation (1.36) gives

$$B(\tau, \eta, \zeta, \varphi) = \frac{\lambda}{4\pi} \int_0^{2\pi} d\varphi' \left[\int_0^1 x(\gamma') \, d\eta' \int_0^\tau B(\tau', \eta', \zeta, \varphi') e^{-(\tau-\tau')/\eta'} \frac{d\tau'}{\eta'} \right.$$

$$\left. - \int_{-1}^0 x(\gamma') \, d\eta' \int_\tau^{\tau_0} B(\tau', \eta', \zeta, \varphi') e^{-(\tau-\tau')/\eta'} \frac{d\tau'}{\eta'} \right] + \frac{\lambda}{4} Sx(\gamma) e^{-\tau/\zeta}. \qquad (1.50)$$

After the determination of $B(\tau, \eta, \zeta, \varphi)$ from the integral equation (1.50), the radiation intensity $I(\tau, \eta, \zeta, \varphi)$ may be found from equations (1.48) and (1.49).

It is possible to obtain integral equations determining the quantities $B^m(\tau, \eta, \zeta)$, the coefficients of the azimuthal expansion of the source function, in a similar manner. From equation (1.45) with the boundary conditions (1.47), we find

$$I^m(\tau, \eta, \zeta) = \int_0^\tau B^m(\tau', \eta, \zeta) e^{-(\tau-\tau')/\eta} \frac{d\tau'}{\eta} \qquad (\eta > 0), \qquad (1.51)$$

$$I^m(\tau, \eta, \zeta) = - \int_\tau^{\tau_0} B^m(\tau', \eta, \zeta) e^{-(\tau-\tau')/\eta} \frac{d\tau'}{\eta} \qquad (\eta < 0). \qquad (1.52)$$

By substituting (1.51) and (1.52) into (1.46), we arrive at the desired integral equation for the function $B^m(\tau, \eta, \zeta)$:

$$B^m(\tau, \eta, \zeta) = \frac{\lambda}{2} \int_0^1 p^m(\eta, \eta') \, d\eta' \int_0^\tau B^m(\tau', \eta', \zeta) e^{-(\tau-\tau')/\eta'} \frac{d\tau'}{\eta'}$$

$$-\frac{\lambda}{2} \int_{-1}^0 p^m(\eta, \eta') \, d\eta' \int_\tau^{\tau_0} B^m(\tau', \eta', \zeta) e^{-(\tau-\tau')/\eta'} \frac{d\tau'}{\eta'} + \frac{\lambda}{4} Sp^m(\eta, \zeta) e^{-\tau/\zeta}. \tag{1.53}$$

After solving equation (1.53) for each m and determining the quantities $I^m(\tau, \eta, \zeta)$ through equations (1.51) and (1.52), the radiation intensity $I(\tau, \eta, \zeta, \varphi)$ may be found from equation (1.44).

The integral equation (1.53) determines $B^m(\tau, \eta, \zeta)$ as a function of two variables: τ and η (ζ is a parameter). However, this equation may be replaced by a system of integral equations that determine certain functions which depend only on τ. Since the quantity $p^m(\eta, \eta')$ appearing in equation (1.53) is given by equation (1.41), we can conclude that $B^m(\tau, \eta, \zeta)$ has the following form:

$$B^m(\tau, \eta, \zeta) = \sum_{i=m}^n c_i^m P_i^m(\eta) B_i^m(\tau, \zeta). \tag{1.54}$$

By substituting expressions (1.41) and (1.54) into equation (1.53), we obtain

$$B_i^m(\tau, \zeta) = \frac{\lambda}{2} \sum_{j=m}^n c_j^m \left[\int_0^1 P_i^m(\eta') P_j^m(\eta') \, d\eta' \int_0^\tau B_j^m(\tau', \zeta) e^{-(\tau-\tau')/\eta'} \frac{d\tau'}{\eta'} \right.$$

$$\left. - \int_{-1}^0 P_i^m(\eta') P_j^m(\eta') \, d\eta' \int_\tau^{\tau_0} B_j^m(\tau', \zeta) e^{-(\tau-\tau')/\eta'} \frac{d\tau'}{\eta'} \right] + \frac{\lambda}{4} SP_i^m(\zeta) e^{-\tau/\zeta}, \tag{1.55}$$

where $i = m, m+1, \ldots n$. In this manner, $B^m(\tau, \eta, \zeta)$ is expressed in terms of the functions $B_i^m(\tau, \zeta)$ with the aid of formula (1.54), while for the determination of $B_i^m(\tau, \zeta)$ we have the system of equations (1.55).

For the complete solution of the problem (i.e. in order to determine the function $B(\tau, \eta, \zeta, \varphi)$), it is necessary to solve the system of equations (1.55) for all m. For given n, the total number of functions $B_i^m(\tau, \zeta)$ to be determined is $(n+1)(n+2)/2$. It is clear that when n is large, solution of the problem by way of finding the functions $B_i^m(\tau, \zeta)$ from equation (1.55) becomes difficult in practice. A much simpler method for determining $B^m(\tau, \eta, \zeta)$ will be presented in Chapter 5.

Let us now write the integral equations for the source functions corresponding to the simplest phase functions.

For the case of isotropic scattering, when $x(\gamma) = 1$, the source function depends only on τ and ζ and not on η and φ. We shall thus denote the source function simply by $B(\tau, \zeta)$.

In this case equation (1.50) takes the following form:

$$B(\tau, \zeta) = \frac{\lambda}{2} \int_0^1 d\eta' \int_0^\tau B(\tau', \zeta) e^{-(\tau-\tau')/\eta'} \frac{d\tau'}{\eta'} - \frac{\lambda}{2} \int_{-1}^0 d\eta' \int_\tau^{\tau_0} B(\tau', \zeta) e^{-(\tau-\tau')/\eta'} \frac{d\tau'}{\eta'} + \frac{\lambda}{4} Se^{-\tau/\zeta}.$$

(1.56)

By interchanging the order of integration we find

$$B(\tau, \zeta) = \frac{\lambda}{2} \int_0^{\tau_0} B(\tau', \zeta) E_1(|\tau-\tau'|) \, d\tau' + \frac{\lambda}{4} Se^{-\tau/\zeta},$$

(1.57)

where $E_1(\tau)$ is the first exponential integral, defined by

$$E_k(\tau) = \int_0^1 e^{-\tau/y} \frac{dy}{y^{2-k}}.$$

(1.58)

Equation (1.57) was first obtained by Professor O. D. Hvolson [6] of the Petersburg University, who studied light scattering in milky glass. We note that if in (1.57) we omit the inhomogeneous term and also set $\tau_0 = \infty$ and $\lambda = 1$, then we arrive at the equation

$$B(\tau) = \frac{1}{2} \int_0^\infty B(\tau') E_1(|\tau-\tau'|) \, d\tau',$$

(1.59)

which is generally called the "Milne equation" [7]. It determines the radiation field in an atmosphere when the sources of radiation are located at an infinitely great optical depth (as in a stellar photosphere).

For the phase function $x(\gamma) = 1 + x_1 \cos \gamma$, the source function takes the following form:

$$B(\tau, \eta, \zeta, \varphi) = B^0(\tau, \eta, \zeta) + 2B^1(\tau, \eta, \zeta) \cos \varphi.$$

(1.60)

For the determination of the quantities $B^0(\tau, \eta, \zeta)$ and $B^1(\tau, \eta, \zeta)$, we obtain the following equations from (1.53) with the aid of (1.41):

$$B^0(\tau, \eta, \zeta) = \frac{\lambda}{2} \int_0^1 (1+x_1\eta\eta') \, d\eta' \int_0^\tau B^0(\tau', \eta', \zeta) e^{-(\tau-\tau')/\eta'} \frac{d\tau'}{\eta'}$$

$$- \frac{\lambda}{2} \int_{-1}^0 (1+x_1\eta\eta') \, d\eta' \int_\tau^{\tau_0} B^0(\tau', \eta', \zeta) e^{-(\tau-\tau')/\eta'} \frac{d\tau'}{\eta'} + \frac{\lambda}{4} S(1+x_1\eta\zeta) e^{-\tau/\zeta}, \qquad (1.61)$$

$$B^1(\tau, \eta, \zeta) = \frac{x_1}{2} \sqrt{1-\eta^2} \left[\frac{\lambda}{2} \int_0^1 \sqrt{1-\eta'^2} \, d\eta' \int_0^\tau B^1(\tau', \eta', \zeta) e^{-(\tau-\tau')/\eta'} \frac{d\tau'}{\eta'} \right.$$

$$\left. - \frac{\lambda}{2} \int_{-1}^0 \sqrt{1-\eta'^2} \, d\eta' \int_\tau^{\tau_0} B^1(\tau', \eta', \zeta) e^{-(\tau-\tau')/\eta'} \frac{d\tau'}{\eta'} + \frac{\lambda}{4} S \sqrt{1-\zeta^2} \, e^{-\tau/\zeta} \right]. \qquad (1.62)$$

In accordance with (1.54) we have

$$B^0(\tau, \eta, \zeta) = B^0_0(\tau, \zeta) + x_1 B^0_1(\tau, \zeta)\eta, \tag{1.63}$$

$$B^1(\tau, \eta, \zeta) = \frac{x_1}{2} B^1_1(\tau, \zeta) \sqrt{1-\eta^2}, \tag{1.64}$$

and from (1.55) it follows that the functions $B^0_0(\tau, \zeta)$ and $B^0_1(\tau, \zeta)$ are determined by the system of equations

$$B^0_0(\tau, \zeta) = \frac{\lambda}{2} \int_0^{\tau_0} B^0_0(\tau', \zeta) E_1(|\tau-\tau'|)\, d\tau' + \frac{\lambda}{2} x_1 \int_0^{\tau} B^0_1(\tau', \zeta) E_2(\tau-\tau')\, d\tau'$$

$$- \frac{\lambda}{2} x_1 \int_{\tau}^{\tau_0} B^0_1(\tau', \zeta) E_2(\tau'-\tau)\, d\tau' + \frac{\lambda}{4} S e^{-\tau/\zeta}, \tag{1.65}$$

and

$$B^0_1(\tau, \zeta) = \frac{\lambda}{2} \int_0^{\tau} B^0_0(\tau', \zeta) E_2(\tau-\tau')\, d\tau' - \frac{\lambda}{2} \int_{\tau}^{\tau_0} B^0_0(\tau', \zeta) E_2(\tau'-\tau)\, d\tau'$$

$$+ \frac{\lambda}{2} x_1 \int_0^{\tau_0} B^0_1(\tau', \zeta) E_3(|\tau-\tau'|)\, d\tau' + \frac{\lambda}{4} S\zeta e^{-\tau/\zeta}, \tag{1.66}$$

while the function $B^1_1(\tau, \zeta)$ satisfies

$$B^1_1(\tau, \zeta) = \frac{\lambda}{4} x_1 \int_0^{\tau_0} B^1_1(\tau', \zeta) \left[E_1(|\tau-\tau'|) - E_3(|\tau-\tau'|) \right] d\tau' + \frac{\lambda}{4} S \sqrt{1-\zeta^2}\, e^{-\tau/\zeta}. \tag{1.67}$$

Equations for the functions $B^m_i(\tau, \zeta)$ for other phase functions represented by formula (1.8) may be written in an analogous manner. In all cases the kernels of these equations are expressed in terms of the functions $E_k(\tau)$.

1.5. The Diffuse Radiation Field

A number of other quantities in addition to the radiation intensity are used to characterize the radiation field. Among the most important are the density of radiant energy u and the flux of radiant energy H, which is commonly referred to simply as the radiation flux. We may write expressions for these quantities in terms of the radiation intensity I.

The *radiant energy density u* is defined such that the quantity $u\, d\nu\, dV$ represents the amount of radiant energy in the frequency interval $d\nu$ within the volume dV. The energy density is thus given by the equation

$$u = \frac{1}{c} \int I\, d\omega, \tag{1.68}$$

where c is the speed of light and the integration is carried out over all directions.

The *radiation flux H* is defined such that $H d\sigma \, dv \, dt$ represents the net amount of energy flowing through an area $d\sigma$ in the frequency interval dv during the time dt. The radiation flux is a vector quantity with direction defined by the normal to $d\sigma$, so that the energy falling on one side makes a positive contribution, while that incident on the other side makes a negative contribution. Thus, we have

$$H = \int I \cos \vartheta \, d\omega, \tag{1.69}$$

where ϑ is the angle between the direction of radiation and the normal to the area. The radiation flux incident on an area from one hemisphere only is called the *illumination E*, so that H equals the difference between the illumination from one side and that from the other.

The density of radiant energy depends only on position in the layer, while the flux depends on position and on direction. In the future, H will represent the flux of radiant energy in the downward direction through an area with normal perpendicular to the plane bounding the atmosphere. Accordingly, ϑ in equation (1.69) will signify the angle between the radiation direction and the normal directed toward increasing depth.

Let us rewrite equations (1.68) and (1.69) in an explicit form applicable to planetary atmospheres. Substituting expression (1.44) into these equations, we obtain

$$u(\tau, \zeta) = \frac{2\pi}{c} \int_{-1}^{1} I^0(\tau, \eta, \zeta) \, d\eta \tag{1.70}$$

and

$$H(\tau, \zeta) = 2\pi \int_{-1}^{1} I^0(\tau, \eta, \zeta) \eta \, d\eta, \tag{1.71}$$

where $I^0(\tau, \eta, \zeta)$ is the radiation intensity averaged over azimuth.

For practical applications, the radiation intensity emerging from the atmosphere and the flux at its boundaries are of special interest. These quantities may be easily expressed in terms of the source function $B(\tau, \eta, \zeta, \varphi)$. Setting $\tau = 0$ in equation (1.49) and substituting $-\eta$ for η, we find

$$I(0, -\eta, \zeta \, \varphi) = \int_0^{\tau_0} B(\tau, -\eta, \zeta, \varphi) \, e^{-\tau/\eta} \, \frac{d\tau}{\eta}. \tag{1.72}$$

The radiation intensity diffusely reflected by the atmosphere is determined by this equation where $\eta > 0$ is now the cosine of the angle of reflection $(\pi - \vartheta)$. Setting $\tau = \tau_0$ in equation (1.48), we obtain the following expression for the diffuse radiation emerging from the bottom of the atmosphere:

$$I(\tau_0, \eta, \zeta, \varphi) = \int_0^{\tau_0} B(\tau, \eta, \zeta, \varphi) \, e^{-(\tau_0 - \tau)/\eta} \, \frac{d\tau}{\eta}, \tag{1.73}$$

where η is the cosine of the transmission angle ϑ. The radiation intensities emerging from the atmosphere are usually represented as

$$I(0, -\eta, \zeta, \varphi) = S\varrho(\eta, \zeta, \varphi) \, \zeta, \qquad \eta > 0 \tag{1.74}$$

$$I(\tau_0, \eta, \zeta, \varphi) = S\sigma(\eta, \zeta, \varphi) \, \zeta, \qquad \eta > 0 \tag{1.75}$$

where the quantity ϱ is called the *reflection coefficient*, the quantity σ is known as the *transmission coefficient*, the incident solar flux through the upper boundary has been taken equal to $\pi S\zeta$, and η represents in each case the cosine of the angle of emergence with respect to the outward normal.

The physical meaning of these coefficients follows from the fact that the quantity $S\zeta$ is the radiation intensity scattered by a perfectly white orthotropic screen that is located horizontally on the upper boundary of the atmosphere (such a screen reflects all radiation incident on it with equal probability in all outward directions). Thus, the reflection coefficient ϱ (or the transmission coefficient σ) represents the ratio of the radiation intensity diffusely reflected (or diffusely transmitted) by the atmosphere to the radiation intensity scattered by such a screen.

From formulas (1.72)–(1.75) we find the following expressions for the reflection and transmission coefficients in terms of the source function $B(\tau, \eta, \zeta, \varphi)$:

$$\varrho(\eta, \zeta, \varphi) = \frac{1}{S} \int_0^{\tau_0} B(\tau, -\eta, \zeta, \varphi) e^{-\tau/\eta} \frac{d\tau}{\eta\zeta}, \tag{1.76}$$

$$\sigma(\eta, \zeta, \varphi) = \frac{1}{S} \int_0^{\tau_0} B(\tau, \eta, \zeta, \varphi) e^{-(\tau_0-\tau)/\eta} \frac{d\tau}{\eta\zeta}. \tag{1.77}$$

An important property of the reflection and transmission coefficients is their symmetry with respect to the variables η and ζ, i.e.

$$\varrho(\eta, \zeta, \varphi) = \varrho(\zeta, \eta, \varphi), \tag{1.78}$$

$$\sigma(\eta, \zeta, \varphi) = \sigma(\zeta, \eta, \varphi). \tag{1.79}$$

These relations express "the principle of reciprocity" for optical phenomena. They may be easily obtained from the basic integral equation (1.50).

In order to determine the radiation flux at the atmospheric boundaries, we must set $\tau = 0$ and $\tau = \tau_0$ in equation (1.71). Since we assume that there is no diffuse radiation incident on the atmosphere from outside, the flux $H(0, \zeta)$ is the illumination of the upper atmospheric boundary from below with negative sign, while the flux $H(\tau_0, \zeta)$ is the illumination of the lower atmospheric boundary from above. Designating these illuminations by $E(0, \zeta)$ and $E(\tau_0, \zeta)$, respectively, we obtain

$$E(0, \zeta) = -H(0, \zeta) = 2\pi \int_0^1 I^0(0, -\eta, \zeta) \eta \, d\eta, \tag{1.80}$$

$$E(\tau_0, \zeta) = H(\tau_0, \zeta) = 2\pi \int_0^1 I^0(\tau_0, \eta, \zeta) \eta \, d\eta. \tag{1.81}$$

Instead of the quantities $I^0(0, -\eta, \zeta)$ and $I^0(\tau_0, \eta, \zeta)$ we introduce the azimuth-averaged reflection and transmission coefficients $\varrho^0(\eta, \zeta)$ and $\sigma^0(\eta, \zeta)$ with the aid of equations analo-

3*

gous to (1.74) and (1.75). Then instead of equations (1.80) and (1.81) we have

$$E(0, \zeta) = 2\pi S\zeta \int_0^1 \varrho^0(\eta, \zeta) \eta \, d\eta, \tag{1.82}$$

$$E(\tau_0, \zeta) = 2\pi S\zeta \int_0^1 \sigma^0(\eta, \zeta) \eta \, d\eta. \tag{1.83}$$

It must be noted that equation (1.83) does not determine the entire illumination of the planetary surface, but only that part corresponding to the diffuse atmospheric radiation. In order to obtain the complete illumination it is necessary to add to expression (1.83) the surface illumination coming from the direct solar radiation transmitted by the atmosphere.

The quantity $E(0, \zeta)$ represents the energy *emerging* from the atmosphere per unit area at its upper boundary per unit time. The energy from the Sun *incident* per unit area on this upper boundary per unit time (i.e. the illumination of this boundary by the Sun's rays) is denoted by $E_0(\zeta)$. It is apparent that

$$E_0(\zeta) = \pi S\zeta. \tag{1.84}$$

The ratio of these energies,

$$A(\zeta) = \frac{E(0, \zeta)}{E_0(\zeta)}, \tag{1.85}$$

is called the *atmospheric albedo*. Substituting expressions (1.82) and (1.84) into (1.85) we find

$$A(\zeta) = 2 \int_0^1 \varrho^0(\eta, \zeta) \eta \, d\eta. \tag{1.86}$$

Equation (1.86) determines the atmospheric albedo at that location on the planet where the incident angle of the solar radiation is arccos ζ. Since locally the atmosphere may be considered as a plane layer, the quantity $A(\zeta)$ is sometimes called the *plane albedo*.

In addition to the plane albedo, the *spherical albedo*, which represents the ratio of the energy reflected by the whole planet to the solar energy falling on it, is used in planetary studies. In order to obtain the spherical albedo, let us consider a planet of radius R (Fig. 1.3). It is apparent that the energy incident per unit time equals $\pi R^2 \pi S$. In order to find the energy reflected by the planet, we introduce on the disc a ring with radius r and width dr. The energy reflected by this ring per unit time will be $A(\zeta) \pi S \, 2\pi r \, dr$. Since $r \, dr = R^2 \zeta \, d\zeta$, then this energy may be rewritten as $2\pi R^2 \pi S A(\zeta) \zeta \, d\zeta$. Thus, the energy reflected by the whole planet will be

$$2\pi R^2 \pi S \int_0^1 A(\zeta) \zeta \, d\zeta.$$

If we denote the spherical albedo by A_s, we obtain

$$A_s = 2 \int_0^1 A(\zeta) \zeta \, d\zeta. \tag{1.87}$$

In this manner, if the plane albedo $A(\zeta)$ is known, it is easy to find the spherical albedo A_s.

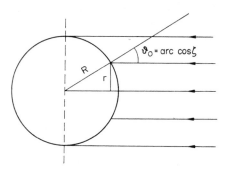

FIG. 1.3. Geometry for the determination of the spherical albedo A_s of a planet.

The equations for the various physical quantities (density of radiant energy, atmospheric albedo, radiation flux, surface illumination) introduced in the previous paragraphs will often be used below. We emphasize that it is not the radiation intensity I that enters into all of these formulas, but only the quantity I^0, the radiation intensity averaged over azimuth. Later, other equations will be derived containing the quantity I^0. For this reason, finding the radiation intensity averaged over azimuth represents an especially important problem.

1.6. The Case of Pure Scattering

Let us examine in more detail the case when pure scattering occurs in the atmosphere. For this case, often encountered in practice, we must set $\lambda = 1$ in the equations derived above. It is important that when pure scattering occurs, the equation of radiative transfer admits two simple integrals. We shall now find these integrals and draw certain conclusions from them.

From equations (1.35) and (1.36) we obtain the following equation of radiative transfer for $\lambda = 1$:

$$\eta \frac{dI}{d\tau} = -I + \int Ix(\gamma') \frac{d\omega'}{4\pi} + \frac{S}{4} x(\gamma) e^{-\tau/\zeta}. \tag{1.88}$$

By multiplying equation (1.88) by $d\omega$, integrating over all directions, and using equation (1.3), we find

$$\frac{dH}{d\tau} = \pi S e^{-\tau/\zeta}, \tag{1.89}$$

where H is the radiation flux determined by equation (1.69). Integration of (1.89) gives

$$H(\tau, \zeta) + \pi S \zeta e^{-\tau/\zeta} = C, \tag{1.90}$$

where C is an arbitrary constant.

Equation (1.90) represents one of the desired integrals, the so-called *flux integral*. The first term is the flux of diffuse radiation, while the second term is the flux of direct solar energy. Equation (1.90) thus expresses the constancy of the total radiation flux in a purely scattering atmosphere.

By multiplying equation (1.88) by $\eta \, d\omega$, integrating over all directions, and using equations (1.10) and (1.69), we obtain

$$\frac{dK}{d\tau} = -\left(1 - \frac{x_1}{3}\right)H + \frac{x_1}{3}\pi S\zeta e^{-\tau/\zeta}, \tag{1.91}$$

where we have introduced the notation

$$K(\tau, \zeta) = \int I \cos^2 \vartheta \, d\omega = 2\pi \int_{-1}^{1} I^0(\tau, \eta, \zeta) \, \eta^2 \, d\eta. \tag{1.92}$$

Substituting H from (1.90) into (1.91) and carrying out the integration, we find

$$K(\tau, \zeta) = -\left(1 - \frac{x_1}{3}\right)C\tau - \pi S\zeta^2 e^{-\tau/\zeta} + C_1, \tag{1.93}$$

where C_1 is a new arbitrary constant. Equation (1.93) gives the other integral we are seeking, which is generally known as the *K-integral*.

Let us apply relations (1.90) and (1.93) to the atmospheric boundary. Setting $\tau = 0$ and $\tau = \tau_0$ in equation (1.90) and using equations (1.80) and (1.81), we obtain

$$-E(0, \zeta) + \pi S\zeta = C, \tag{1.94}$$

$$E(\tau_0, \zeta) + \pi S\zeta e^{-\tau_0/\zeta} = C. \tag{1.95}$$

It follows that

$$E(0, \zeta) + E(\tau_0, \zeta) = \pi S\zeta(1 - e^{-\tau_0/\zeta}). \tag{1.96}$$

This equation expresses the fact that for $\lambda = 1$, all energy which interacts with the atmosphere emerges again following multiple scatterings.

With the aid of equations (1.82) and (1.83), equation (1.96) may be rewritten as

$$2\int_0^1 \varrho^0(\eta, \zeta) \, \eta \, d\eta + 2\int_0^1 \sigma^0(\eta, \zeta) \, \eta \, d\eta = 1 - e^{-\tau_0/\zeta}. \tag{1.97}$$

Setting $\tau = 0$ and $\tau = \tau_0$ in equation (1.93) and eliminating the constant C_1 from the resulting equations, we find

$$K(0, \zeta) - K(\tau_0, \zeta) = \left(1 - \frac{x_1}{3}\right)C\tau_0 - \pi S\zeta^2(1 - e^{-\tau_0/\zeta}). \tag{1.98}$$

Introducing constant C from (1.90) into (1.98) and applying equations (1.82) and (1.92), we have

$$2\int_0^1 \varrho^0(\eta, \zeta) \, \eta^2 \, d\eta - 2\int_0^1 \sigma^0(\eta, \zeta) \, \eta^2 \, d\eta$$

$$= \left(1 - \frac{x_1}{3}\right)\tau_0\left[1 - 2\int_0^1 \varrho^0(\eta, \zeta) \, \eta \, d\eta\right] - \zeta(1 - e^{-\tau_0/\eta}). \tag{1.99}$$

Equations (1.97) and (1.99) may be used in determining the quantities $\varrho^0(\eta, \zeta)$ and $\sigma^0(\eta, \zeta)$, the azimuth-averaged reflection and transmission coefficients of the atmosphere.

1.7. Methods for Solving the Problem

The system of equations (1.35) and (1.36) or its equivalent, equation (1.50), is very difficult to solve. As a result, the search for effective methods of solution has gone on for many years. A number of scholars from various fields (astrophysics, geophysics, physics, and mathematics) have taken part in this work. The characteristics of several of the proposed methods will now be discussed briefly.

The application of exact methods to the solution of light scattering problems in planetary atmospheres began with the work of V. A. Ambartsumyan (see his collected works [8]). He considered the problem of finding directly the equations determining the radiation intensity emerging from the atmosphere. In order to obtain these equations, he proposed "the method of combination of layers". The following "invariance principle" may be used to illustrate this method. The reflection properties of a semi-infinite medium will not change if we add to it a layer with optical properties the same as those of the whole medium. Considering the optical thickness of the additional layer to be very small, accounting for all of the changes in the intensity of the reflected radiation brought about by this layer, and equating these changes to zero, we arrive at the desired equation for the reflection coefficient of the semi-infinite atmosphere. From this equation it is possible to determine the structure of the reflection coefficient, i.e. to express it through certain auxiliary functions $\varphi_i^m(\eta)$ which depend on one argument only. The same results may be obtained with another method proposed by Ambartsumyan based on the use of the integral equation (1.50). In the case of an atmosphere with finite optical thickness τ_0, the reflection and transmission coefficients are expressed through auxiliary functions $\varphi_i^m(\eta, \tau_0)$ and $\psi_i^m(\eta, \tau_0)$. Both of these methods are applied in Chapters 2 and 3.

The application of various "invariance principles" in radiative transfer theory was subsequently continued by Chandrasekhar [9]. He also introduced into the theory the important functions $H^m(\eta)$, $X^m(\eta, \tau_0)$ and $Y^m(\eta, \tau_0)$. For the simplest phase functions he succeeded in expressing the auxiliary functions $\varphi_i^m(\eta)$ in terms of $H^m(\eta)$, and the auxiliary functions $\varphi_i^m(\eta, \tau_0)$ and $\psi_i^m(\eta, \tau_0)$ in terms of $X^m(\eta, \tau_0)$ and $Y^m(\eta, \tau_0)$. The properties of the functions $H^m(\eta)$, $X^m(\eta, \tau_0)$ and $Y^m(\eta, \tau_0)$ are studied in detail in the book by Busbridge [10].

In the works of V. V. Sobolev, described in Chapters 5 and 6 of the present book, the fundamental functions $\Phi^m(\tau)$ (for semi-infinite atmospheres) and $\Phi^m(\tau, \tau_0)$ (for atmospheres of optical thickness τ_0) were introduced into the theory. Knowledge of these functions allows one to find the radiation intensity at any optical depth in the atmosphere. In particular, it is easy to express the function $H^m(\eta)$ in terms of $\Phi^m(\tau)$ and the functions $X^m(\eta, \tau_0)$ and $Y^m(\eta, \tau_0)$ in terms of $\Phi^m(\tau, \tau_0)$. Moreover, a general method is given (for arbitrary phase function) for the determination of $\varphi_i^m(\eta)$ if $H^m(\eta)$ is known, and for $\varphi_i^m(\eta, \tau_0)$ and $\psi_i^m(\eta, \tau_0)$ if $X^m(\eta, \tau_0)$ and $Y^m(\eta, \tau_0)$ are known. The function $\Phi^m(\tau, \tau_0)$ plays this important role because the resolvent of the integral equation (1.53) may be expressed in terms of it. Consequently, the determination of the function $\Phi^m(\tau, \tau_0)$ allows one to find the radiation field in the atmosphere for an arbitrary distribution of energy sources.

If we are only interested in the reflection and transmission coefficients, then it is possible to determine them not only from the nonlinear integral equations derived by the methods of Ambartsumyan, but also from linear integral equations. These equations may be derived from the system of equations (1.45) and (1.46). In the case of a semi-infinite atmosphere, they may be solved explicitly. Linear integral equations which determine the reflection and transmission coefficients, as well as the auxiliary functions $\varphi_i^m(\eta, \tau_0)$ and $\psi_i^m(\eta, \tau_0)$, are examined in Chapter 7.

Approximate methods of solution for equations (1.35) and (1.36) play an important role in various applications of the theory. The method based on averaging the radiation intensity over angle is one example. It represents a generalization of the well-known Eddington method used in the theory of stellar atmospheres for isotropic scattering. Approximate formulas obtained with its aid prove to be very useful in practice. This approximate method (and a number of others) are presented in Chapter 8.

Other methods of considerable interest for the solution of radiative transfer problems have also been proposed, but will not be considered in detail here. We shall only mention them, referring to the works in which they appear and are applied.

Chandrasekhar developed a "discrete ordinate method", replacing the integral term of the transfer equation by a Gaussian sum. This procedure is used in the book by Chandrasekhar [9], as well as in the works of many other authors.

It is well known that the integral equations of radiative transfer theory have kernels which depend on the absolute value of the difference of two arguments. Solution of these equations for $\tau_0 = \infty$ may be obtained by the Wiener–Hopf method, based on the application of the Laplace transform. This particular procedure is explained in the Busbridge book [10].

If it is necessary to determine the radiation intensity emerging from the atmosphere for different internal source distributions, it is expedient to find the probability that a photon originating at a given depth will emerge from the atmosphere in a particular direction (generally speaking, after multiple scattering). This method was proposed by Sobolev (see [11], chapter VI) and is often applied for isotropic scattering of light. I. N. Minin generalized the method to the case of anisotropic scattering (see, particularly, [12]).

The "invariance principles" proposed by Ambartsumyan and Chandrasekhar are widely used to obtain various equations determining the intensity of the diffuse radiation. Developing these principles, Bellman and his colleagues arrived at the "invariant imbedding" method. Equations determining the changes of the reflection and transmission coefficients as a function of optical thickness of the atmosphere were found with the aid of this procedure. Equations for the determination of the radiation intensity within the atmosphere were also obtained. A review of the work in this direction is presented in the articles by Bellman [13] and Kagiwada and Kalaba [14].

The method of Case merits careful attention. It is based on the expansion of the quantities being sought in eigenfunctions of the integro-differential equation of transfer. This approach and its application to a number of problems is described in the book by Case and Zweifel [15].

In recent years, electronic computers have facilitated the wide use of numerical methods for the solution of radiative transfer problems. The method of "layer doubling" proposed by van de Hulst [16] is an example of one of these. It utilizes equations which relate the radiation intensity in a layer with optical thickness τ_0 to the radiation intensity in a layer with optical thickness $2\tau_0$. These computations begin with very small values of τ_0. Hunt and Grant [17] have developed a related computational method based on the invariance principles; it may be used in an inhomogeneous atmosphere for an arbitrary phase function and for arbitrarily distributed radiation sources.

It should be pointed out that the theory of radiative transfer is close to the theory of neutron diffusion in the methods it employs. For this reason, research in each of these two areas has substantially influenced the other. The theory of neutron diffusion and the methods applicable to it may be found in the book by Davison [18].

References

1. RAYLEIGH, On the scattering of light by small particles, *Phil. Mag.* **41**, 102, 447 (1871).
2. G. MIE, Beiträge zur Optik trüben Medien, Speziell kolloidaler Mettallösungen, *Ann. d. Phys.* **25**, 377 (1908).
3. K. S. SHIFRIN, *Rasseyanie Sveta v Mutnoi Srede*, Gostekhizdat, Moscow, 1951.
4. H. C. VAN DE HULST, *Light Scattering by Small Particles*, John Wiley & Sons, Inc., New York, 1957.
5. D. DEIRMENDJIAN, *Electromagnetic Scattering on Spherical Polydispersions*, American Elsevier Pub. Co., Inc., New York, 1969.
6. O. D. HVOLSON, Grundzüge einer mathematischen Theorie der inneren Diffusion des Lichtes, *Izv. Petersburg. Akad. Nauk.* **33**, 221 (1890).
7. E. A. MILNE, Radiative equilibrium in the outer layers of a star, *M.N.R.A.S.* **81**, 361 (1921).
8. V. A. AMBARTSUMYAN, *Nauchnye Trudy*, Vol. 1, Izd. Akad. Nauk Armen. SSR, Yerevan, 1960.
9. S. CHANDRASEKHAR, *Radiative Transfer*, Oxford University Press, 1950.
10. I. W. BUSBRIDGE, *The Mathematics of Radiative Transfer*, Cambridge University Press, 1960.
11. V. V. SOBOLEV, *Perenos Luchistoi Energii v Atmosferakh Zvezd i Planet*, Gostekhizdat, Moscow, 1956 [*A Treatise on Radiative Transfer*, Van Nostrand, Princeton, N.J., 1963].
12. I. N. MININ, Diffusion of radiation in a plane layer for anisotropic scattering, *Astron. Zh.* **43**, 1244 (1966); **45**, 264 (1968). [*Sov. Astron.—A.J.* **10**, 995 (1967); **12**, 209 (1968)].
13. R. BELLMAN, Invariant imbedding and computational methods in radiative transfer, *Transport Theory*, Amer. Math. Soc., Providence, R.I., 1969.
14. H. H. KAGIWADA and R. E. KALABA, Direct and inverse problems for integral equation via initial-value methods, *Transport Theory*, Amer. Math. Soc., Providence, R.I., 1969.
15. K. M. CASE and P. F. ZWEIFEL, *Linear Transport Theory*, Addison-Wesley Pub. Co., Reading, Massachusetts, 1967.
16. H. C. VAN DE HULST, *A New Look at Multiple Scattering*, NASA Institute for Space Studies, New York, 1963.
17. G. E. HUNT and I. P. GRANT, Discrete space theory of radiative transfer and its application to problems in planetary atmospheres, *J. Atmos. Sci.* **26**, 963 (1969).
18. B. DAVISON, *Neutron Transport Theory*, Clarendon Press, Oxford, 1957.

Chapter 2

SEMI-INFINITE ATMOSPHERES

WE SHALL begin closer study of the equations derived in the previous chapter by considering the problem of determining the radiation field in a semi-infinite atmosphere (i.e. an atmosphere for which $\tau_0 = \infty$). This problem has great intrinsic interest both analytically and for application to planets with very deep atmospheres such as Venus and Jupiter. In addition, its solution is useful in the determination of the radiation field in an atmosphere of finite optical thickness τ_0.

In this chapter we shall be principally concerned with the problem of diffuse reflection and transmission of light by a semi-infinite atmosphere. In the case of transmission, we seek the relative angular distribution of diffusely transmitted radiation as $\tau_0 \to \infty$. We shall also determine the radiation field in deep layers of the atmosphere ($\tau \gg 1$). It turns out that knowledge of the relative intensity of radiation in deep layers significantly simplifies the determination of the intensity of diffusely reflected and transmitted light. At the end of the chapter we examine the case of small true absorption in the atmosphere ($1 - \lambda \ll 1$); this situation is important in many applications. Asymptotic formulas expressing various characteristics of the radiation field will then be found; these become more exact as $(1 - \lambda)$ decreases. The complete solution of the problem of radiative transfer in a semi-infinite atmosphere will be given later, in Chapter 5.

2.1. The Radiation Field in Deep Layers (Relative Intensity of Radiation)

Under conditions of anisotropic scattering, the radiation field in an atmosphere is generally quite complicated. Deep in a semi-infinite atmosphere, however, the character of the radiation field gradually simplifies. At great optical depths ($\tau \gg 1$), it assumes an asymptotic form characterized by rather simple properties.

We shall now consider the determination of the intensity of radiation in the deep layers of a semi-infinite medium. This problem is important not only in the study of atmospheres, but also in other areas (for example, the study of oceans). As will become apparent below, the solution for the deep layers enables us to also solve the problem of diffuse reflection and transmission of light by an atmosphere of large optical thickness.

The following properties of the radiation field in deep layers may be shown to follow from physical considerations: (1) the role of direct (unscattered) radiation is negligible compared with the role of diffuse radiation, (2) the intensity of radiation does not depend upon azi-

24

muth, (3) the relative angular distribution of the intensity does not depend on the optical depth. All these assertions may be proven, but we shall not do so here.

It follows from our assertions that we need be interested only in the quantities B^0 and I^0, the source function and radiation intensity averaged over azimuth. For the determination of these quantities, we use equations (1.45) and (1.46) with $m = 0$ and neglect the term representing direct radiation. These equations then take the form:

$$\eta \frac{dI^0(\tau, \eta, \zeta)}{d\tau} = -I^0(\tau, \eta, \zeta) + B^0(\tau, \eta, \zeta), \tag{2.1}$$

$$B^0(\tau, \eta, \zeta) = \frac{\lambda}{2} \int_{-1}^{1} p^0(\eta, \eta') \, I^0(\tau, \eta', \zeta) \, d\eta', \tag{2.2}$$

where

$$p^0(\eta, \eta') = \frac{1}{2\pi} \int_{0}^{2\pi} x(\gamma') \, d\varphi, \tag{2.3}$$

and γ' is determined by equation (1.37).

From a position within the deep layers of a semi-infinite medium, we may consider that the medium extends without limit in all directions. The solution of equations (2.1) and (2.2) may then be sought in the form

$$B^0(\tau, \eta, \zeta) = Sc(\zeta) \, b(\eta) \, e^{-k\tau}, \tag{2.4}$$

$$I^0(\tau, \eta, \zeta) = Sc(\zeta) \, i(\eta) \, e^{-k\tau}, \tag{2.5}$$

where $c(\zeta)$, $b(\eta)$, $i(\eta)$ are unknown functions and k is an unknown constant. Substituting equations (2.4) and (2.5) into (2.1), we obtain

$$i(\eta) = \frac{b(\eta)}{1 - k\eta}. \tag{2.6}$$

The substitution of (2.4) and (2.5) into (2.2) gives, with the aid of (2.6),

$$b(\eta) = \frac{\lambda}{2} \int_{-1}^{1} p^0(\eta, \eta') \frac{b(\eta')}{1 - k\eta'} \, d\eta'. \tag{2.7}$$

Thus, the function $b(\eta)$ is determined from the integral equation (2.7), and the constant k from the condition that this equation has a solution. If the function $b(\eta)$ is found, then the function $i(\eta)$ is determined from equation (2.6).

It is clear that the function $c(\zeta)$, which depends on the angle of incidence of the radiation coming from outside the atmosphere, cannot be found from equations (2.1) and (2.2). It will be determined in Section 2.4.

Equation (2.7) was first obtained by V. A. Ambartsumyan [1]. In a series of papers [1–4], methods of solution for this equation were proposed, and numerical results for the function $b(\eta)$ and the parameter k were given for various phase functions $x(\gamma)$ and different values of λ. The most complete mathematical investigation of equation (2.7) has been carried out by M. V. Maslennikov [5].

Before considering the solution of equation (2.7), we note that it determines $b(\eta)$ only to within a multiplicative constant. This raises the question of normalization for this function. In what follows we shall always assume that it is normalized according to the condition

$$\frac{1}{2} \int_{-1}^{1} b(\eta) \, d\eta = 1. \tag{2.8}$$

Equation (2.7) may be solved quite easily if the phase function is represented by a sum of a finite number of Legendre polynomials, as in equation (1.8). In this case we find, on the basis of (1.41) and (1.42), that

$$p^0(\eta, \eta') = \sum_{i=0}^{n} x_i P_i(\eta) P_i(\eta'). \tag{2.9}$$

Substituting (2.9) into (2.7) we conclude that the function $b(\eta)$ has the form

$$b(\eta) = \sum_{i=0}^{n} x_i b_i P_i(\eta), \tag{2.10}$$

where

$$b_i = \frac{\lambda}{2} \int_{-1}^{1} \frac{b(\eta) P_i(\eta)}{1 - k\eta} \, d\eta. \tag{2.11}$$

The normalization condition (2.8) gives $b_0 = 1$. To determine the other coefficients b_i and the parameter k, we substitute (2.10) into (2.11) and obtain the system of equations

$$b_i = \frac{\lambda}{2} \sum_{j=0}^{n} b_j \int_{-1}^{1} \frac{P_j(\eta) P_i(\eta)}{1 - k\eta} \, d\eta. \tag{2.12}$$

In the case of isotropic scattering $(x(\gamma) = 1)$, we have $b(\eta) = b_0 = 1$. As a result, on the basis of (2.6),

$$i(\eta) = \frac{1}{1 - k\eta}. \tag{2.13}$$

From (2.12) it follows that in this case the parameter k is determined by the equation

$$\frac{\lambda}{2k} \ln \frac{1+k}{1-k} = 1. \tag{2.14}$$

Table 2.1 lists values of the parameter k obtained from the solution of equation (2.14) for various values of λ.

TABLE 2.1. RELATION BETWEEN λ AND k FOR ISOTROPIC SCATTERING

λ	k	λ	k	λ	k	λ	k
0	1.0000	0.4	0.9856	0.8	0.7104	0.96	0.3408
0.1	1.0000	0.5	0.9575	0.9	0.5254	0.98	0.2430
0.2	0.9999	0.6	0.9073	0.92	0.4740	0.99	0.1725
0.3	0.9974	0.7	0.8286	0.94	0.4140	1.00	0.0000

In the case of the simplest anisotropic phase function, $x(\gamma) = 1 + x_1 \cos \gamma$, equation (2.10) takes the form

$$b(\eta) = 1 + x_1 b_1 \eta, \tag{2.15}$$

and we obtain from (2.12),

$$\frac{\lambda}{2k} \ln \frac{1+k}{1-k} + x_1 b_1 \frac{\lambda}{k} \left(\frac{1}{2k} \ln \frac{1+k}{1-k} - 1 \right) = 1, \tag{2.16}$$

$$b_1 = \frac{1-\lambda}{k}. \tag{2.17}$$

Elimination of b_1 from (2.16) and (2.17) yields

$$\frac{\lambda}{2k} \left(1 + x_1 \frac{1-\lambda}{k^2} \right) \ln \frac{1+k}{1-k} - \lambda x_1 \frac{1-\lambda}{k^2} = 1. \tag{2.18}$$

After the determination of k from equation (2.18), the constant b_1 is found from equation (2.17).

If the number of terms in the expansion (2.9) is large, then the coefficients b_i may be conveniently determined from a recursion relation. This is obtained from equation (2.11) by using the recursion relation for Legendre polynomials:

$$(i+1) P_{i+1}(\eta) + i P_{i-1}(\eta) = (2i+1) \eta P_i(\eta). \tag{2.19}$$

The recursion formula for the coefficients b_i then has the form

$$(i+1) b_{i+1} + i b_{i-1} = (2i+1-\lambda x_i) b_i/k \tag{2.20}$$

where $b_0 = 1$, $b_1 = (1-\lambda)/k$.

In order to find the cofficients b_i from equation (2.20), it is necessary to know the value of the parameter k. This may be determined from the equation

$$\frac{\lambda}{2} \sum_{j=0}^{n} b_j \int_{-1}^{1} \frac{P_j(\eta)}{1-k\eta} d\eta = 1, \tag{2.21}$$

which is equation (2.12) for the case $i = 0$. Finding the quantities b_i from equation (2.20) for various k and substituting these into equation (2.21), we may find the necessary value of k to reduce the latter equation to an identity.

An alternative method of determining the parameter k utilizes a relation between k and λ which does not involve the quantities b_i. This relation may be obtained from equation (2.7) in the form of the following continuing fraction:

$$1 - \lambda = \cfrac{k^2}{3 - \lambda x_1 - \cfrac{4k^2}{5 - \lambda x_2 - \cfrac{9k^2}{7 - \lambda x_3 - \ldots}}}. \tag{2.22}$$

The simplest means of solving the problem under consideration would seem to be the determination of the parameter k from equation (2.22) and the quantities b_i from equation (2.20).

As an example we present the results of a calculation of the radiation field in the deep layers of a medium which scatters according to the Henyey–Greenstein phase function [equation (1.16)]. These computations were performed by V. M. Loskutov [4] utilizing equation (2.22) and the recursion relation (2.20).

TABLE 2.2. VALUES OF SINGLE SCATTERING ALBEDO λ
AS A FUNCTION OF g AND k

k	g						
	0.50	0.60	0.70	0.80	0.85	0.90	0.95
0	1.000	1.000	1.000	1.000	1.000	1.000	1.000
0.10	0.993	0.992	0.989	0.984	0.979	0.971	0.954
0.20	0.973	0.967	0.957	0.940	0.925	0.905	0.873
0.30	0.940	0.926	0.906	0.875	0.852	0.823	0.782
0.40	0.892	0.870	0.839	0.795	0.767	0.731	0.685
0.50	0.830	0.799	0.758	0.704	0.671	0.633	0.585
0.60	0.753	0.712	0.663	0.603	0.568	0.528	0.481
0.70	0.657	0.609	0.554	0.491	0.456	0.417	0.373
0.80	0.539	0.485	0.428	0.366	0.334	0.299	0.260
0.90	0.394	0.331	0.278	0.225	0.198	0.171	0.142

Table 2.2 contains values of the single scattering albedo λ which were computed for various values of g and k. It is clear that as g increases (i.e. as the phase function becomes more elongated), a given value of λ corresponds to decreasing values of k. It should be remembered that the degree of elongation of the phase function is ordinarily characterized by the para-

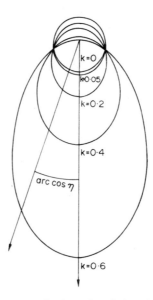

FIG. 2.1. Polar diagram of the angular distribution of radiation deep within a semi-infinite layer for a Henyey–Greenstein phase function with $g = 0.7$. The parameter k characterizes absorption, and is related to the single scattering albedo λ by equation (2.22).

meter x_1, which is always less than 3. In the present case $x_1 = 3g$, so that the last columns of Table 2.2 represent very elongated phase functions.

Figure 2.1 illustrates the angular distribution of radiation for $g=0.7$ and various k. When $k=0$, the intensity of radiation does not depend upon direction. As k increases, so that the role of true absorption increases, the fraction of radiation directed downwards increases.

It is clear from equations (2.6), (2.7), and (2.8) that for the case of pure scattering ($\lambda = 1$)

$$k = 0, \quad b(\eta) = 1, \quad i(\eta) = 1, \tag{2.23}$$

for arbitrary phase function.

When the role of true absorption is small ($1-\lambda \ll 1$), then asymptotic formulas may be obtained for the quantities k, $b(\eta)$ and $i(\eta)$. These formulas become more exact as $1-\lambda$ decreases. They may be found by expanding the desired quantities in a power series in $\sqrt{1-\lambda}$ and retaining only the leading terms. Limiting ourselves to terms of order $\sqrt{1-\lambda}$, we obtain from equation (2.22)

$$k = \sqrt{(1-\lambda)(3-x_1)}. \tag{2.24}$$

Asymptotic formulas for $i(\eta)$ and $b(\eta)$ may be obtained with the aid of equations (2.6), (2.10), and (2.20). Retaining only terms of order $\sqrt{(1-\lambda)}$, we find

$$b(\eta) = 1+x_1 \sqrt{\frac{1-\lambda}{3-x_1}} \, \eta, \tag{2.25}$$

$$i(\eta) = 1+3 \sqrt{\frac{1-\lambda}{3-x_1}} \, \eta. \tag{2.26}$$

If in the asymptotic formulas we retain also terms of order $1-\lambda$, then expression (2.24) for the quantity k does not change, because it is correct to terms of order $(1-\lambda)^{3/2}$. For the quantities $b(\eta)$ and $i(\eta)$ we find the following relations

$$b(\eta) = 1+x_1 \sqrt{\frac{1-\lambda}{3-x_1}} \, \eta + x_2 \frac{2(1-\lambda)}{5-x_2} P_2(\eta), \tag{2.27}$$

$$i(\eta) = 1+3 \sqrt{\frac{1-\lambda}{3-x_1}} \, \eta + (1-\lambda)\left[1+\frac{10}{5-x_2} P_2(\eta)\right]. \tag{2.28}$$

Because the case of small true absorption often occurs in real atmospheres, equations (2.24)–(2.28) are very important. From them we may determine immediately the radiation field deep inside the atmosphere. In addition, we shall use these equations below (Sections 2.5 and 2.6) to solve other problems for the case when $1-\lambda \ll 1$.

2.2. Diffuse Reflection of Light

Most of our knowledge about the atmospheres of the planets is obtained from an analysis of the radiation diffusely reflected by the planetary atmosphere. We shall now consider this problem for the case of an atmosphere of infinitely large optical thickness ($\tau_0 = \infty$).

As previously, we shall assume that the phase function may be represented by a finite sum of Legendre polynomials, i.e. by equation (1.8). It then follows from equations (1.44) and (1.74) that the reflection coefficient $\varrho(\eta, \zeta, \varphi)$ may be expressed as

$$\varrho(\eta, \zeta, \varphi) = \varrho^0(\eta, \zeta)+2 \sum_{m=1}^{n} \varrho^m(\eta, \zeta) \cos m\varphi, \tag{2.29}$$

where ζ is the cosine of the angle of incidence of the solar radiation on the atmosphere, η is the cosine of the angle of reflection, and φ is the difference in azimuthal angles for the reflected and incident rays.

Substituting expressions (2.29) and (1.43) into equation (1.76), we obtain the following expression for the quantity $\varrho^m(\eta, \zeta)$ in terms of the function $B^m(\tau, \eta, \zeta)$:

$$\varrho^m(\eta, \zeta) = \frac{1}{S} \int_0^\infty B^m(\tau, -\eta, \zeta) e^{-\tau/\eta} \frac{d\tau}{\eta\zeta}. \tag{2.30}$$

It then follows from (1.53) that the function $B^m(\tau, \eta, \zeta)$ for $\tau_0 = \infty$ is determined by the integral equation

$$B^m(\tau, \eta, \zeta) = \frac{\lambda}{2} \int_0^1 p^m(\eta, \eta')\, d\eta' \int_0^\tau B^m(\tau'\, \eta', \zeta) e^{-(\tau-\tau')/\eta'} \frac{d\tau'}{\eta'}$$

$$+ \frac{\lambda}{2} \int_0^1 p^m(\eta, -\eta')\, d\eta' \int_\tau^\infty B^m(\tau', -\eta', \zeta) e^{-(\tau'-\tau)/\eta'} \frac{d\tau'}{\eta'} + \frac{\lambda}{4} Sp^m(\eta, \zeta) e^{-\tau/\zeta}, \tag{2.31}$$

where $p^m(\eta, \eta')$ is given by equation (1.41).

If equation (2.31) has been solved, then $\varrho^m(\eta, \zeta)$ may be found from equation (2.30). It is important to note, however, that one may obtain an equation which determines $\varrho^m(\eta, \zeta)$ directly. This possibility was first pointed out by V. A. Ambartsumyan, who proposed two derivations of this equation. The first procedure is based on the utilization of the "principle of invariance", formulated in Section 1.7. The second procedure involves derivation of an equation for the quantity $\varrho^m(\eta, \zeta)$ from equation (2.31), and we shall use this method.

We begin by rewriting equation (2.31) in the form

$$B^m(\tau, \eta, \zeta) = \frac{\lambda}{2} \int_0^1 p^m(\eta, \eta')\, d\eta' \int_0^\tau B^m(\tau-\alpha, \eta', \zeta) e^{-\alpha/\eta'} \frac{d\alpha}{\eta'}$$

$$+ \frac{\lambda}{2} \int_0^1 p^m(\eta, -\eta')\, d\eta' \int_0^\infty B^m(\tau+\alpha, -\eta', \zeta) e^{-\alpha/\eta'} \frac{d\alpha}{\eta'} + \frac{\lambda}{4} Sp^m(\eta, \zeta) e^{-\tau/\zeta}. \tag{2.32}$$

Differentiating equation (2.32) with respect to τ and carrying out a transformation which is the inverse of that which converted (2.31) into (2.32), we find

$$\frac{dB^m(\tau, \eta, \zeta)}{d\tau} = \frac{\lambda}{2} \int_0^1 p^m(\eta, \eta')\, d\eta' \int_0^\tau \frac{dB^m(\tau', \eta', \zeta)}{d\tau'} e^{-(\tau-\tau')/\eta'} \frac{d\tau'}{\eta'}$$

$$+ \frac{\lambda}{2} \int_0^1 p^m(\eta, -\eta')\, d\eta' \int_\tau^\infty \frac{dB^m(\tau', -\eta', \zeta)}{d\tau'} e^{-(\tau'-\tau)/\eta'} \frac{d\tau'}{\eta'}$$

$$+ \frac{\lambda}{2} \int_0^1 p^m(\eta, \eta') B^m(0, \eta', \zeta) e^{-\tau/\eta'} \frac{d\eta'}{\eta'} - \frac{\lambda}{4\zeta} Sp^m(\eta, \zeta) e^{-\tau/\zeta}. \tag{2.33}$$

We see that equation (2.33) for

$$\frac{dB^m(\tau, \eta, \zeta)}{d\tau}$$

differs from equation (2.31) for $B^m(\tau, \eta, \zeta)$ only in the inhomogeneous term. Furthermore, the inhomogeneous term in (2.33) is a superposition of functions of the same type as those which form the inhomogeneous term of (2.31). It follows from the linearity of the equations being considered that the solution of equation (2.33) will be a superposition of solutions of equation (2.31), namely

$$\frac{dB^m(\tau, \eta, \zeta)}{d\tau} = -\frac{1}{\zeta} B^m(\tau, \eta, \zeta) + \frac{2}{S} \int_0^1 B^m(0, \eta', \zeta) B^m(\tau, \eta, \eta') \frac{d\eta'}{\eta'}. \tag{2.34}$$

Multiplying equation (2.34) by $e^{-\tau/\eta}$, integrating over τ from 0 to infinity and using (2.30), we obtain

$$S(\eta+\zeta) \varrho^m(\eta, \zeta) = B^m(0, -\eta, \zeta) + 2\eta \int_0^1 B^m(0, \eta', \zeta) \varrho^m(\eta, \eta') d\eta'. \tag{2.35}$$

For the determination of the quantity $B^m(0, \eta, \zeta)$ appearing in (2.35), it is sufficient to set $\tau = 0$ in equation (2.31). With the aid of (2.30), we then find

$$B^m(0, \eta, \zeta) = \frac{\lambda}{4} S \left[p^m(\eta, \zeta) + 2\zeta \int_0^1 p^m(\eta, -\eta') \varrho^m(\eta', \zeta) d\eta' \right]. \tag{2.36}$$

Substitution of (2.36) into (2.35) gives the desired integral equation for the function $\varrho^m(\eta, \zeta)$:

$$(\eta+\zeta) \varrho^m(\eta, \zeta) = \frac{\lambda}{4} p^m(-\eta, \zeta) + \frac{\lambda}{2} \zeta \int_0^1 p^m(\eta, \eta') \varrho^m(\eta', \zeta) d\eta'$$

$$+\frac{\lambda}{2} \eta \int_0^1 p^m(\eta', \zeta) \varrho^m(\eta, \eta') d\eta' + \lambda\eta\zeta \int_0^1 \varrho^m(\eta, \eta') d\eta' \int_0^1 p^m(\eta', -\eta'') \varrho^m(\eta'', \zeta) d\eta''. \tag{2.37}$$

The function $\varrho^m(\eta, \zeta)$, which depends on two arguments, may be expressed in terms of certain auxiliary functions which depend on only one argument. For this purpose we utilize expression (1.41) for the quantity $p^m(\eta, \eta')$. Substituting into equation (2.36) we find

$$B^m(0, \eta, \zeta) = \frac{\lambda}{4} S \sum_{i=m}^n c_i^m P_i^m(\eta) \varphi_i^m(\zeta), \tag{2.38}$$

where we have introduced the notation

$$\varphi_i^m(\zeta) = P_i^m(\zeta) + 2\zeta \int_0^1 P_i^m(-\eta') \varrho^m(\eta', \zeta) d\eta'. \tag{2.39}$$

By substituting equation (2.38) into (2.35), we obtain

$$(\eta+\zeta)\,\varrho^m(\eta,\,\zeta) = \frac{\lambda}{4} \sum_{i=m}^{n} c_i^m \varphi_i^m(\zeta) \left[P_i^m(-\eta) + 2\eta \int_0^1 P_i^m(\eta')\,\varrho^m(\eta,\,\eta')\,d\eta' \right]. \qquad (2.40)$$

Using the symmetry of the function $\varrho^m(\eta,\,\zeta)$, which follows from (1.78), and recalling (2.39), we find

$$\varrho^m(\eta,\,\zeta) = \frac{\lambda}{4} \sum_{i=m}^{n} c_i^m (-1)^{i+m}\, \frac{\varphi_i^m(\eta)\varphi_i^m(\zeta)}{\eta+\zeta}. \qquad (2.41)$$

This equation expresses the quantity $\varrho^m(\eta,\,\zeta)$ in terms of the auxiliary functions $\varphi_i^m(\eta)$.

Substitution of (2.41) into (2.39) gives a system of nonlinear integral equations for the determination of the functions $\varphi_i^m(\eta)$:

$$\varphi_i^m(\zeta) = P_i^m(\zeta) + \frac{\lambda}{2} \sum_{j=m}^{n} c_j^m (-1)^{i+j} \varphi_j^m(\zeta) \int_0^1 \frac{P_i^m(\eta)\varphi_j^m(\eta)}{\eta+\zeta}\, d\eta. \qquad (2.42)$$

The functions $\varphi_i^m(\eta)$ are usually called the Ambartsumyan functions, since he first introduced them into the theory of light scattering [6]. In order to find the quantity $\varrho^m(\eta,\,\zeta)$, it is necessary to know $(n-m+1)$ such functions. The number of functions $\varphi_i^m(\eta)$ necessary for the determination of the entire reflection coefficient $\varrho(\eta,\,\zeta,\,\varphi)$ is just $\frac{1}{2}(n+1)(n+2)$.

We shall now apply these results to the case of the two simplest phase functions. For isotropic scattering $x(\gamma) = 1$, the reflection coefficient is

$$\varrho(\eta,\,\zeta) = \frac{\lambda}{4}\,\frac{\varphi(\eta)\varphi(\zeta)}{\eta+\zeta}, \qquad (2.43)$$

and the function $\varphi(\eta)$, according to (2.42), is determined by the equation

$$\varphi(\eta) = 1 + \frac{\lambda}{2}\,\eta\varphi(\eta) \int_0^1 \frac{\varphi(\zeta)}{\eta+\zeta}\, d\zeta. \qquad (2.44)$$

Here we have written simply $\varphi(\eta)$ in place of $\varphi_0^0(\eta)$.

The function $\varphi(\eta)$ has been well studied in a number of papers. In particular, expressions for this function in explicit form have been found (cf. [7]). In the solution of some problems, one encounters the moments of the function $\varphi(\eta)$, that is, the quantities

$$\alpha_k = \int_0^1 \varphi(\eta)\eta^k\, d\eta. \qquad (2.45)$$

The zeroth moment is easily obtained from equation (2.44). It is

$$\alpha_0 = \frac{2}{\lambda}\,\{1 - \sqrt{1-\lambda}\}. \qquad (2.46)$$

Table 2.3 gives values of the function $\varphi(\eta)$ and its first three moments.

TABLE 2.3. THE FUNCTION $\varphi(\eta)$ AND ITS MOMENTS

η	λ							
	0.5	0.7	0.8	0.9	0.950	0.975	0.995	1.000
0	1.000	1.000	1.000	1.000	1.000	1.000	1.000	1.000
0.1	1.072	1.113	1.139	1.172	1.195	1.211	1.232	1.247
0.2	1.113	1.183	1.229	1.291	1.337	1.370	1.415	1.450
0.3	1.144	1.236	1.301	1.391	1.461	1.512	1.583	1.643
0.4	1.168	1.281	1.361	1.478	1.571	1.641	1.742	1.829
0.5	1.188	1.318	1.413	1.556	1.672	1.762	1.895	2.013
0.6	1.204	1.350	1.459	1.626	1.765	1.875	2.042	2.194
0.7	1.219	1.378	1.500	1.689	1.851	1.982	2.184	2.374
0.8	1.231	1.403	1.536	1.747	1.931	2.083	2.322	2.553
0.9	1.242	1.453	1.568	1.801	2.007	2.178	2.456	2.731
1.0	1.251	1.445	1.598	1.850	2.077	2.271	2.587	2.908
α_0	1.172	1.292	1.382	1.520	1.634	1.727	1.868	2.000
α_1	0.603	0.679	0.736	0.825	0.902	0.964	1.062	1.155
α_2	0.407	0.461	0.503	0.569	0.627	0.674	0.748	0.820

For the simplest anisotropic phase function $x(\gamma) = 1 + x_1 \cos \gamma$, the reflection coefficient is given by the formula

$$\varrho(\eta, \zeta, \varphi) = \varrho^0(\eta, \zeta) + 2\varrho^1(\eta, \zeta) \cos \varphi, \tag{2.47}$$

and the quantities $\varrho^0(\eta, \zeta)$ and $\varrho^1(\eta, \zeta)$ have, according to (2.41), the form

$$\varrho^0(\eta, \zeta) = \frac{\lambda}{4} \frac{\varphi_0^0(\eta) \varphi_0^0(\zeta) - x_1 \varphi_1^0(\eta) \varphi_1^0(\zeta)}{\eta + \zeta}, \tag{2.48}$$

$$\varrho^1(\eta, \zeta) = \frac{\lambda}{8} x_1 \frac{\varphi_1^1(\eta) \varphi_1^1(\zeta)}{\eta + \zeta}. \tag{2.49}$$

It follows from (2.42) that the functions $\varphi_0^0(\eta)$ and $\varphi_1^0(\eta)$ are determined by the system of equations

$$\varphi_0^0(\eta) = 1 + \frac{\lambda}{2} \eta \varphi_0^0(\eta) \int_0^1 \frac{\varphi_0^0(\zeta)}{\eta + \zeta} \, d\zeta - \frac{\lambda}{2} x_1 \eta \varphi_1^0(\eta) \int_0^1 \frac{\varphi_1^0(\zeta)}{\eta + \zeta} \, d\zeta, \tag{2.50}$$

$$\varphi_1^0(\eta) = \eta - \frac{\lambda}{2} \eta \varphi_0^0(\eta) \int_0^1 \frac{\varphi_0^0(\zeta)}{\eta + \zeta} \zeta \, d\zeta + \frac{\lambda}{2} x_1 \eta \varphi_1^0(\eta) \int_0^1 \frac{\varphi_1^0(\zeta)}{\eta + \zeta} \zeta \, d\zeta, \tag{2.51}$$

and the function $\varphi_1^1(\eta)$ from the equation

$$\varphi_1^1(\eta) = \sqrt{1 - \eta^2} + \frac{\lambda}{4} x_1 \eta \varphi_1^1(\eta) \int_0^1 \frac{\varphi_1^1(\zeta)}{\eta + \zeta} \sqrt{1 - \zeta^2} \, d\zeta. \tag{2.52}$$

4*

We note that the following inter-dependence between the functions $\varphi_0^0(\eta)$ and $\varphi_1^0(\eta)$ results from equations (2.50) and (2.51):

$$\varphi_1^0(\eta) = \frac{\left(1 - \frac{\lambda}{2}\alpha\right)\eta\varphi_0^0(\eta)}{1 - \frac{\lambda}{2}x_1\beta\eta}, \qquad (2.53)$$

where

$$\alpha = \int_0^1 \varphi_0^0(\eta)\,d\eta, \quad \beta = \int_0^1 \varphi_1^0(\eta)\,d\eta. \qquad (2.54)$$

In the case of pure scattering, it is easy to see that equations (2.50) and (2.51) have the solution

$$\varphi_1^0(\eta) = 0, \quad \varphi_0^0(\eta) = \varphi(\eta), \qquad (2.55)$$

where $\varphi(\eta)$ is determined by equation (2.44) for $\lambda = 1$. Consequently, in this case the azimuth-averaged reflection coefficient $\varrho^0(\eta, \zeta)$ is for arbitrary x_1 the same as in the case of isotropic scattering. Detailed tables of the functions $\varphi_0^0(\eta)$, and $\varphi_1^0(\eta)$ for various values of the parameters λ and x_1 have been obtained [8] by numerical solution of equations (2.50)–(2.52).

In the investigation of light diffusely reflected from planetary atmospheres it is often not important to know the intensity of the reflected radiation, but only the entire amount of energy emerging through the upper boundary of the atmosphere (per unit area per unit time). As is well known, the ratio of this quantity to the illumination of the upper surface of the atmosphere by the Sun is called the albedo of the atmosphere. In Section 1.5 it was shown that the plane albedo is determined by the equation

$$A(\zeta) = 2\int_0^1 \varrho^0(\eta, \zeta)\eta\,d\eta. \qquad (2.56)$$

It is easy to see that the albedo may be very simply expressed through the auxiliary function $\varphi_1^0(\eta)$. Since, on the basis of equation (2.39),

$$\varphi_1^0(\zeta) = \zeta - 2\zeta \int_0^1 \varrho^0(\eta, \zeta)\eta\,d\eta, \qquad (2.57)$$

we have for the albedo

$$A(\zeta) = 1 - \frac{1}{\zeta}\varphi_1^0(\zeta). \qquad (2.58)$$

Using equation (1.87), we find that the spherical albedo of the planet is just

$$A_s = 1 - 2\int_0^1 \varphi_1^0(\zeta)\,d\zeta. \qquad (2.59)$$

Equations (2.58) and (2.59) are valid for any phase function. For the case of the simplest anisotropic phase function $1 + x_1 \cos \gamma$, we may use expression (2.53) to rewrite equation

(2.58) in the form

$$A(\zeta) = 1 - \frac{1 - \dfrac{\lambda}{2}\alpha}{1 - \dfrac{\lambda}{2}x_1\beta\zeta}\; \varphi_0^0(\zeta). \tag{2.60}$$

For isotropic scattering, we take $x_1 = 0$ in (2.60) and use equation (2.46) to obtain

$$A(\zeta) = 1 - \varphi(\zeta)\sqrt{1-\lambda}, \tag{2.61}$$

where the function $\varphi(\eta)$ is determined by equation (2.44).

In this section we have seen that the reflection coefficient for a semi-infinite atmosphere may be expressed in terms of the auxiliary functions $\varphi_i^m(\eta)$. Moreover, the role of $\varphi_i^m(\eta)$ is not restricted to this application. It is possible to show that the intensity emergent from an atmosphere containing an arbitrary distribution of internal energy sources may also be expressed in terms of these functions. The appropriate equations for the case of isotropic scattering have been presented in the book *TRT* (Chapter VI, § 3), and for the case of anisotropic scattering in the work of I. N. Minin [9].

2.3. Diffuse Transmission of Light

The radiation falling on an atmosphere of finite optical thickness τ_0 is partly reflected and partly transmitted. As τ_0 increases, the intensity of diffusely transmitted radiation decreases. In addition, the relative angular distribution of this radiation tends toward some limiting distribution. We shall speak of this distribution as representing the transmission by an atmosphere of infinitely large optical thickness ($\tau_0 = \infty$). This limiting intensity represents the solution to the problem of the distribution of brightness over a cloudy sky.

The radiation emerging from a semi-infinite atmosphere for an arbitrary distribution of energy sources located at infinitely great optical depth possesses a similar angular distribution. The radiative transfer problem in this case is called the "Milne problem", and was first encountered in the study of stellar photospheres. Thus, part of the Milne problem consists in finding this same limiting intensity distribution.

It is obvious that the intensity of radiation diffusely transmitted by an atmosphere of infinitely great optical thickness will not depend on azimuth since it cannot depend on the conditions of incidence at the upper boundary. We will designate this relative intensity by $u(\eta)$, where η is the cosine of the angle of transmission. We shall determine the normalization of the function $u(\eta)$ later.

The function $u(\eta)$ may be easily found if the solution to the diffuse reflection problem is known. Specifically, $u(\eta)$ is expressed in terms of the auxiliary functions $\varphi_i^0(\eta)$. In order to obtain this expression, we shall use three different methods. Each method is interesting for its own sake, and may in addition be used for the solution of a variety of other problems.

In order to simplify notation, we shall in the future omit the index "0" for all quantities averaged over azimuth; that is, we shall write simply $\varphi_i(\eta)$ instead of $\varphi_i^0(\eta)$, $\varrho(\eta, \zeta)$ instead of $\varrho^0(\eta, \zeta)$, etc.

First method

Let us add a thin layer of optical thickness $\Delta\tau$ to the lower atmospheric boundary and find the resultant change in intensity of the diffusely transmitted radiation $u(\eta)$. First of all, the layer produces an attenuation of the radiation traveling in the specified direction. The intensity of the attenuated radiation is

$$ u(\eta) \left(1 - \frac{\Delta\tau}{\eta} \right). $$

In addition, the layer will scatter radiation emerging from the previous boundary into the given direction. The intensity of this scattered radiation may be represented as

$$ \frac{\lambda}{2} \frac{\Delta\tau}{\eta} \int_0^1 u(\eta') \, p(\eta, \eta') \, d\eta', $$

where the quantity $p(\eta, \eta')$ is determined by equation (2.3).

Finally, the layer scatters radiation back towards the previous boundary, and part of this radiation is diffusely reflected at an angle arccos η to the normal. The corresponding intensity is

$$ \lambda\Delta\tau \int_0^1 \varrho(\eta, \eta'') \, d\eta'' \int_0^1 u(\eta') p(-\eta'', \eta') \, d\eta'. $$

Adding the last three expressions, we obtain the radiation intensity emerging through the new atmospheric boundary. Alternatively, this intensity may be written as $u(\eta)(1-k\Delta\tau)$, where k is a constant. This latter expression follows from the fact that the relative angular distribution of radiation diffusely transmitted by an atmosphere of infinitely great optical thickness will not change with the addition of the layer (cf. eqn. (2.5)).

On the basis of the previous discussion we have

$$ u(\eta)(1-k\Delta\tau) = u(\eta) \left(1 - \frac{\Delta\tau}{\eta} \right) + \frac{\lambda}{2} \frac{\Delta\tau}{\eta} \int_0^1 u(\eta') p(\eta, \eta') \, d\eta' $$

$$ + \lambda\Delta\tau \int_0^1 \varrho(\eta, \eta'') \, d\eta'' \int_0^1 u(\eta') \, p(-\eta'', \eta') \, d\eta'. \qquad (2.62) $$

It then follows that

$$ u(\eta)(1-k\eta) = \frac{\lambda}{2} \int_0^1 u(\eta') \, p(\eta, \eta') \, d\eta' $$

$$ + \lambda\eta \int_0^1 \varrho(\eta, \eta'') \, d\eta'' \int_0^1 u(\eta') p(-\eta'', \eta') \, d\eta'. \qquad (2.63) $$

Introducing the notation

$$K(\eta, \eta') = p(\eta, \eta') + 2\eta \int_0^1 \varrho(\eta, \eta'') p(-\eta'', \eta') \, d\eta'', \qquad (2.64)$$

we obtain in place of (2.63)

$$u(\eta)(1 - k\eta) = \frac{\lambda}{2} \int_0^1 u(\eta') K(\eta, \eta') \, d\eta'. \qquad (2.65)$$

The function $u(\eta)$ may thus be found from the integral equation (2.65), whose kernel $K(\eta, \eta')$ is expressed in terms of the reflection coefficient $\varrho(\eta, \eta')$ by equation (2.64). The constant k is determined by the condition that equation (2.65) possess a solution.

Comparing equations (2.64) and (2.36), we find

$$B(0, \eta, \zeta) = \frac{\lambda}{4} SK(\zeta, \eta). \qquad (2.66)$$

Recalling equation (2.38), we have

$$K(\eta, \eta') = \sum_{i=0}^n x_i \varphi_i(\eta) P_i(\eta'), \qquad (2.67)$$

since, according to (1.42), $c_i^0 = x_i$.

We see that substitution of (2.67) into (2.65) permits us to express the function $u(\eta)$ in terms of the functions $\varphi_i(\eta)$. Carrying out this substitution, we obtain

$$u(\eta) = \frac{\lambda}{2} \sum_{i=0}^n \frac{x_i a_i \varphi_i(\eta)}{1 - k\eta}, \qquad (2.68)$$

where

$$a_i = \int_0^1 u(\eta) P_i(\eta) \, d\eta. \qquad (2.69)$$

Substitution of (2.68) into (2.69) leads to a system of homogeneous algebraic equations for the determination of the coefficients a_i. The quantity k may then be found from the condition that the determinant of this system vanishes. Equation (2.68) was first derived by V. A. Ambartsumyan [10] using the method described here.

Second method

Let us consider that the energy sources within a semi-infinite atmosphere are located at an infinitely large optical depth. The determination of the radiation field in this case is known as the "Milne problem". This problem may be reduced to the solution of the following integral equation for the source function $B(\tau, \eta)$:

$$B(\tau, \eta) = \frac{\lambda}{2} \int_0^1 p(\eta, \eta') \, d\eta' \int_0^\tau B(\tau, \eta') e^{-(\tau - \tau')/\eta'} \frac{d\tau'}{\eta'}$$

$$+ \frac{\lambda}{2} \int_0^1 p(\eta, -\eta') \, d\eta' \int_\tau^\infty B(\tau' - \eta') e^{-(\tau' - \tau)/\eta'} \frac{d\tau'}{\eta'}. \qquad (2.70)$$

This equation may be obtained from equation (2.31) by setting $m = 0$ (since the source function in the present case does not depend on azimuth) and by dropping the inhomogeneous term, which represents radiation incident on the atmosphere from outside.

At the present time we are not interested in the complete solution of the Milne problem, but only in finding the radiation intensity emerging from the atmosphere; that is, the quantity

$$u(\eta) = \int_0^\infty B(\tau, -\eta) e^{-\tau/\eta} \frac{d\tau}{\eta}. \tag{2.71}$$

In order to find an equation for the function $u(\eta)$, we proceed in the same manner as in solving the problem of diffuse reflection of light. Differentiating (2.70) with respect to τ, we find

$$\frac{dB(\tau, \eta)}{d\tau} = \frac{\lambda}{2} \int_0^1 p(\eta, \eta') \, d\eta' \int_0^\tau \frac{dB(\tau', \eta')}{d\tau'} e^{-(\tau - \tau')/\eta'} \frac{d\tau'}{\eta'}$$

$$+ \frac{\lambda}{2} \int_0^1 p(\eta, -\eta') \, d\eta' \int_\tau^\infty \frac{dB(\tau', -\eta')}{d\tau'} e^{-(\tau' - \tau)/\eta'} \frac{d\tau'}{\eta'} \tag{2.72}$$

$$+ \frac{\lambda}{2} \int_0^1 p(\eta, \eta') B(0, \eta') e^{-\tau/\eta'} \frac{d\eta'}{\eta'}.$$

A comparison of (2.72) with (2.31) gives

$$\frac{dB(\tau, \eta)}{d\tau} = kB(\tau, \eta) + \frac{2}{S} \int_0^1 B(0, \eta') B(\tau, \eta, \eta') \frac{d\eta'}{\eta'}, \tag{2.73}$$

where k is some constant. The term $kB(\tau, \eta)$ represents a solution of equation (2.70), which is the homogeneous form of (2.31) and (2.72). We did not consider this term in our earlier examination of the diffuse reflection of light since it is identically zero in the absence of radiation sources at infinitely great optical depths.

Multiplying equation (2.73) by $e^{-\tau/\eta}$, integrating over τ from 0 to infinity, and using equations (2.30) and (2.71), we obtain

$$u(\eta)(1 - k\eta) = B(0, -\eta) + 2\eta \int_0^1 B(0, \eta') \varrho(\eta, \eta') \, d\eta'. \tag{2.74}$$

Setting $\tau = 0$ in equation (2.70) and utilizing equations (2.30) and (2.71), we find

$$B(0, \eta) = \frac{\lambda}{2} \int_0^1 p(\eta, -\eta') u(\eta') \, d\eta'. \tag{2.75}$$

Substitution of (2.75) into (2.74) and use of equation (2.64) leads us once again to equation (2.65). From this equation, as we have already seen, we may derive the equation (2.68) which we are seeking.

Third method

Knowledge of the radiation field in the deep layers of a semi-infinite atmosphere (see Section 2.1) may be used to determine the function $u(\eta)$. Let us consider a plane placed parallel to the atmospheric boundary and located at a large optical depth, and then examine radiation passing through this plane. We designate by $i(\eta)$ the radiation intensity traveling downward (in relative units) and by $i(-\eta)$ the upward directed intensity (in both cases $\eta > 0$). We may consider the radiation traveling upward to result from the diffuse reflection of the radiation $i(\eta)$ by a semi-infinite atmosphere located below our hypothetical plane. In consequence, we have

$$i(-\eta) = 2 \int_0^1 i(\eta') \varrho(\eta, \eta') \eta' \, d\eta'. \tag{2.76}$$

The downward directed intensity consists of two parts. The first part represents the radiation diffusely transmitted by that layer of the atmosphere lying above the plane of interest. The second part results from diffuse reflection by that layer of the upward directed radiation. Since the atmospheric layer may be considered as semi-infinite, we obtain

$$i(\eta) = Mu(\eta) + 2 \int_{-0}^1 i(-\eta') \varrho(\eta, \eta') \eta' \, d\eta', \tag{2.77}$$

where M is some constant.

Of course, if the function $u(\eta)$ is not normalized, then it is not necessary to introduce the constant M, and equations (2.76) and (2.77) may be used for the normalization. For future use, however, it is convenient to normalize $u(\eta)$ as follows:

$$2 \int_0^1 u(\eta) \, i(\eta) \, \eta d\eta = 1. \tag{2.78}$$

We shall henceforth always use this convention. The function $i(\eta)$ will in turn be normalized according to the condition

$$\frac{\lambda}{2} \int_0^1 i(\eta) \, d\eta = 1, \tag{2.79}$$

which follows from equations (2.6)–(2.8).

In order to find the constant M, we multiply (2.77) by $i(\eta)\eta$ and integrate over η from 0 to 1. Using equations (2.76) and (2.78), we obtain

$$M = 2 \int_{-1}^1 i^2(\eta) \, \eta d\eta. \tag{2.80}$$

We see that the function $u(\eta)$ which we are seeking may be expressed in terms of $i(\eta)$ and $\varrho(\eta, \zeta)$ with the aid of equations (2.77) and (2.80).

Since the quantity $i(\eta)$ may, with the aid of equations (2.6) and (2.10), be expressed as

$$i(\eta) = \frac{1}{1-k\eta} \sum_{i=0}^n x_i b_i P_i(\eta), \tag{2.81}$$

we have in place of (2.77)

$$Mu(\eta) = \sum_{i=0}^{n} x_i b_i \left[\frac{P_i(\eta)}{1-k\eta} - 2 \int_0^1 \varrho(\eta, \eta') \frac{P_i(-\eta')}{1+k\eta'} \eta' d\eta' \right].$$ (2.82)

From equation (2.41) for $m = 0$ we obtain the following expression for the function $\varrho(\eta, \zeta)$:

$$\varrho(\eta, \zeta) = \frac{\lambda}{4} \sum_{i=0}^{n} x_i (-1)^i \frac{\varphi_i(\eta)\varphi_i(\zeta)}{\eta+\zeta}.$$ (2.83)

Likewise, from (2.39) we find

$$\varphi_i(\eta) = P_i(\eta) + 2\eta \int_0^1 P_i(-\eta') \varrho(\eta, \eta') d\eta'.$$ (2.84)

Substituting (2.83) into (2.82), breaking the integral into two parts with the aid of the identity

$$\frac{\eta'}{(\eta+\eta')(1+k\eta')} = \frac{1}{1-k\eta} \left(\frac{1}{1+k\eta'} - \frac{\eta}{\eta+\eta'} \right)$$

and using equations (2.81) and (2.84), we again obtain equation (2.68) for $u(\eta)$, with the coefficients a_i determined by the expression

$$Ma_i = \frac{2}{\lambda} x_i b_i - (-1)^i \int_0^1 \varphi_i(\eta) i(-\eta) d\eta.$$ (2.85)

It is possible to give an alternative expression for the coefficients a_i. Multiplying (2.84) by $i(\eta)$, integrating over η between 0 and 1, and using (2.76), we find

$$\int_{-1}^{1} i(\eta) P_i(\eta) d\eta = \int_0^1 i(\eta) \varphi_i(\eta) d\eta.$$ (2.86)

Relations (2.6), (2.11), and (2.86) allow us to write

$$a_i = \frac{1}{M} \int_0^1 \varphi_i(\eta) [i(\eta) - (-1)^i i(-\eta)] d\eta$$ (2.87)

in place of (2.85). It should be noted that the function $u(\eta)$ as determined by equation (2.68) with the coefficients a_i computed according to (2.87) satisfies the normalization condition (2.78) (since this condition was used in the derivation of these equations).

Thus, we have obtained one and the same formula (2.68) for $u(\eta)$ by three methods. The third method, however, allowed us to obtain two new results:

  1. It was proven that the quantity k which enters equation (2.68) has for any phase function the same value as in the equations characterizing the radiation field in deep layers of the medium. It follows that k may be determined, for example, from equation (2.22).
  2. Equation (2.87) expressing the coefficients a_i explicitly in terms of the functions $i(\eta)$ and $\varphi_i(\eta)$ was obtained.

In order to use these results to find the function $u(\eta)$, it is first necessary to determine the radiation field in the deep layers (more precisely, the quantities k and $i(\eta)$). It is obviously much easier to do this than to find k and a_i from the system of equations obtained from substituting (2.68) into (2.69).

As a sample of the application of this procedure, let us consider the case of light scattering according to the phase function $x(\gamma) = 1 + x_1 \cos \gamma$. Equation (2.68) then becomes

$$u(\eta) = \frac{\lambda}{2} \frac{a_0 \varphi_0(\eta) + x_1 a_1 \varphi_1(\eta)}{1 - k\eta}. \tag{2.88}$$

In order to find the coefficients a_0 and a_1, it is necessary to use the expression for $i(\eta)$ which may be obtained from (2.6), (2.15) and (2.17). Substituting this expression into (2.87) and (2.80), we obtain

$$a_0 = \frac{2}{M} \left(k + x_1 \frac{1-\lambda}{k} \right) \int_0^1 \frac{\varphi_0(\eta)\eta \, d\eta}{1 - k^2 \eta^2}, \tag{2.89}$$

$$a_1 = \frac{2}{M} \int_0^1 \frac{\varphi_1(\eta)}{1 - k^2 \eta^2} [1 + x_1(1-\lambda)\eta^2] \, d\eta, \tag{2.90}$$

where

$$M = \frac{4}{k} \left[\left(1 + x_1 \frac{1-\lambda}{k^2} \right)^2 \frac{k^2}{1-k^2} + \left(1 + 3x_1 \frac{1-\lambda}{k^2} \right) \left(1 - \frac{1}{\lambda} \right) \right]. \tag{2.91}$$

In the derivation of equation (2.91) we have used equation (2.18), which serves to determine k.

For pure scattering ($\lambda = 1$, $k = 0$), $\varphi_1(\eta) = 0$ and $\varphi_0(\eta)$ coincides with the analogous function in the case of isotropic scattering (see Section 2.2). Equation (2.88) thus shows that the function $u(\eta)$ also takes on the same value as for isotropic scattering. In other words, $u(\eta)$ for the phase function $x(\gamma) = 1 + x_1 \cos \gamma$ does not depend on the parameter x_1 when $\lambda = 1$.

We shall see in what follows that the function $u(\eta)$ introduced in the present section occurs in the solution of many problems. We shall give below other equations for the determination of this function which differ from (2.68) (exact expressions in Chapter 7, and approximate ones in Chapter 8).

2.4. The Radiation Field in Deep Layers (Absolute Intensity)

Let us return to the consideration of the radiation field in deep layers of a semi-infinite medium. The relative intensity of radiation in these layers was determined in Section 2.1. Now we shall find the radiation intensity in absolute units, using several of the results obtained above.

In Section 2.1 we pointed out that the source function $B(\tau, \eta, \zeta)$ and the radiation intensity $I(\tau, \eta, \zeta)$ in the deep layers are independent of azimuth and have the form given by equations (2.4) and (2.5). We also showed how the quantities k, $b(\eta)$ and $i(\eta)$ may be determined. In this section we shall obtain an expression for the function $c(\zeta)$, which depends on the angle arccos ζ at which radiation is incident on the atmosphere.

We begin with the basic integral equation for the source function $B(\tau, \eta, \zeta)$. It may be obtained from (2.31) for $m = 0$ and has the form

$$B(\tau, \eta, \zeta) = \frac{\lambda}{2} \int_0^1 p(\eta, \eta')\, d\eta' \int_0^\tau B(\tau', \eta', \zeta) e^{-(\tau-\tau')/\eta'} \frac{d\tau'}{\eta'}$$

$$+ \frac{\lambda}{2} \int_0^1 p(\eta, -\eta')\, d\eta' \int_\tau^\infty B(\tau', -\eta', \zeta) e^{-(\tau'-\tau)/\eta'} \frac{d\tau'}{\eta'} + \frac{\lambda}{4} Sp(\eta, \zeta) e^{-\tau/\zeta}. \tag{2.92}$$

From (2.92) it is possible to find the following relation between the function $B(\tau, \eta, \zeta)$ and its derivative with respect to τ:

$$\frac{dB(\tau, \eta, \zeta)}{d\tau} = -\frac{1}{\zeta} B(\tau, \eta, \zeta) + \frac{2}{S} \int_0^1 B(0, \eta', \zeta) B(\tau, \eta, \eta') \frac{d\eta'}{\eta'}. \tag{2.93}$$

This relation is just equation (2.34) for $m = 0$.

Substituting equation (2.4) into (2.93), we obtain an integral equation for the function $c(\zeta)$:

$$c(\zeta) = \frac{\zeta}{1-k\zeta} \frac{2}{S} \int_0^1 c(\eta) B(0, \eta, \zeta) \frac{d\eta}{\eta}. \tag{2.94}$$

Recalling (2.66), we may replace (2.94) by

$$c(\zeta)(1-k\zeta) = \zeta \frac{\lambda}{2} \int_0^1 c(\eta) K(\zeta, \eta) \frac{d\eta}{\eta}, \tag{2.95}$$

where $K(\zeta, \eta)$ is given by equation (2.67). Comparison of (2.94) with (2.65) allows us to conclude that

$$c(\zeta) = Cu(\zeta)\zeta, \tag{2.96}$$

where C is some constant.

For the determination of C we use the following approach. We rewrite equation (2.92) in order to apply it to the deep layers of the medium. Dropping the inhomogeneous term and replacing 0 by $-\infty$ as the lower limit of the integration over τ', we have

$$B(\tau, \eta, \zeta) = \frac{\lambda}{2} \int_0^1 p(\eta, \eta')\, d\eta' \int_{-\infty}^\tau B(\tau', \eta', \zeta) e^{-(\tau-\tau')/\eta'} \frac{d\tau'}{\eta'}$$

$$+ \frac{\lambda}{2} \int_0^1 p(\eta, -\eta')\, d\eta' \int_\tau^\infty B(\tau', -\eta', \zeta) e^{-(\tau'-\tau)/\eta'} \frac{d\tau'}{\eta'}. \tag{2.97}$$

The solution of equation (2.97) is just the function $B(\tau, \eta, \zeta)$ given by equation (2.4). Substituting (2.4) into (2.97), we obtain the identity

$$b(\eta)e^{-k\tau} = \frac{\lambda}{2} \int_0^1 p(\eta, \eta')\, d\eta' \int_0^\tau b(\eta')e^{-k\tau'-(\tau-\tau')/\eta'}\, \frac{d\tau'}{\eta'}$$

$$+ \frac{\lambda}{2} \int_0^1 p(\eta, -\eta')\, d\eta' \int_\tau^\infty b(-\eta')e^{-k\tau'-(\tau'-\tau)/\eta'}\, \frac{d\tau'}{\eta'} + \frac{\lambda}{2} \int_0^1 p(\eta, \eta')e^{-\tau/\eta'}i(\eta')\, d\eta'. \quad (2.98)$$

Comparing (2.98) with (2.92) and using the superposition theorem for solutions of linear equations, we find

$$b(\eta)e^{-k\tau} = \frac{2}{S} \int_0^1 B(\tau, \eta, \eta')i(\eta')\, d\eta'. \quad (2.99)$$

Equation (2.99) is valid for any τ. Substituting equation (2.4) into it and using equation (2.96), we obtain

$$2C \int_0^1 u(\zeta)i(\zeta)\zeta\, d\zeta = 1. \quad (2.100)$$

Since we have stipulated that the function $u(\zeta)$ is normalized according to equation (2.78), it follows from relation (2.100) that $C = 1$. This determines the function $c(\zeta)$ completely. Substitution of equation (2.96) with $C = 1$ into equations (2.4) and (2.5) gives

$$B(\tau, \eta, \zeta) = Su(\zeta)\zeta b(\eta)e^{-k\tau}, \quad (2.101)$$
$$I(\tau, \eta, \zeta) = Su(\zeta)\zeta i(\eta)e^{-k\tau}, \quad (2.102)$$

where $b(\eta)$ and $i(\eta)$ are related by equation (2.6). Equations (2.101) and (2.102) give the complete solution to the problem of finding the radiation field in deep layers of the medium. We note that it follows from (2.102) that

$$\frac{I(\tau, \eta, \zeta_2)}{I(\tau, \eta, \zeta_1)} = \frac{u(\zeta_2)\zeta_2}{u(\zeta_1)\zeta_1}. \quad (2.103)$$

This equation relates the relative intensity of radiation at a given place and in a given direction for various angles of incidence of the external radiation to the relative intensity of the radiation diffusely transmitted in different directions. With the aid of equation (2.103), one of these quantities may be found if the other has been measured.

2.5. The Atmospheric Albedo for Small True Absorption

A particularly important case for practical applications occurs when the coefficient of true absorption is much less than the scattering coefficient; that is, when $1 - \lambda \ll 1$. We have previously obtained (Section 2.1) asymptotic formulas which determine the radiation intensity in the deep layers with increasing accuracy for decreasing values of $1 - \lambda$. Similar formulas may be obtained for other quantities characterizing the radiation field in a semi-infinite atmosphere (albedo, reflection coefficient, etc.). Such formulas express these quantities for small true absorption in terms of the corresponding quantities for the case of pure scattering ($\lambda = 1$).

In the present section we shall find asymptotic formulas for the albedo of a semi-infinite atmosphere. Similar results for other characteristics of the radiation field will be obtained in the following section. The derivation of these formulas is based on the asymptotic relations (2.26) and (2.28) for the radiation intensity in the deep layers of the medium, and on equations (2.76) and (2.77) which relate this intensity to the reflection and transmission coefficients.

The plane albedo $A(\zeta)$ may be found from equation (1.86) if the reflection coefficient $\varrho(\eta, \zeta)$ is known. As with equation (2.26), which may be rewritten in the form

$$i(\eta) = 1 + \frac{3k}{3 - x_1}\,\eta, \tag{2.104}$$

we may express $\varrho(\eta, \zeta)$ in the form

$$\varrho(\eta, \zeta) = \varrho_0(\eta, \zeta) - k\varrho_1(\eta, \zeta). \tag{2.105}$$

In these expressions k is determined by equation (2.24) and $\varrho_0(\eta, \zeta)$ is the reflection coefficient for the case of pure scattering ($k = 0$).

We recall that for pure scattering the albedo $A(\zeta)$ is unity, so that

$$2 \int_0^1 \varrho_0(\eta, \zeta)\eta\, d\eta = 1. \tag{2.106}$$

Then, substituting (2.105) into (1.86), we obtain

$$A(\zeta) = 1 - 2k \int_0^1 \varrho_1(\eta, \zeta)\eta\, d\eta. \tag{2.107}$$

We may now make use of equation (2.76). Introducing into it equations (2.104) and (2.105) and taking account of (2.106) and (2.107), we find

$$A(\zeta) = 1 - \frac{3k}{3 - x_1}\left[\zeta + 2\int_0^1 \varrho_0(\eta, \zeta)\,\eta^2\, d\eta\right]. \tag{2.108}$$

This equation gives the desired expression for the plane albedo in terms of the function $\varrho_0(\eta, \zeta)$.

The plane albedo may, however, also be expressed in terms of the quantity $u_0(\eta)$, which is the transmission function $u(\eta)$ for the case $k = 0$. In order to show this, we use equation (2.77). It follows from (2.80) and (2.104) that the constant M entering this equation is, for small k, just

$$M = \frac{8k}{3 - x_1}. \tag{2.109}$$

We then find from (2.77) and (2.104)–(2.108):

$$u_0(\zeta) = \tfrac{3}{4}\left[\zeta + 2\int_0^1 \varrho_0(\eta,\,\zeta)\,\eta^2\,d\eta\right]. \tag{2.110}$$

This important equation establishes for the case of pure scattering the relation between the reflection coefficient and the function $u_0(\zeta)$ characterizing the transmission.

Comparison of (2.108) with (2.110) then gives the following simple formula for the plane albedo in terms of the function $u_0(\zeta)$:

$$A(\zeta) = 1 - \frac{4k}{3-x_1}\,u_0(\zeta). \tag{2.111}$$

Since $i(\eta) = 1$ in the case of pure scattering, the normalization condition (2.78) has the form

$$2\int_0^1 u_0(\eta)\eta\,d\eta = 1. \tag{2.112}$$

Once the plane albedo $A(\zeta)$ is known, we may determine the spherical albedo A_s from equation (1.87). Substituting equation (2.111) into (1.87) and using (2.112), we obtain

$$A_s = 1 - \frac{4k}{3-x_1}. \tag{2.113}$$

This expression may be rewritten with the aid of (2.24) as

$$A_s = 1 - 4\sqrt{\frac{1-\lambda}{3-x_1}}. \tag{2.114}$$

Thus, calculation of the spherical albedo for small values of $1-\lambda$ does not require knowledge of any auxiliary functions.

The asymptotic formulas which we have obtained for the quantities $A(\zeta)$ and A_s represent the zeroth- and first-order terms in their expansions in powers of $\sqrt{1-\lambda}$. It is possible to find also the next term in each of these expansions, which is of order $1-\lambda$. For this purpose we may again use equations (2.76) and (2.77), into which it is necessary to substitute equation (2.28) for $i(\eta)$, equation (2.139) for $u(\eta)$ (derived below), and the expression for $\varrho(\eta,\,\zeta)$ corresponding to (2.105) but with terms of order k^2. After a short computation (for more details see [11]), we obtain

$$A(\zeta) = 1 - 4\sqrt{\frac{1-\lambda}{3-x_1}}\,u_0(\zeta) + \left[\frac{15}{5-x_2}\,v_0(\zeta) + \frac{D}{3-x_1}\,u_0(\zeta)\right](1-\lambda), \tag{2.115}$$

where

$$D = 24\int_0^1 u_0(\eta)\eta^2\,d\eta \tag{2.116}$$

and

$$v_0(\zeta) = \zeta^2 - 2\int_0^1 \varrho_0(\eta,\,\zeta)\eta^3\,d\eta. \tag{2.117}$$

It is easy to see that the function $v_0(\zeta)$ satisfies the condition

$$\int_0^1 v_0(\zeta)\zeta \, d\zeta = 0. \tag{2.118}$$

Substituting equation (2.115) into (1.87) and using (2.112) and (2.118), we find the following asymptotic formula for the spherical albedo:

$$A_s = 1 - 4\sqrt{\left(\frac{1-\lambda}{3-x_1}\right)} + D\frac{1-\lambda}{3-x_1}. \tag{2.119}$$

We see that, if we include only the first two terms in the expansion in powers of $\sqrt{1-\lambda}$, the spherical albedo does not depend on the entire phase function but only on the parameter x_1. This property is approximately true even to order $1-\lambda$, since the quantity D depends only weakly on the phase function. We may in practice use the value of this quantity for isotropic scattering, $D = 8.5$.

Table 2.4 gives as an example values of the plane albedo $A(\zeta)$ for the phase function $x(\gamma) = 1 + \cos\gamma + P_2(\cos\gamma)$ for $\lambda = 0.99$. The approximate values were found first from equation (2.111) and then from equation (2.115), and the exact values from the results obtained in Section 7.4.

TABLE 2.4. VALUES OF THE ALBEDO $A(\zeta)$ FOR THE PHASE FUNCTION
$x(\gamma) = 1 + \cos\gamma + P_2(\cos\gamma)$ and $\lambda = 0.99$

ζ	Approximate		Exact	ζ	Approximate		Exact
0	0.886	0.889	0.889	0.6	0.731	0.767	0.764
0.1	0.854	0.860	0.860	0.7	0.709	0.752	0.748
0.2	0.827	0.838	0.838	0.8	0.686	0.738	0.732
0.3	0.803	0.819	0.818	0.9	0.663	0.725	0.717
0.4	0.779	0.800	0.799	1.0	0.641	0.712	0.702
0.5	0.755	0.783	0.781				

It is clear from Table 2.4 that inclusion of the term of order $1-\lambda$ significantly increases the precision of the determination of $A(\zeta)$ and A_s. For very small values of $1-\lambda$, however, it is sufficient to use equations (2.111) and (2.114) to determine these quantities.

2.6. Other Quantities in the Case of Small True Absorption

In the preceding section we obtained asymptotic expressions for the atmospheric albedo in the case of small true absorption. We shall now find corresponding expressions for other quantities: the reflection coefficient $\varrho(\eta, \zeta)$, the function $B(0, \eta, \zeta)$, the transmission function $u(\eta)$, and the auxiliary functions $\varphi_i(\eta)$. In this process we shall expand the desired quantities in powers of $\sqrt{1-\lambda}$ and retain only terms of zeroth and first order.

Reflection coefficient

Setting $m = 0$ in (2.35) and (2.36), we obtain the following system of equations for the determination of $\varrho(\eta, \zeta)$ and $B(0, \eta, \zeta)$:

$$S(\eta+\zeta)\varrho(\eta, \zeta) = B(0, -\eta, \zeta)+2\eta \int_0^1 B(0, \eta', \zeta)\varrho(\eta, \eta')\, d\eta', \qquad (2.120)$$

$$B(0, \eta, \zeta) = \frac{\lambda}{4} S\left[p(\eta, \zeta)+2\zeta \int_0^1 p(\eta, -\eta')\, \varrho(\eta', \zeta)\, d\eta'\right]. \qquad (2.121)$$

We shall represent the function $\varrho(\eta, \zeta)$ by equation (2.105), and the function $B(0, \eta, \zeta)$ by the expression

$$B(0, \eta, \zeta) = B_0(0, \eta, \zeta)-kB_1(0, \eta, \zeta). \qquad (2.122)$$

Our problem is to determine $\varrho_1(\eta, \zeta)$ and $B_1(0, \eta, \zeta)$ in terms of $\varrho_0(\eta, \zeta)$ and $B_0(0, \eta, \zeta)$ (or in terms of other quantities which are known for $\lambda = 1$).

Substituting equations (2.105) and (2.122) into (2.120) and (2.121) and recalling that $k \ll 1$, we find

$$S(\eta+\zeta)\varrho_1(\eta, \zeta) = B_1(0, -\eta, \zeta)+2\eta \int_0^1 [\varrho_0(\eta, \eta')\, B_1(0, \eta', \zeta)+\varrho_1(\eta, \eta')\, B_0(0, \eta', \zeta)]\, d\eta',$$
$$\qquad (2.123)$$

$$B_1(0, \eta, \zeta) = \tfrac{1}{2}S\zeta \int_0^1 p(\eta, -\eta')\varrho_1(\eta', \zeta)\, d\eta'. \qquad (2.124)$$

Substituting (2.124) into (2.123) and using (2.121) with $\lambda = 1$, we obtain

$$S(\eta+\zeta)\varrho_1(\eta, \zeta) = 2\eta \int_0^1 B_0(0, \eta', \zeta)\, \varrho_1(\eta, \eta')\, d\eta'+2\zeta \int_0^1 B_0(0, \eta', \eta)\varrho_1(\eta', \zeta)\, d\eta'. \quad (2.125)$$

On the other hand, setting $k = 0$ in (2.65) and recalling (2.66), we have

$$Su_0(\eta) = 2 \int_0^1 u_0(\eta')B_0(0, \eta', \eta)\, d\eta'. \qquad (2.126)$$

Comparison of (2.125) and (2.126) shows that the function $\varrho_1(\eta, \zeta)$ has the form

$$\varrho_1(\eta, \zeta) = C_1 u_0(\eta)u_0(\zeta), \qquad (2.127)$$

where C_1 is some constant. In order to find C_1, we substitute (2.127) into (2.107). With the aid of (2.111) and (2.112), we then find

$$C_1 = \frac{4}{3-x_1}. \qquad (2.128)$$

In this way the reflection coefficient $\varrho(\eta, \zeta)$ for small k is determined to be

$$\varrho(\eta, \zeta) = \varrho_0(\eta, \zeta)-\frac{4k}{3-x_1} u_0(\eta)\, u_0(\zeta). \qquad (2.129)$$

Equations (2.129) and (2.110) allow us to find the function $\varrho(\eta, \zeta)$ if we know only the single function $\varrho_0(\eta, \zeta)$.

5

The function $B(0, \eta, \zeta)$

The function $B(0, \eta, \zeta)$ for small k may be easily determined after the quantity $\varrho_1(\eta, \zeta)$ has been found. Substituting (2.127) into (2.124) we obtain

$$B_1(0, \eta, \zeta) = \frac{2}{3-x_1} S\zeta u_0(\zeta) \int_0^1 p(\eta, -\eta')u_0(\eta')\, d\eta'. \tag{2.130}$$

But it follows from (2.75) that

$$B_0(0, \eta) = \frac{1}{2} \int_0^1 p(\eta, -\eta')u_0(\eta')\, d\eta', \tag{2.131}$$

where $B_0(0, \eta)$ is the value of the source function at the boundary for the Milne problem with $\lambda = 1$. Thus, introducing (2.130) into (2.122) and making use of (2.131), we find

$$B(0, \eta, \zeta) = B_0(0, \eta, \zeta) - \frac{4k}{3-x_1} S\zeta u_0(\zeta)B_0(0, \eta). \tag{2.132}$$

The fact that $B_0(0, \eta)$ may be expressed quite simply in terms of $B_0(0, \eta, \zeta)$ deserves attention. In order to obtain this expression, we must substitute (2.110) into (2.131) and then use (2.121). This yields the result

$$4B_0(0, \eta) = \frac{6}{S} \int_0^1 B_0(0, \eta, \zeta)\zeta\, d\zeta - x_1\eta. \tag{2.133}$$

Transmission function

In analogy with the quantities discussed above, we write the function $u(\eta)$ as

$$u(\eta) = u_0(\eta) - ku_1(\eta). \tag{2.134}$$

In order to determine $u_1(\eta)$, we use the relation

$$Su(\eta)(1-k\eta) = 2 \int_0^1 u(\eta')B(0, \eta', \eta)\, d\eta', \tag{2.135}$$

which may be derived from (2.65) and (2.66). Substitution of equations (2.132) and (2.134) into (2.135) then yields

$$S[\eta u_0(\eta) + u_1(\eta)] = 2 \int_0^1 u_1(\eta')\, B_0(0, \eta', \eta)\, d\eta' + \frac{8}{3-x_1} S\eta u_0(\eta) \int_0^1 u_0(\eta')B_0(0, \eta')\, d\eta'. \tag{2.136}$$

Introducing (2.133) into (2.122) and then utilizing (2.126) and (2.112), it is easy to show that

$$\frac{8}{3-x_1} \int_0^1 u_0(\eta')B_0(0, \eta')\, d\eta' = 1. \tag{2.137}$$

Making use of (2.137), we may write (2.136) as

$$Su_1(\eta) = 2 \int_0^1 u_1(\eta')B_0(0, \eta', \eta) \, d\eta'. \tag{2.138}$$

By comparing equations (2.126) and (2.138), we conclude that $u_1(\eta)$ is proportional to $u_0(\eta)$. Consequently,

$$u(\eta) = u_0(\eta)(1 - k D_1), \tag{2.139}$$

where D_1 is some constant. This constant is determined by the normalization condition (2.78). Substituting equations (2.104) and (2.139) into equation (2.78), we find

$$D_1 = \frac{6}{3 - x_1} \int_0^1 u_0(\eta)\eta^2 \, d\eta. \tag{2.140}$$

We may draw the following conclusion from equation (2.139): the radiation intensity diffusely transmitted by a semi-infinite atmosphere in the case of small true absorption has the same relative angular distribution as in the case of pure scattering. This conclusion holds to terms of order k^2.

Because of the importance of the function $u_0(\eta)$, we shall derive expressions for it in terms of the auxiliary functions $\varphi_{i0}(\eta)$, which we here define as the functions $\varphi_i(\eta)$ for the case $k = 0$. Substituting equation (2.83) for $k = 0$ into (2.110), we find

$$u_0(\eta) = \frac{3}{4}\left[\eta + \frac{1}{2}\sum_{i=0}^{n} x_i(-1)^i \alpha_{i1}\varphi_{i0}(\eta) - 2\eta \int_0^1 \varrho_0(\eta, \zeta)\zeta \, d\zeta\right], \tag{2.141}$$

where we have set

$$\alpha_{i1} = \int_0^1 \varphi_{i0}(\zeta)\zeta \, d\zeta. \tag{2.142}$$

Using (2.106) we may rewrite (2.141) as

$$u_0(\eta) = \frac{3}{8}\sum_{i=0}^{n} x_i(-1)^i \alpha_{i1}\varphi_{i0}(\eta). \tag{2.143}$$

We may obtain a much simpler expression for $u_0(\eta)$ if we compare equations (2.110) and (2.84) for $i = 0$ and $i = 2$. From this comparison it follows that

$$2\eta u_0(\eta) = \tfrac{1}{2}\varphi_{00}(\eta) + \varphi_{20}(\eta). \tag{2.144}$$

The functions $\varphi_i(\eta)$

Formulas for the auxiliary functions $\varphi_i(\eta)$ for small k may be obtained with the aid of the relation

$$B(0, \eta, \zeta) = \frac{\lambda}{4}S\sum_{i=0}^{n} x_i P_i(\eta)\varphi_i(\zeta), \tag{2.145}$$

which follows from (2.38) for $m = 0$. Substituting (2.145) for $k = 0$ into (2.133) gives

$$4B_0(0, \eta) = \tfrac{3}{2}\sum_{i=0}^{n} x_i \alpha_{i1} P_i(\eta) - x_1\eta. \tag{2.146}$$

Introducing (2.145) and (2.146) into (2.132), we find

$$\varphi_i(\eta) = \varphi_{i0}(\eta) - \frac{6k}{3-x_1}\alpha_{i1}u_0(\eta)\eta \qquad (i \neq 1), \tag{2.147}$$

$$\varphi_1(\eta) = \frac{4k}{3-x_1}u_0(\eta)\eta. \tag{2.148}$$

In the derivation of equation (2.148) we have used the fact that always

$$\varphi_{10}(\eta) = 0, \tag{2.149}$$

as follows from (2.84) and (2.106).

Let us apply the equations which we have obtained to the two simplest phase functions. In the case of isotropic scattering, there is only one auxiliary function $\varphi_0(\eta)$. For pure scattering we denote it by $\varphi_{00}(\eta)$. The first moment of $\varphi_{00}(\eta)$ is just $\alpha_{01} = 2/\sqrt{3}$, so that on the basis of (2.143) we have

$$u_0(\eta) = \frac{\sqrt{3}}{4}\varphi_{00}(\eta). \tag{2.150}$$

In consequence, equation (2.147) takes the form

$$\varphi_0(\eta) = \varphi_{00}(\eta)\left[1 - \eta\sqrt{3(1-\lambda)}\right]. \tag{2.151}$$

For the simplest anisotropic phase function $x(\gamma) = 1 + x_1\cos\gamma$, the radiation intensity averaged over azimuth is expressed in terms of the functions $\varphi_0(\eta)$ and $\varphi_1(\eta)$. For pure scattering $\varphi_{10}(\eta) = 0$ and $\varphi_{00}(\eta)$ has the same value as for isotropic scattering. Equation (2.150) thus remains valid. Substituting it and (2.24) into equations (2.147) and (2.148), we obtain

$$\varphi_0(\eta) = \varphi_{00}(\eta)\left[1 - 3\eta\sqrt{\frac{1-\lambda}{3-x_1}}\right], \tag{2.152}$$

$$\varphi_1(\eta) = \varphi_{00}(\eta)\eta\sqrt{\frac{3(1-\lambda)}{3-x_1}}. \tag{2.153}$$

As we have previously stated, both the intensity emerging from the atmosphere for the case of external illumination and the intensity due to internal energy sources are expressed in terms of the functions $\varphi_i(\eta)$. Equations (2.147) and (2.148) are thus important in both these cases.

The derivation of the asymptotic expressions given in the last two sections was based on the work of V. V. Sobolev [12]. Equation (2.111) for the plane albedo and equation (2.129) for the reflection coefficient, as well as a number of other relations, have been obtained by van de Hulst from physical considerations (cf. reference following Chapter 3).

Recently, van de Hulst [13] has examined the problem of transforming from an expansion of the reflection coefficient in powers of k to its expansion in powers of $\sqrt{1-\lambda}$, and hence to an expansion in powers of λ. He found an expression for the coefficient of the λ^n term with a relative error of order n^{-2}. It is apparent that the term containing λ^n represents scattering of the nth order. Determining the role of various orders of scattering is of considerable interest for certain problems.

References

1. V. A. AMBARTSUMYAN, A new method for computing light scattering in turbid media, *Izv. Akad. Nauk SSSR, ser. geogr. i geofiz.* **3**, 97 (1942).
2. V. V. SOBOLEV, On the light regime in deep layers of a turbid medium, *Izv. Akad. Nauk SSSR, ser. geogr. i geofiz.* **8**, 273 (1944).
3. L. M. ROMANOVA, Application of Kellog's method for the computation of the radiation intensity in the depths of a scattering and absorbing medium, *Izv. Akad. Nauk SSSR, ser. geofiz.*, No. 9, 1294 (1962) [*Bull. (Izv.) Acad. Sci. USSR, Geophys. Ser.*, No. 9, 807 (1962)].
4. V. M. LOSKUTOV, The light regime in deep layers of a turbid medium for a strongly elongated phase function, *Vest. Leningrad Univ.*, No. 13 (1969).
5. M. V. MASLENNIKOV, The Milne problem with anisotropic scattering, *Trudy Matem. Inst. Akad. Nauk SSSR*, No. 97 (1968).
6. V. A. AMBARTSUMYAN, On the problem of diffuse reflection of light, *J. Phys. (USSR)* **8**, 65 (1944).
7. V. V. SOBOLEV, On certain functions in the theory of light scattering, *Astrofiz.* **3**, 433 (1967) [*Astrophysics* **3**, 205 (1967)].
8. I. N. MININ, A. G. PILIPOSYAN and N. A. SHIDLOVSKAYA, Tables of the Ambartsumyan functions for anisotropic scattering, *Trudy Astron. Obs. Leningrad Gos. Univ.* **20** (1963).
9. I. N. MININ, Diffuse reflection from a semi-infinite medium for anisotropic scattering. I, *Vest. Leningrad Univ.*, No. 1 (1961).
10. V. A. AMBARTSUMYAN, Diffusion of light through a scattering medium of large optical thickness, *Doklady Akad. Nauk SSSR* **43**, 102 (1944).
11. V. V. SOBOLEV, The albedo of a planetary atmosphere, *Doklady Akad. Nauk SSSR* **184,** 318 (1969) [*Sov. Phys.—Doklady* **14,** 1 (1969)].
12. V. V. SOBOLEV, Anisotropic scattering of light in a semi-infinite atmosphere. I, *Astron. Zh.* **45,** 254 (1968) [*Sov. Astron.—A.J.* **12,** 202 (1968)].
13. H. C. VAN DE HULST, High-order scattering in diffuse reflection from a semi-infinite atmosphere, *Astronomy and Astrophysics* **9**, 374 (1971).

Chapter 3

ATMOSPHERES OF FINITE OPTICAL THICKNESS

IN THIS chapter we shall examine the problem o diff use reflection and transmission of light by an atmosphere of finite optical thickness τ_0. It is very important to note that this problem may be solved without explicit knowledge of the source function (to whose determination we shall return in Chapter 6).

We may obtain equations of two types for the determination of the reflection and transmission coefficients. We shall first give equations which include an integration over angle. They represent a generalization of those equations which were considered in the previous chapter for the case $\tau_0 = \infty$. Next we derive equations containing derivatives with respect to the optical thickness τ_0. These equations solve the problem by considering a series of atmospheres with gradually increasing τ_0.

Special attention is given to the case of atmospheres with large optical thickness ($\tau_0 \gg 1$), which is of particular practical interest. We obtain in this case asymptotic formulas for the reflection and transmission coefficients and also for the auxiliary functions in terms of which these coefficients may be expressed. In this process we assume that the corresponding quantities for a semi-infinite atmosphere are known. The resulting equations become more exact as τ_0 increases. At the end of the chapter we briefly examine the problem of the radiation from an inhomogeneous atmosphere in which the single scattering albedo λ and the phase function $x(\gamma)$ depend on optical depth τ.

3.1. Diffuse Reflection and Transmission of Light

Let us consider a plane-parallel atmosphere of optical thickness τ_0 illuminated by parallel radiation incident at an angle arccos ζ to the normal. We set the flux through a horizontal area at the surface equal to $\pi S\zeta$. We wish to find the intensity of radiation diffusely reflected and diffusely transmitted by the atmosphere.

As previously, we shall assume that the phase function may be expanded as a finite sum of Legendre polynomials. The source function B and the radiation intensity I are then given by equations (1.43) and (1.44). The coefficients B^m which enter the first of these equations are determined by the integral equations (1.53), which may be rewritten in the form

$$B^m(\tau, \eta, \zeta, \tau_0) = \frac{\lambda}{2} \int_0^1 p^m(\eta, \eta') \, d\eta' \int_0^\tau B^m(\tau', \eta', \zeta, \tau_0) e^{-(\tau-\tau')/\eta'} \frac{d\tau'}{\eta'} +$$

$$+\frac{\lambda}{2}\int_0^1 p^m(\eta, -\eta')\, d\eta' \int_\tau^{\tau_0} B^m(\tau', -\eta', \zeta, \tau_0)e^{-(\tau'-\tau)/\eta'}\frac{d\tau'}{\eta'}+\frac{\lambda}{4}Sp^m(\eta, \zeta)e^{-\tau/\zeta}, \quad (3.1)$$

where we have indicated explicitly the dependence of B^m on τ_0.

If the functions B^m have been found, then it is easy to determine the radiation intensity at arbitrary optical depth. In particular, the two quantities of interest to us, the reflection coefficient ϱ and transmission coefficient σ, may be obtained. These quantities are related to the intensity emerging from the atmosphere by equations (1.74) and (1.75). They may be expanded in a form similar to that used for the intensity, so that

$$\varrho(\eta, \zeta, \varphi, \tau_0) = \varrho^0(\eta, \zeta, \tau_0)+2\sum_{m=1}^n \varrho^m(\eta, \zeta, \tau_0)\cos m\varphi, \quad (3.2)$$

$$\sigma(\eta, \zeta, \varphi, \tau_0) = \sigma^0(\eta, \zeta, \tau_0)+2\sum_{m=1}^n \sigma^m(\eta, \zeta, \tau_0)\cos m\varphi. \quad (3.3)$$

The coefficients ϱ^m and σ^m are determined by the equations

$$S\varrho^m(\eta, \zeta, \tau_0)\,\zeta = \int_0^{\tau_0} B^m(\tau, -\eta, \zeta, \tau_0)e^{-\tau/\eta}\frac{d\tau}{\eta}, \quad (3.4)$$

$$S\sigma^m(\eta, \zeta, \tau_0)\zeta = \int_0^{\tau_0} B^m(\tau_0-\tau, \eta, \zeta, \tau_0)e^{-\tau/\eta}\frac{d\tau}{\eta}. \quad (3.5)$$

It is possible, however, to obtain equations which determine ϱ^m and σ^m directly, without use of the function B^m. We shall now do this, using the same procedure as in the solution of the problem of diffuse reflection by a semi-infinite atmosphere (see Section 2.2).

Differentiating (3.1) with respect to τ, we find

$$\frac{dB^m(\tau, \eta, \zeta, \tau_0)}{d\tau} = \frac{\lambda}{2}\int_0^1 p^m(\eta, \eta')\, d\eta' \int_0^\tau \frac{dB^m(\tau', \eta', \zeta, \tau_0)}{d\tau'}e^{-(\tau-\tau')/\eta'}\frac{d\tau'}{\eta'}$$

$$+\frac{\lambda}{2}\int_0^1 p^m(\eta, -\eta')\, d\eta' \int_\tau^{\tau_0} \frac{dB^m(\tau', -\eta', \zeta, \tau_0)}{d\tau'}e^{-(\tau'-\tau)/\eta'}\frac{d\tau'}{\eta'}$$

$$+\frac{\lambda}{2}\int_0^1 p^m(\eta, \eta')B^m(0, \eta', \zeta, \tau_0)e^{-\tau/\eta}\frac{d\eta'}{\eta'}$$

$$-\frac{\lambda}{2}\int_0^1 p^m(\eta, -\eta')B^m(\tau_0, -\eta', \zeta, \tau_0)e^{-(\tau_0-\tau)/\eta}\frac{d\eta'}{\eta'}-\frac{\lambda}{4\zeta}Sp^m(\eta, \zeta)e^{-\tau/\zeta}. \quad (3.6)$$

Comparing (3.6) with (3.1) we see that

$$\frac{dB^m(\tau, \eta, \zeta, \tau_0)}{d\tau} = -\frac{1}{\zeta} B^m(\tau, \eta, \zeta, \tau_0) + \frac{2}{S} \int_0^1 B^m(0, \eta', \zeta, \tau_0) B^m(\tau, \eta, \eta', \tau_0) \frac{d\eta'}{\eta'}$$

$$-\frac{2}{S} \int_0^1 B^m(\tau_0, -\eta', \zeta, \tau_0) B^m(\tau_0-\tau, -\eta, \eta', \tau_0) \frac{d\eta'}{\eta'}. \qquad (3.7)$$

In equation (3.7) we now set η equal to $-\eta$, multiply by $e^{-\tau/\eta}$, and integrate over τ between 0 and τ_0. Using equations (3.4) and (3.5), we obtain

$$S\varrho^m(\eta, \zeta, \tau_0)(\eta + \zeta) = B^m(0, -\eta, \zeta, \tau_0)$$

$$+ 2\eta \int_0^1 B^m(0, \eta', \zeta, \tau_0) \varrho^m(\eta, \eta', \tau_0) \, d\eta' - B^m(\tau_0, -\eta, \zeta, \tau_0) e^{-\tau_0/\eta}$$

$$- 2\eta \int_0^1 B^m(\tau_0, -\eta', \zeta, \tau_0) \, \sigma^m(\eta, \eta', \tau_0) \, d\eta'. \qquad (3.8)$$

In the same manner, we multiply (3.7) by $e^{-(\tau_0-\tau)/\eta}$, integrate, and find

$$S\sigma^m(\eta, \zeta, \tau_0)(\eta - \zeta) = B^m(0, \eta, \zeta, \tau_0) e^{-\tau_0/\eta}$$

$$+ 2\eta \int_0^1 B^m(0, \eta', \zeta, \tau_0) \, \sigma^m(\eta, \eta', \tau_0) \, d\eta' - B^m(\tau_0, \eta, \zeta, \tau_0) \qquad (3.9)$$

$$- 2\eta \int_0^1 B^m(\tau_0, -\eta', \zeta, \tau_0) \varrho^m(\eta, \eta', \tau_0) \, d\eta'.$$

On the other hand, setting in equation (3.1) first $\tau = 0$ and then $\tau = \tau_0$ we obtain

$$B^m(0, \eta, \zeta, \tau_0) = \frac{\lambda}{4} S \left[p^m(\eta, \zeta) + 2\zeta \int_0^1 p^m(\eta, -\eta') \varrho^m(\eta', \zeta, \tau_0) \, d\eta' \right], \qquad (3.10)$$

$$B^m(\tau_0, \eta, \zeta, \tau_0) = \frac{\lambda}{4} S \left[p^m(\eta, \zeta) e^{-\tau_0/\zeta} + 2\zeta \int_0^1 p^m(\eta, \eta') \sigma^m(\eta', \zeta, \tau_0) \, d\eta' \right]. \qquad (3.11)$$

The relations (3.8)–(3.11) may be considered as a system of four equations for the determination of four unknown functions: $\varrho^m(\eta, \zeta, \tau_0)$, $\sigma^m(\eta, \zeta, \tau_0)$, $B^m(0, \eta, \zeta, \tau_0)$ and $B^m(\tau_0, \eta, \zeta, \tau_0)$. These equations are the generalization to an atmosphere of finite optical thickness of equations (2.35) and (2.36) which were obtained previously for a semi-infinite medium.

The optical thickness τ_0 enters equations (3.8)–(3.11) as a parameter. The functions entering these equations thus depend only on two variables (η and ζ). It is possible, moreover, to express these functions in terms of other functions which depend on only a single variable, and to obtain equations for the determination of these latter functions.

With this goal in mind, let us use equation (1.41) for the quantity $p^m(\eta, \zeta)$, and substitute this into equations (3.10) and (3.11). We find

$$B^m(0, \eta, \zeta, \tau_0) = \frac{\lambda}{4} S \sum_{i=m}^{n} c_i^m P_i^m(\eta) \varphi_i^m(\zeta, \tau_0), \tag{3.12}$$

$$B^m(\tau_0, \eta, \zeta, \tau_0) = \frac{\lambda}{4} S \sum_{i=m}^{n} c_i^m P_i^m(\eta) \psi_i^m(\zeta, \tau_0), \tag{3.13}$$

where we have set

$$\varphi_i^m(\zeta, \tau_0) = P_i^m(\zeta) + 2\zeta \int_0^1 P_i^m(-\eta) \varrho^m(\eta, \zeta, \tau_0)\, d\eta, \tag{3.14}$$

$$\psi_i^m(\zeta, \tau_0) = P_i^m(\zeta) e^{-\tau_0/\zeta} + 2\zeta \int_0^1 P_i^m(\eta) \sigma^m(\eta, \zeta, \tau_0)\, d\eta. \tag{3.15}$$

Substituting (3.12) and (3.13) into equations (3.8) and (3.9) and using (3.14) and (3.15) we obtain

$$\varrho^m(\eta, \zeta, \tau_0) = \frac{\lambda}{4} \sum_{i=m}^{n} c_i^m (-1)^{i-m} \frac{\varphi_i^m(\eta, \tau_0)\varphi_i^m(\zeta, \tau_0) - \psi_i^m(\eta, \tau_0)\psi_i^m(\zeta, \tau_0)}{\eta + \zeta}, \tag{3.16}$$

$$\sigma^m(\eta, \zeta, \tau_0) = \frac{\lambda}{4} \sum_{i=m}^{n} c_i^m \frac{\varphi_i^m(\zeta, \tau_0)\psi_i^m(\eta, \tau_0) - \varphi_i^m(\eta, \tau_0)\psi_i^m(\zeta, \tau_0)}{\eta - \zeta}. \tag{3.17}$$

Equations (3.16) and (3.17) thus express the reflection and transmission coefficients in terms of the auxiliary functions $\varphi_i^m(\eta, \tau_0)$ and $\psi_i^m(\eta, \tau_0)$. In order to find equations determining the auxiliary functions, it is sufficient to substitute (3.16) and (3.17) into (3.14) and (3.15). We obtain

$$\varphi_i^m(\zeta, \tau_0) = P_i^m(\zeta) + \frac{\lambda}{2} \zeta \sum_{j=m}^{n} c_j^m (-1)^{j+m}$$

$$\times \int_0^1 P_i^m(-\eta) \frac{\varphi_j^m(\eta, \tau_0)\varphi_j^m(\zeta, \tau_0) - \psi_j^m(\eta, \tau_0)\psi_j^m(\zeta, \tau_0)}{\eta + \zeta}\, d\eta, \tag{3.18}$$

$$\psi_i^m(\zeta, \tau_0) = P_i^m(\zeta) e^{-\tau_0/\zeta} + \frac{\lambda}{2} \zeta \sum_{j=m}^{n} c_j^m \int_0^1 P_i^m(\eta) \frac{\varphi_j^m(\zeta, \tau_0)\psi_j^m(\eta, \tau_0) - \varphi_j^m(\eta, \tau_0)\psi_j^m(\zeta, \tau_0)}{\eta - \zeta}\, d\eta. \tag{3.19}$$

Thus, to find the quantities $\varrho^m(\eta, \zeta, \tau_0)$ and $\sigma^m(\eta, \zeta, \tau_0)$ it is necessary to determine $2(n-m+1)$ auxiliary functions from equations (3.18) and (3.19). Once this has been done, the complete reflection and transmission coefficients are found from equations (3.2) and (3.3). The total number of auxiliary functions for given n is $(n+1)(n+2)$.

In the case of isotropic scattering, solution of the problem of diffuse reflection and transmission requires knowledge of two auxiliary functions $\varphi_0^0(\eta, \tau_0)$ and $\psi_0^0(\eta, \tau_0)$. We shall designate these functions by $\varphi(\eta, \tau_0)$ and $\psi(\eta, \tau_0)$ for simplicity. According to equations

(3.16) and (3.17), the reflection and transmission coefficients in this case are just

$$\varrho(\eta, \zeta, \tau_0) = \frac{\lambda}{4} \frac{\varphi(\eta, \tau_0)\varphi(\zeta, \tau_0) - \psi(\eta, \tau_0)\psi(\zeta, \tau_0)}{\eta + \zeta},\tag{3.20}$$

$$\sigma(\eta, \zeta, \tau_0) = \frac{\lambda}{4} \frac{\varphi(\zeta, \tau_0)\psi(\eta, \tau_0) - \varphi(\eta, \tau_0)\psi(\zeta, \tau_0)}{\eta - \zeta},\tag{3.21}$$

where the functions $\varphi(\eta, \tau_0)$ and $\psi(\eta, \tau_0)$ are, according to (3.18) and (3.19), determined by the equations

$$\varphi(\zeta, \tau_0) = 1 + \frac{\lambda}{2}\zeta \int_0^1 \frac{\varphi(\eta, \tau_0)\,\varphi(\zeta, \tau_0) - \psi(\eta, \tau_0)\psi(\zeta, \tau_0)}{\eta + \zeta}\,d\eta,\tag{3.22}$$

$$\psi(\zeta, \tau_0) = e^{-\tau_0/\zeta} + \frac{\lambda}{2}\zeta \int_0^1 \frac{\varphi(\zeta, \tau_0)\psi(\eta, \tau_0) - \varphi(\eta, \tau_0)\psi(\zeta, \tau_0)}{\eta - \zeta}\,d\eta.\tag{3.23}$$

We note that a simple relation exists between the zeroth moments of $\varphi(\eta, \tau_0)$ and $\psi(\eta, \tau_0)$. Setting

$$\alpha_0 = \int_0^1 \varphi(\eta, \tau_0)\,d\eta, \quad \beta_0 = \int_0^1 \psi(\eta, \tau_0)\,d\eta,\tag{3.24}$$

it follows from equation (3.22) that

$$\alpha_0 = 1 + \frac{\lambda}{4}(\alpha_0^2 - \beta_0^2).\tag{3.25}$$

When $\lambda = 1$ the last equation yields

$$\alpha_0 + \beta_0 = 2.\tag{3.26}$$

Equations (3.22) and (3.23) for the functions $\varphi(\eta, \tau_0)$ and $\psi(\eta, \tau_0)$ and the expressions (3.20) and (3.21) for $\varrho(\eta, \zeta, \tau_0)$ and $\sigma(\eta, \zeta, \tau_0)$ were first obtained by V. A. Ambartsumyan [1]. He also examined the problem of diffuse reflection by a semi-infinite atmosphere for anisotropic scattering, and introduced the functions $\varphi_i^m(\eta)$ which we considered in Chapter 2. The generalization of these results to atmospheres of finite optical thickness was made by Chandrasekhar [2]. In addition to the system of nonlinear integral equations (3.18) and (3.19) for $\varphi_i^m(\eta, \tau_0)$ and $\psi_i^m(\eta, \tau_0)$, linear integral equations for these functions have been obtained (see Section 7.6).

It should be noted that some of the functions $\varphi_i^m(\eta, \tau_0)$ and $\psi_i^m(\eta, \tau_0)$ play a special role in the theory of radiative transfer. For example, the albedo of an atmosphere and the illumination of an adjoining surface are specified by the functions $\varphi_1^0(\eta, \tau_0)$ and $\psi_1^0(\eta, \tau_0)$. For the atmospheric albedo, in analogy with the semi-infinite case (Section 2.2), we have

$$A(\zeta, \tau_0) = 1 - \frac{1}{\zeta}\varphi_1^0(\zeta, \tau_0).\tag{3.27}$$

We may express the total surface illumination (including both diffuse radiation and direct, attenuated solar radiation) in the form $\pi S \zeta V(\zeta, \tau_0)$ where $\pi S \zeta$ is the incident solar flux at the upper boundary of the atmosphere. From equation (1.83) we obtain

$$V(\zeta, \tau_0) = e^{-\tau_0/\zeta} + 2 \int_0^1 \sigma^0(\eta, \zeta, \tau_0) \eta \, d\eta, \tag{3.28}$$

while from (3.15) it follows that

$$\psi_1^0(\zeta, \tau_0) = \zeta e^{-\tau_0/\zeta} + 2\zeta \int_0^1 \sigma^0(\eta, \zeta, \tau_0) \eta \, d\eta. \tag{3.29}$$

We thus find

$$V(\zeta, \tau_0) = \frac{1}{\zeta} \psi_1^0(\zeta, \tau_0). \tag{3.30}$$

Equations (3.27) and (3.30) will be used below.

3.2. Dependence of the Reflection and Transmission Coefficients on Optical Thickness

In the equations introduced in the previous section, integration is carried out over the cosine of the polar angle, and the optical thickness τ_0 enters as a parameter. It is possible, however, to obtain instead equations for the reflection and transmission coefficients and auxiliary functions which contain integrals or derivatives with respect to τ_0.

To find such equations we return again to the basic integral equation (3.1) which determines the source function. We shall also need the following equation, which can be obtained from (3.1) by replacing τ by $\tau_0 - \tau$ and τ' by $\tau_0 - \tau'$:

$$B^m(\tau_0 - \tau, \eta, \zeta, \tau_0) = \frac{\lambda}{2} \int_0^1 p^m(\eta, \eta',) \, d\eta' \int_\tau^{\tau_0} B^m(\tau_0 - \tau', \eta', \zeta, \tau_0) e^{-(\tau'-\tau)/\eta'} \frac{d\tau'}{\eta'}$$

$$+ \frac{\lambda}{2} \int_0^1 p^m(\eta, -\eta') \, d\eta' \int_0^\tau B^m(\tau_0 - \tau', -\eta', \zeta, \tau_0) e^{-(\tau-\tau')/\eta'} \frac{d\tau'}{\eta'} + \frac{\lambda}{4} S p^m(\eta, \zeta) e^{-(\tau_0-\tau)/\zeta}. \tag{3.31}$$

By comparing equations (3.1) and (3.31) with the derivatives of those equations with respect to τ_0, we find

$$\frac{\partial B^m(\tau, \eta, \zeta, \tau_0)}{\partial \tau_0} = \frac{2}{S} \int_0^1 B^m(\tau_0, -\eta', \zeta, \tau_0) B^m(\tau_0 - \tau, -\eta, \eta', \tau_0) \frac{d\eta'}{\eta'}, \tag{3.32}$$

$$\frac{\partial B^m(\tau_0 - \tau, \eta, \zeta, \tau_0)}{\partial \tau_0} = -\frac{1}{\zeta} B^m(\tau_0 - \tau, \eta, \zeta, \tau_0)$$

$$+ \frac{2}{S} \int_0^1 B^m(0, \eta', \zeta, \tau_0) B^m(\tau_0 - \tau, \eta, \eta', \tau_0) \frac{d\eta'}{\eta'}. \tag{3.33}$$

Differentiating (3.4) and (3.5) with respect to τ_0, we obtain

$$S\zeta \frac{\partial \varrho^m(\eta, \zeta, \tau_0)}{\partial \tau_0} = \int_0^{\tau_0} \frac{\partial B^m(\tau, -\eta, \zeta, \tau_0)}{\partial \tau_0} e^{-\tau/\eta} \frac{d\tau}{\eta} + \frac{1}{\eta} B^m(\tau_0, -\eta, \zeta, \tau_0) e^{-\tau_0/\eta}, \quad (3.34)$$

$$S\zeta \frac{\partial \sigma^m(\eta, \zeta, \tau_0)}{\partial \tau_0} = \int_0^{\tau_0} \frac{\partial B^m(\tau_0-\tau, \eta, \zeta, \tau_0)}{\partial \tau_0} e^{-\tau/\eta} \frac{d\tau}{\eta} + \frac{1}{\eta} B^m(0, \eta, \zeta, \tau_0) e^{-\tau_0/\eta}. \quad (3.35)$$

Substituting (3.32) into (3.34) and (3.33) into (3.35) gives, following use of equation (3.5),

$$S\zeta \frac{\partial \varrho^m(\eta, \zeta, \tau_0)}{\partial \tau_0} = \frac{1}{\eta} B^m(\tau_0, -\eta, \zeta, \tau_0) e^{-\tau_0/\eta} + 2 \int_0^1 B^m(\tau_0, -\eta', \zeta, \tau_0) \sigma^m(\eta, \eta', \tau_0) \, d\eta', \quad (3.36)$$

$$S\zeta \frac{\partial \sigma^m(\eta, \zeta, \tau_0)}{\partial \tau_0} = -S\sigma^m(\eta, \zeta, \tau_0) + \frac{1}{\eta} B^m(0, \eta, \zeta, \tau_0) e^{-\tau_0/\eta}$$

$$+ 2 \int_0^1 B^m(0, \eta', \zeta, \tau_0) \sigma^m(\eta, \eta', \tau_0) \, d\eta'. \quad (3.37)$$

If we substitute equations (3.10) and (3.11) into (3.36) and (3.37), we arrive at a system of two equations for the determination of the functions $\varrho^m(\eta, \zeta, \tau_0)$ and $\sigma^m(\eta, \zeta, \tau_0)$.

Another approach used to determine these functions consists in first finding the quantities $B^m(0, \eta, \zeta, \tau_0)$ and $B^m(\tau_0, \eta, \zeta, \tau_0)$ from the equations

$$\frac{\partial B^m(0, \eta, \zeta, \tau_0)}{\partial \tau_0} = \frac{2}{S} \int_0^1 B^m(\tau_0, -\eta', \zeta, \tau_0) B^m(\tau_0, -\eta, \eta', \tau_0) \frac{d\eta'}{\eta'}, \quad (3.38)$$

$$\frac{\partial B^m(\tau_0, \eta, \zeta, \tau_0)}{\partial \tau_0} = -\frac{1}{\zeta} B^m(\tau_0, \eta, \zeta, \tau_0) + \frac{2}{S} \int_0^1 B^m(0, \eta', \zeta, \tau_0) B^m(\tau_0, \eta, \eta', \tau_0) \frac{d\eta'}{\eta'}, \quad (3.39)$$

which follow from (3.32) and (3.33) for $\tau = 0$. After the solution of these equations, the reflection and transmission coefficients may be found from equations (3.36) and (3.37) or (3.8) and (3.9).

In a previous section we expressed the quantities $\varrho^m(\eta, \zeta, \tau_0)$ and $\sigma^m(\eta, \zeta, \tau_0)$ by means of equations (3.16) and (3.17) in terms of the auxiliary functions $\varphi_i^m(\eta, \tau_0)$ and $\psi_i^m(\eta, \tau_0)$, which were determined by equations (3.18) and (3.19). We may now use the relations (3.36)–(3.39) to find alternative expressions for $\varrho^m(\eta, \zeta, \tau_0)$ and $\sigma^m(\eta, \zeta, \tau_0)$ and alternative equations for the auxiliary functions. Substituting expressions (3.12) and (3.13) into (3.36) and (3.37) and using equations (3.14) and (3.15), we obtain

$$\eta\zeta \frac{\partial \varrho^m(\eta, \zeta, \tau_0)}{\partial \tau_0} = \frac{\lambda}{4} \sum_{i=m}^n c_i^m (-1)^{i+m} \psi_i^m(\eta, \tau_0) \psi_i^m(\zeta, \tau_0), \quad (3.40)$$

$$\eta\zeta \frac{\partial \sigma^m(\eta, \zeta, \tau_0)}{\partial \tau_0} = -\eta\sigma^m(\eta, \zeta, \tau_0) + \frac{\lambda}{4} \sum_{i=m}^n c_i^m \psi_i^m(\eta, \tau_0) \varphi_i^m(\zeta, \tau_0). \quad (3.41)$$

Introducing equations (3.12) and (3.13) into (3.38) and (3.39), we find

$$\frac{\partial \varphi_i^m(\zeta, \tau_0)}{\partial \tau_0} = \frac{\lambda}{2} \sum_{j=m}^{n} c_j^m (-1)^{i+j} \psi_j^m(\zeta, \tau_0) \int_0^1 P_i^m(\eta) \psi_i^m(\eta, \tau_0) \frac{d\eta}{\eta}, \tag{3.42}$$

$$\frac{\partial \psi_i^m(\zeta, \tau_0)}{\partial \tau_0} = -\frac{1}{\zeta} \psi_i^m(\zeta, \tau_0) + \frac{\lambda}{2} \sum_{j=m}^{n} c_j^m \varphi_j^m(\zeta, \tau_0) \int_0^1 P_i^m(\eta) \psi_i^m(\eta, \tau_0) \frac{d\eta}{\eta}. \tag{3.43}$$

The system of equations (3.42) and (3.43) for the determination of $\varphi_i^m(\eta, \tau_0)$ and $\psi_i^m(\eta, \tau_0)$ must be solved subject to the initial conditions

$$\varphi_i^m(\zeta, 0) = P_i^m(\zeta), \quad \psi_i^m(\zeta, 0) = P_i^m(\zeta). \tag{3.44}$$

The quantities $\varrho^m(\eta, \zeta, \tau_0)$ and $\sigma^m(\eta, \zeta, \tau_0)$ may then be found by integrating equations (3.40) and (3.41) with the conditions that $\varrho^m(\eta, \zeta, 0) = 0$ and $\sigma^m(\eta, \zeta, 0) = 0$.

In the case of isotropic scattering, equations (3.40) and (3.41) give

$$\eta \zeta \varrho(\eta, \zeta, \tau_0) = \frac{\lambda}{4} \int_0^{\tau_0} \psi(\eta, \tau) \psi(\zeta, \tau) \, d\tau, \tag{3.45}$$

$$\eta \zeta \sigma(\eta, \zeta, \tau_0) = \frac{\lambda}{4} \int_0^{\tau_0} e^{-(\tau_0-\tau)/\zeta} \varphi(\zeta, \tau) \psi(\eta, \tau) \, d\tau, \tag{3.46}$$

while we obtain the following integral equations for the determination of $\varphi(\zeta, \tau_0)$ and $\psi(\zeta, \tau_0)$ from (3.42) and (3.43):

$$\varphi(\zeta, \tau_0) = 1 + \frac{\lambda}{2} \int_0^{\tau_0} \psi(\zeta, \tau) \, d\tau \int_0^1 \psi(\eta, \tau) \frac{d\eta}{\eta}, \tag{3.47}$$

$$\psi(\zeta, \tau_0) = e^{-\tau_0/\zeta} + \frac{\lambda}{2} \int_0^{\tau_0} e^{-(\tau_0-\tau)/\zeta} \varphi(\zeta, \tau) \, d\tau \int_0^1 \psi(\eta, \tau) \frac{d\eta}{\eta}. \tag{3.48}$$

TABLE 3.1. FUNCTION $\varphi(\eta, \tau_0)$ FOR $\lambda = 1$

η	τ_0								
	0	0.2	0.4	0.6	0.8	1.0	2.0	3.0	∞
0	1	1.000	1.000	1.000	1.000	1.000	1.000	1.000	1.000
0.1	1	1.147	1.171	1.182	1.189	1.194	1.211	1.219	1.247
0.2	1	1.198	1.261	1.293	1.313	1.327	1.365	1.385	1.450
0.3	1	1.222	1.314	1.368	1.404	1.429	1.498	1.531	1.643
0.4	1	1.236	1.349	1.421	1.472	1.510	1.613	1.663	1.829
0.5	1	1.245	1.373	1.460	1.524	1.574	1.715	1.785	2.013
0.6	1	1.251	1.390	1.489	1.565	1.626	1.805	1.895	2.194
0.7	1	1.256	1.403	1.512	1.598	1.668	1.884	1.997	2.374
0.8	1	1.259	1.413	1.531	1.625	1.703	1.954	2.090	2.553
0.9	1	1.262	1.422	1.546	1.647	1.732	2.016	2.175	2.731
1.0	1	1.265	1.429	1.558	1.666	1.757	2.071	2.254	2.908

Tables 3.1 and 3.2 present values of the functions $\varphi(\eta, \tau_0)$ and $\psi(\eta, \tau_0)$ for the case of pure scattering ($\lambda = 1$). More complete tables of these functions for various λ are given by Carlstedt and Mullikin (see ref. 4, Chapter 6).

TABLE 3.2. FUNCTION $\psi(\eta, \tau_0)$ FOR $\lambda = 1$

η	τ_0								
	0	0.2	0.4	0.6	0.8	1.0	2.0	3.0	∞
0	1	0	0	0	0	0	0	0	0
0.1	1	0.241	0.106	0.075	0.063	0.056	0.037	0.028	0
0.2	1	0.534	0.314	0.213	0.164	0.137	0.086	0.066	0
0.3	1	0.711	0.506	0.379	0.300	0.250	0.149	0.112	0
0.4	1	0.823	0.654	0.530	0.441	0.376	0.224	0.167	0
0.5	1	0.899	0.767	0.657	0.569	0.500	0.311	0.231	0
0.6	1	0.953	0.854	0.762	0.683	0.615	0.404	0.303	0
0.7	1	0.995	0.924	0.850	0.781	0.719	0.501	0.381	0
0.8	1	1.027	0.980	0.923	0.866	0.811	0.597	0.464	0
0.9	1	1.053	1.027	0.986	0.940	0.893	0.691	0.549	0
1.0	1	1.074	1.066	1.039	1.004	0.966	0.780	0.636	0

3.3. Atmospheres of Large Optical Thickness

The equations obtained in the two previous sections apply to atmospheres of any finite optical thickness. The solution of these equations becomes increasingly difficult as τ_0 increases. Consequently, it is very important that simple asymptotic expressions for the various quantities characterizing the radiation field may be found for atmospheres of large optical thickness ($\tau_0 \gg 1$). These expressions become more exact as τ_0 increases. They express the desired quantities in terms of those functions which characterize the radiation field in a semi-infinite atmosphere.

In this section we shall find the asymptotic expressions for the reflection and transmission coefficients. We limit ourselves to the case of quantities averaged over azimuth. This is reasonable because the dependence of the intensity on azimuth rapidly decreases with increasing optical depth (see Section 5.4). Therefore, for $\tau_0 \gg 1$, the transmission coefficient may be considered independent of azimuth, and the azimuth-dependent terms in the reflection coefficient are equal to the corresponding terms in the reflection coefficient of a semi-infinite atmosphere. In the derivation of these asymptotic expressions, we shall use the absolute intensity of radiation in the deep layers of a semi-infinite atmosphere, which was determined in Section 2.4.

Let us consider a semi-infinite atmosphere which is divided in two parts by an imaginary plane placed at a large optical depth τ_0. We may then consider the outer portion of the atmosphere as a layer of optical thickness τ_0 illuminated from above by parallel radiation with an angle of incidence arccos ζ and from below by diffuse radiation whose intensity is given by equation (2.102) for $\tau = \tau_0$ and $\eta < 0$. The reflection and transmission coefficients of the layer of optical thickness τ_0 will be designated $\rho(\eta, \zeta, \tau_0)$ and $\sigma(\eta, \zeta, \tau_0)$, respectively, and the reflection coefficient of the semi-infinite atmosphere by $\rho(\eta, \zeta)$. These quantities are averaged over azimuth, but for simplicity we omit the index "0".

Bearing in mind that the layer under consideration is illuminated from above and from below, we obtain the following two relations:

$$\varrho(\eta, \zeta) = \varrho(\eta, \zeta, \tau_0) + 2u(\zeta) \, e^{-k\tau_0} \int_0^1 \sigma(\eta, \eta', \tau_0) \, i(-\eta') \, \eta' \, d\eta',$$ (3.49)

$$u(\zeta) \, i(\eta) \, e^{-k\tau_0} = \sigma(\eta, \zeta, \tau_0) + 2u(\zeta) \, e^{-k\tau_0} \int_0^1 \varrho(\eta, \eta', \tau_0) \, i(-\eta') \, \eta' \, d\eta'.$$ (3.50)

These two equations give the intensity of radiation emerging from the layer through its upper and through its lower boundary, respectively.

In order to find the integrals entering equations (3.49) and (3.50), we multiply these equations by $i(-\zeta) \, \zeta$ and integrate over ζ between 0 and 1. We obtain

$$2 \int_0^1 \sigma(\eta, \zeta, \tau_0) \, i(-\zeta) \, \zeta \, d\zeta = \left[i(\eta) - 2 \int_0^1 \varrho(\eta, \zeta) \, i(-\zeta) \, \zeta \, d\zeta \right] \frac{N e^{-k\tau_0}}{1 - N^2 e^{-2k\tau_0}},$$ (3.51)

where we have defined

$$N = 2 \int_0^1 u(\zeta) \, i(-\zeta) \zeta \, d\zeta.$$ (3.52)

From the symmetry of the function $\varrho(\eta, \zeta, \tau_0)$ it follows that

$$i(\eta) - 2 \int_0^1 \varrho(\eta, \zeta) \, i(-\zeta) \, \zeta \, d\zeta = Mu(\eta),$$ (3.53)

where M is some constant. Equation (3.53) has already been obtained in Section 2.3, where it was shown that the constant is

$$M = 2 \int_{-1}^1 i^2(\eta) \, \eta \, d\eta,$$ (3.54)

and the function $i(\eta)$ is normalized in the usual way (see equation (2.79)).

Substituting (3.53) into (3.51) and the resulting expression into (3.49), we obtain

$$\varrho(\eta, \zeta, \tau_0) = \varrho(\eta, \zeta) - f(\tau_0) \, u(\eta) \, u(\zeta),$$ (3.55)

where

$$f(\tau_0) = \frac{M N e^{-2k\tau_0}}{1 - N^2 e^{-2k\tau_0}}.$$ (3.56)

From (3.50) we find, with the aid of (3.55),

$$\sigma(\eta, \zeta, \tau_0) = g(\tau_0) \, u(\eta) \, u(\zeta),$$ (3.57)

where

$$g(\tau_0) = \frac{M e^{-k\tau_0}}{1 - N^2 e^{-2k\tau_0}}.$$ (3.58)

Equations (3.55) and (3.57) give the desired asymptotic expressions for the reflection and transmission coefficients for an atmosphere of large optical thickness. We recall that the constant k which enters equations (3.56) and (3.58) was introduced in the study of the radiation field in deep layers and is given by equation (2.22), while the function $u(\eta)$ describes the relative intensity of the transmitted radiation for a semi-infinite atmosphere and is normalized according to equation (2.78).

Because of the importance of equations (3.55) and (3.57) for many applications, we shall examine them in more detail. In the derivation of these equations it was assumed that $\tau_0 \gg 1$. The constant k, however, may take any value between 0 and 1. As a result we meet in practice different particular cases according to the values of the quantities k and $k\tau_0$. Let us deduce the asymptotic expressions for the reflection and transmission coefficients for several particular cases.

1. First, we consider the case of small true absorption in the atmosphere ($k \ll 1$). Then, with the aid of equations (2.104) and (2.139), we obtain for the constants M and N determined by equations (3.54) and (3.52)

$$M = \frac{8k}{3-x_1},$$
(3.59)

and

$$N = 1 - \frac{3k\delta}{3-x_1},$$
(3.60)

where

$$\delta = 4 \int_0^1 u_0(\eta)\, \eta^2\, d\eta$$
(3.61)

and $u_0(\eta)$ is the transmission function of a semi-infinite atmosphere for the case of pure scattering. Substitution of (3.59) and (3.60) into equations (3.56) and (3.58) yields

$$f(\tau_0) = \frac{8k}{(3-x_1)(e^{2k\tau_0}-1)+6\delta k},$$
(3.62)

$$g(\tau_0) = \frac{8ke^{k\tau_0}}{(3-x_1)(e^{2k\tau_0}-1)+6\delta k}.$$
(3.63)

The reflection coefficient of a semi-infinite atmosphere $\varrho(\eta, \zeta)$ is given by equation (2.129) when $k \ll 1$. As a result, if we define

$$h(\tau_0) = \frac{4k}{3-x_1} + f(\tau_0),$$
(3.64)

we may replace equation (3.55) by

$$\varrho(\eta, \zeta, \tau_0) = \varrho_0(\eta, \zeta) - h(\tau_0)\, u_0(\eta)\, u_0(\zeta),$$
(3.65)

where $\varrho_0(\eta, \zeta)$ is the reflection coefficient of a semi-infinite atmosphere for $\lambda = 1$. Equation (3.57) in this case takes the form

$$\sigma(\eta, \zeta, \tau_0) = g(\tau_0)\, u_0(\eta)\, u_0(\zeta).$$
(3.66)

The reflection and transmission coefficients of an atmosphere with $\tau_0 \gg 1$ and small true absorption are thus determined by the asymptotic expressions (3.65) and (3.66). The functions $\varrho_0(\eta, \zeta)$ and $u_0(\eta)$ entering these equations depend only on the phase function.

2. Let us next consider a purely scattering atmosphere ($\lambda = 1$). To obtain the asymptotic expressions for the reflection and transmission coefficients in this case it is necessary to let $k \to 0$ in the equations derived above. Doing this in equations (3.62)–(3.64), we find

$$g(\tau_0) = h(\tau_0) = \frac{4}{(3-x_1)\tau_0+3\delta} . \tag{3.67}$$

Substitution of (3.67) into (3.65) and (3.66) yields

$$\varrho(\eta, \zeta, \tau_0) = \varrho_0(\eta, \zeta) - \frac{4u_0(\eta)\, u_0(\zeta)}{(3-x_1)\,\tau_0+3\delta}, \tag{3.68}$$

$$\sigma(\eta, \zeta, \tau_0) = \frac{4u_0(\eta)\, u_0(\zeta)}{(3-x_1)\tau_0+3\delta} . \tag{3.69}$$

It is clear that equations (3.68) and (3.69) give asymptotic expressions for $\varrho(\eta, \zeta, \tau_0)$ and $\sigma(\eta, \zeta, \tau_0)$ not only for the case of pure scattering, but whenever $k\tau_0 \ll 1$. According to these equations, the atmosphere absorbs no energy, but instead entirely reflects or transmits it. In actual fact, when $k\tau_0 \ll 1$ the atmosphere will absorb a certain portion of the energy falling on it, but this portion will be very small. We thus speak of an atmosphere for which $k\tau_0 \ll 1$ as weakly absorbing (in contrast with a strongly absorbing atmosphere, for which $k\tau_0 \gg 1$).

3. In the case of a strongly absorbing atmosphere for small true absorption (i.e. when $k\tau_0 \gg 1$ and $k \ll 1$), it is easy to see that equations (3.65) and (3.66) become

$$\varrho(\eta, \zeta, \tau_0) = \varrho_0(\eta, \zeta) - \frac{4k}{3-x_1}\, u_0(\eta)\, u_0(\zeta), \tag{3.70}$$

$$\sigma(\eta, \zeta, \tau_0) = \frac{8k}{3-x_1}\, e^{-k\tau_0} u_0(\eta)\, u_0(\zeta). \tag{3.71}$$

In this case the reflection coefficient coincides with that for a semi-infinite atmosphere, and the transmission coefficient is very small.

4. A more general case than the previous one is that of a strongly absorbing atmosphere for arbitrary values of k. In order to find the asymptotic expressions for $\varrho(\eta, \zeta, \tau_0)$ and $\sigma(\eta, \zeta, \tau_0)$ in this case, it is necessary to take $k\tau_0 \gg 1$ in expressions (3.56) and (3.58) and then substitute them into equations (3.55) and (3.57). We obtain as a result

$$\varrho(\eta, \zeta, \tau_0) = \varrho(\eta, \zeta) - MNe^{-2k\tau_0} u(\eta)\, u(\zeta), \tag{3.72}$$

$$\sigma(\eta, \zeta, \tau_0) = Me^{-k\tau_0} u(\eta)\, u(\zeta), \tag{3.73}$$

where the constants M and N are determined by equations (3.54) and (3.52) If we take $k \ll 1$ in equations (3.72) and (3.73), these expressions reduce to equations (3.70) and (3.71).

Let us apply the asymptotic expressions (3.55) and (3.57) to the case of isotropic scattering. The function $\varrho(\eta, \zeta)$ is then given by equation (2.43), while an expression for $u(\eta)$ may be

6

obtained from (2.88) for $x_1 = 0$. The latter equation takes the form

$$u(\eta) = \frac{\lambda}{2} a_0 \frac{\varphi(\eta)}{1-k\eta},$$

(3.74)

where, as follows from (2.89) and (2.91),

$$a_0 = \frac{2k}{M} \int_0^1 \frac{\varphi(\eta)}{1-k^2\eta^2} \eta \, d\eta$$

(3.75)

and

$$M = \frac{4}{k}\left(\frac{1}{1-k^2}-\frac{1}{\lambda}\right).$$

(3.76)

We thus obtain

$$\varrho(\eta, \zeta, \tau_0) = \frac{\lambda}{4} \frac{\varphi(\eta)\varphi(\zeta)}{\eta+\zeta} - CN\frac{e^{-2k\tau_0}}{1-N^2 e^{-2k\tau_0}}\frac{\varphi(\eta)}{1-k\eta}\frac{\varphi(\zeta)}{1-k\zeta},$$

(3.77)

and

$$\sigma(\eta, \zeta, \tau_0) = C\frac{e^{-k\tau_0}}{1-N^2 e^{-2k\tau_0}}\frac{\varphi(\eta)}{1-k\eta}\frac{\varphi(\zeta)}{1-k\zeta},$$

(3.78)

where

$$C = \left(\frac{\lambda}{2} a_0\right)^2 M,$$

(3.79)

and

$$N = \lambda a_0 \int_0^1 \frac{\varphi(\eta)}{1-k^2\eta^2} \eta \, d\eta.$$

(3.80)

The last equation follows from (3.52), (3.74) and (2.13).

Other expressions may also be found for the determination of C and N. For this purpose we shall use the expression

$$\lambda a_0 \int_0^1 \frac{\varphi(\eta)}{(1-k\eta)^2} \eta \, d\eta = 1,$$

(3.81)

which is obtained by substitution of (3.74) and (2.13) into (2.78). From equations (3.80) and (3.81) it follows that

$$N \int_0^1 \frac{\varphi(\eta)}{(1-k\eta)^2} \eta \, d\eta = \int_0^1 \frac{\varphi(\eta)}{1-k^2\eta^2} \eta \, d\eta,$$

(3.82)

while substitution of the expressions for a_0 from (3.75) and (3.81) into equation (3.79) gives, after use of (3.82),

$$C = \frac{\lambda}{2} kN.$$

(3.83)

For pure scattering ($\lambda = 1$, $k = 0$), we find from (3.77) and (3.78)

$$\varrho(\eta, \zeta, \tau_0) = \tfrac{1}{4}\, \varphi(\eta)\varphi(\zeta) \left(\frac{1}{\eta+\zeta} - \frac{1}{\tau_0+\delta}\right), \tag{3.84}$$

$$\sigma(\eta, \zeta, \tau_0) = \frac{1}{4}\, \frac{\varphi(\eta)\,\varphi(\zeta)}{\tau_0+\delta}, \tag{3.85}$$

where

$$\delta = \sqrt{3} \int_0^1 \varphi(\eta)\, \eta^2\, d\eta = 1.42\ldots \tag{3.86}$$

The asymptotic expressions for the reflection and transmission coefficients of an atmosphere of large optical thickness for isotropic scattering have been known for a long time (*TRT*, p. 123). Equations (3.55)–(3.58) were obtained by van de Hulst [3] from physical considerations and by V. V. Sobolev [4] by the method presented here. Expressions found earlier by T. A. Germogenova [5] may be put in the same form.

3.4. Asymptotic Formulas for the Auxiliary Functions

In the previous section, we obtained asymptotic expressions for the reflection and transmission coefficients when $\tau_0 \gg 1$. With their aid we shall find asymptotic formulas for the auxiliary functions $\varphi_i(\eta, \tau_0)$ and $\psi_i(\eta, \tau_0)$.

For this purpose we shall use equations (3.14) and (3.15) with $m = 0$. Substituting (3.55) into (3.14) and recalling (2.84), we obtain

$$\varphi_i(\eta, \tau_0) = \varphi_i(\eta) - 2a_i(-1)^i\, \frac{u(\eta)\eta}{1 - N^2 e^{-2k\tau_0}}\, MN e^{-2k\tau_0}, \tag{3.87}$$

where $\varphi_i(\eta)$ is the auxiliary function for $\tau_0 = \infty$. Introducing equation (3.57) into (3.15) and neglecting terms containing $e^{-\tau_0/\zeta}$, we find

$$\psi_i(\eta, \tau_0) = 2a_i\, \frac{u(\eta)\,\eta}{1 - N^2 e^{-2k\tau_0}}\, M e^{-2k\tau_0}. \tag{3.88}$$

The coefficients a_i entering equations (3.87), (3.79), and (3.88) are determined by (2.69). They may also be computed from (2.87).

These asymptotic expressions for the functions $\varphi_i(\eta, \tau_0)$ and $\psi_i(\eta, \tau_0)$ when $\tau_0 \gg 1$ are valid for arbitrary values of the parameters k and $k\tau_0$. It is also possible to obtain the particular forms of these expressions corresponding to the cases considered in the previous section. We shall now do this for some of those special cases.

For small true absorption ($k \ll 1$), we find in place of (3.87) and (3.88)

$$\varphi_i(\eta, \tau_0) = \varphi_{i0}(\eta) - 2h(\tau_0)(-1)^i a_{i0}u_0(\eta)\,\eta \tag{3.89}$$

and

$$\psi_i(\eta, \tau_0) = 2g(\tau_0)\, a_{i0}\, u_0(\eta)\,\eta, \tag{3.90}$$

6*

where $g(\tau_0)$ and $h(\tau_0)$ are given by equations (3.63) and (3.64) and $\varphi_{i0}(\eta)$, $u_0(\eta)$, and a_{i0} signify the quantities $\varphi_i(\eta)$, $u(\eta)$, and a_i for $\lambda = 1$. Since we have from (2.69) that

$$a_{i0} = \int_0^1 u_0(\eta)\, P_i(\eta)\, d\eta, \tag{3.91}$$

it follows by substitution of (2.110) into equation (3.91) that

$$a_{i0} = \tfrac{3}{4}(-1)^i \int_0^1 \varphi_{i0}(\zeta)\,\zeta\, d\zeta \qquad (i \neq 1) \tag{3.92}$$

and $a_{10} = \tfrac{1}{2}$. We have used the fact that always $\varphi_{10}(\eta) = 0$.

When the inequality $k\tau_0 \ll 1$ holds (so that the atmosphere is weakly absorbing or purely scattering), the quantities $g(\tau_0)$ and $h(\tau_0)$ are given by equation (3.67). We then find, instead of (3.89) and (3.90), that

$$\varphi_i(\eta, \tau_0) = \varphi_{i0}(\eta) - (-1)^i \frac{8a_{i0}u_0(\eta)\eta}{(3-x_1)\tau_0 + 3\delta}, \tag{3.93}$$

$$\psi_i(\eta, \tau_0) = \frac{8a_{i0}u_0(\eta)\,\eta}{(3-x_1)\,\tau_0 + 3\delta}, \tag{3.94}$$

where δ is given by equation (3.61).

For isotropic scattering, we have only the two auxiliary functions which we have denoted by $\varphi(\eta, \tau_0)$ and $\psi(\eta, \tau_0)$. It follows from (3.87) and (3.88) that the asymptotic formulas for these functions take the form

$$\varphi(\eta, \tau_0) = \varphi(\eta) - 2kN^2 \frac{e^{-2k\tau_0}}{1 - N^2 e^{-2k\tau_0}} \frac{\varphi(\eta)\eta}{1 - k\eta}, \tag{3.95}$$

$$\psi(\eta, \tau_0) = 2kN \frac{e^{-k\tau_0}}{1 - N^2 e^{-2k\tau_0}} \frac{\varphi(\eta)\eta}{1 - k\eta}, \tag{3.96}$$

where N is given by equation (3.82).

For a weakly absorbing atmosphere ($k\tau_0 \ll 1$), we obtain from equations (3.95) and (3.96)

$$\varphi(\eta, \tau_0) = \varphi(\eta) - \frac{\varphi(\eta)\eta}{\tau_0 + \delta}, \tag{3.97}$$

$$\psi(\eta, \tau_0) = \frac{\varphi(\eta)\eta}{\tau_0 + \delta}, \tag{3.98}$$

where δ has the value given by equation (3.86).

The asymptotic expressions for the auxiliary functions when $\tau_0 \gg 1$ were initially found for the simplest cases, for isotropic scattering [6] and for the phase function $x(\gamma) = 1 + x_1 \cos \gamma$ [7], and subsequently in the general case [8].

3.5. Inhomogeneous Atmospheres

Up to this point we have considered homogeneous atmospheres, for which the single scattering albedo λ and the phase function $x(\gamma)$ do not depend on depth. In real atmospheres, however, both these quantities may depend very strongly upon depth. It thus becomes necessary to consider the problem of light scattering in inhomogeneous atmospheres, in which λ and $x(\gamma)$ are functions of optical depth τ.

Before turning to this problem, however, we emphasize that the theory of light scattering in homogeneous atmospheres is nonetheless of great importance. It clearly represents the first approximation to reality. Moreover, the methods and equations obtained in this case may be generalized without particular difficulty to inhomogeneous atmospheres. In addition, the results obtained for homogeneous atmospheres may be applied to other media of practical importance (cf. the Concluding Remarks to this book).

We begin by considering the problem of diffuse reflection and transmission of light by the simplest inhomogeneous atmosphere, in which λ depends on τ and the phase function is isotropic. The integral equation determining the source function in such an atmosphere is obtained from equation (1.57) by replacing λ by $\lambda(\tau)$. Thus,

$$B(\tau, \zeta, \tau_0) = \frac{\lambda(\tau)}{2} \int_0^{\tau_0} E_1(|\tau - t|) B(t, \zeta, \tau_0) \, dt + \frac{\lambda(\tau)}{4} Se^{-\tau/\zeta}. \tag{3.99}$$

The reflection and transmission coefficients are expressed in terms of the source function by the equations

$$\left.\begin{aligned}
\varrho(\eta, \zeta, \tau_0) &= \frac{1}{S} \int_0^{\tau_0} B(\tau, \zeta, \tau_0) e^{-\tau/\eta} \frac{d\tau}{\eta\zeta}, \\[2mm]
\sigma(\eta, \zeta, \tau_0) &= \frac{1}{S} \int_0^{\tau_0} B(\tau, \zeta, \tau_0) e^{-(\tau_0 - \tau)/\eta} \frac{d\tau}{\eta\zeta}.
\end{aligned}\right\} \tag{3.100}$$

Equation (3.99) determines the source function in an atmosphere illuminated by parallel radiation from above (on the side where $\tau = 0$). In order to find the reflection and transmission coefficients in an inhomogeneous atmosphere, however, it is necessary to consider simultaneously the equation which determines the source function in an atmosphere illuminated by parallel rays from below (from the side where $\tau = \tau_0$). We shall designate this source function by $B^*(\tau, \zeta, \tau_0)$ and the corresponding reflection and transmission coefficients by $\varrho^*(\eta, \zeta, \tau_0)$ and $\sigma^*(\eta, \zeta, \tau_0)$. The function B^* is obviously determined by the equation

$$B^*(\tau, \zeta, \tau_0) = \frac{\lambda(\tau)}{2} \int_0^{\tau_0} E_1(|\tau - t|) B^*(t, \zeta, \tau_0) \, dt + \frac{\lambda(\tau)}{4} Se^{-(\tau_0 - \tau)/\zeta}, \tag{3.101}$$

while the quantities ϱ^* and σ^* are given by

$$\left.\begin{aligned}
\varrho^*(\eta, \zeta, \tau_0) &= \frac{1}{S} \int_0^{\tau_0} B^*(\tau, \zeta, \tau_0) e^{-(\tau_0 - \tau)/\eta} \frac{d\tau}{\eta\zeta}, \\[2mm]
\sigma^*(\eta, \zeta, \tau_0) &= \frac{1}{S} \int_0^{\tau_0} B^*(\tau, \zeta, \tau_0) e^{-\tau/\eta} \frac{d\tau}{\eta\zeta}.
\end{aligned}\right\} \tag{3.102}$$

The reflection and transmission coefficients satisfy certain relations which express the "principle of reciprocity". These relations take the form

$$\varrho(\eta, \zeta, \tau_0) = \varrho(\zeta, \eta, \tau_0), \tag{3.103}$$

$$\varrho^*(\eta, \zeta, \tau_0) = \varrho^*(\zeta, \eta, \tau_0), \tag{3.104}$$

$$\sigma(\eta, \zeta, \tau_0) = \sigma^*(\zeta, \eta, \tau_0). \tag{3.105}$$

They may be easily obtained from equations (3.99) and (3.101) with the aid of (3.100) and (3.102). For example, to obtain equation (3.103) we multiply equation (3.99) by $B(\tau, \eta, \tau_0)/\lambda(\tau)$ and integrate over τ between 0 and τ_0.

We see that each of the reflection coefficients is a symmetric function of η and ζ. The transmission coefficients, however, do not possess this property. This may be explained by the fact that the atmosphere when observed from below is not identical to the atmosphere observed from above.

It should be noted that equations (3.103)–(3.105) remain valid when the phase function depends on τ. They are the generalization of equations (1.78) and (1.79), which were obtained earlier for a homogeneous atmosphere.

In order to find equations for the reflection and transmission coefficients, we shall use the same method as in the case of a homogeneous atmosphere (Section 3.2). Differentiating equations (3.99) and (3.101) with respect to τ_0, we have

$$\frac{\partial B(\tau, \zeta, \tau_0)}{\partial \tau_0} = \frac{\lambda(\tau)}{2} \int_0^{\tau_0} E_1(|\tau - t|) \frac{\partial B(t, \zeta, \tau_0)}{\partial \tau_0} dt + \frac{\lambda(\tau)}{2} E_1(\tau_0 - \tau) B(\tau_0, \zeta, \tau_0), \tag{3.106}$$

$$\frac{\partial B^*(\tau, \zeta, \tau_0)}{\partial \tau_0} = \frac{\lambda(\tau)}{2} \int_0^{\tau_0} E_1(|\tau - t|) \frac{\partial B(t, \zeta, \tau_0)}{\partial \tau_0} dt$$

$$+ \frac{\lambda(\tau)}{2} E_1(\tau_0 - \tau) B^*(\tau_0, \zeta, \tau_0) - \frac{\lambda(\tau)}{4\zeta} S e^{-(\tau_0 - \tau)/\zeta}. \tag{3.107}$$

Comparison of these equations with (3.99) and (3.101) yields

$$\frac{\partial B(\tau, \zeta, \tau_0)}{\partial \tau_0} = \frac{2}{S} B(\tau_0, \zeta, \tau_0) \int_0^1 B^*(\tau, \zeta', \tau_0) \frac{d\zeta'}{\zeta'}, \tag{3.108}$$

$$\frac{\partial B^*(\tau, \zeta, \tau_0)}{\partial \tau_0} = -\frac{1}{\zeta} B^*(\tau, \zeta, \tau_0) + \frac{2}{S} B^*(\tau_0, \zeta, \tau_0) \int_0^1 B^*(\tau, \zeta', \tau_0) \frac{d\zeta'}{\zeta'}. \tag{3.109}$$

We now differentiate equations (3.100) and (3.102) with respect to τ_0. Using equations (3.108), (3.109), (3.100), and (3.102), we obtain

$$S\eta\zeta \frac{\partial \varrho}{\partial \tau_0} = B(\tau_0, \zeta, \tau_0) \left[e^{-\tau_0/\eta} + 2\eta \int_0^1 \sigma^*(\eta, \zeta', \tau_0) d\zeta' \right], \tag{3.110}$$

$$S\zeta \left(\eta \frac{\partial \sigma}{\partial \tau_0} + \sigma \right) = B(\tau_0, \zeta, \tau_0) \left[1 + 2\eta \int_0^1 \varrho^*(\eta, \zeta', \tau_0) \, d\zeta' \right], \qquad (3.111)$$

$$S\eta\zeta \frac{\partial \varrho^*}{\partial \tau_0} = -S(\eta + \zeta) \varrho^* + B^*(\tau_0, \zeta, \tau_0) \left[1 + 2\eta \int_0^1 \varrho^*(\eta, \zeta', \tau_0) \, d\zeta' \right], \qquad (3.112)$$

$$S\eta \left(\zeta \frac{\partial \sigma^*}{\partial \tau_0} + \sigma^* \right) = B^*(\tau_0, \zeta, \tau_0) \left[e^{-\tau_0/\eta} + 2\eta \int_0^1 \sigma^*(\eta, \zeta', \tau_0) \, d\zeta' \right]. \qquad (3.113)$$

The quantities $B(\tau_0, \zeta, \tau_0)$ and $B^*(\tau_0, \zeta, \tau_0)$ entering equations (3.110)–(3.113) are determined by equations (3.99) and (3.101). Setting $\tau = \tau_0$ in the latter equations and using (3.100) and (3.102), we find

$$B(\tau_0, \zeta, \tau_0) = \frac{\lambda(\tau_0)}{4} S \left[e^{-\tau_0/\zeta} + 2\zeta \int_0^1 \sigma(\eta, \zeta, \tau_0) \, d\eta \right], \qquad (3.114)$$

$$B^*(\tau_0, \zeta, \tau_0) = \frac{\lambda(\tau_0)}{4} S \left[1 + 2\zeta \int_0^1 \varrho^*(\eta, \zeta, \tau_0) \, d\eta \right]. \qquad (3.115)$$

Substitution of (3.114) and (3.115) into (3.110)–(3.113) provides us with a system of equations for the determination of the quantities $\varrho, \sigma, \varrho^*$ and σ^*. It is easy to see, however, that all these quantities may be expressed in terms of the two auxiliary functions

$$\varphi(\eta, \tau_0) = 1 + 2\eta \int_0^1 \varrho^*(\eta, \zeta, \tau_0) \, d\zeta, \qquad (3.116)$$

$$\psi(\eta, \tau_0) = e^{-\tau_0/\eta} + 2\eta \int_0^1 \sigma^*(\eta, \zeta, \tau_0) \, d\zeta. \qquad (3.117)$$

Thus, using the symmetry relations (3.104) and (3.105) and the definitions (3.116) and (3.117), we may write in place of (3.114) and (3.115)

$$B(\tau_0, \zeta, \tau_0) = \frac{\lambda(\tau_0)}{4} S\psi(\zeta, \tau_0), \qquad (3.118)$$

$$B^*(\tau_0, \zeta, \tau_0) = \frac{\lambda(\tau_0)}{4} S\varphi(\zeta, \tau_0). \qquad (3.119)$$

Substituting (3.116)–(3.119) into (3.110)–(3.113), we then obtain

$$\eta\zeta \frac{\partial \varrho}{\partial \tau_0} = \frac{\lambda(\tau_0)}{4} \psi(\eta, \tau_0)\psi(\zeta, \tau_0), \qquad (3.120)$$

$$\eta\zeta \frac{\partial \sigma}{\partial \tau_0} = -\zeta\sigma + \frac{\lambda(\tau_0)}{4} \varphi(\eta, \tau_0)\psi(\zeta, \tau_0), \qquad (3.121)$$

$$\eta\zeta\frac{\partial\varrho^*}{\partial\tau_0} = -(\eta+\zeta)\varrho^* + \frac{\lambda(\tau_0)}{4}\,\varphi(\eta,\,\tau_0)\varphi(\zeta,\,\tau_0), \qquad (3.122)$$

$$\eta\zeta\frac{\partial\sigma^*}{\partial\tau_0} = -\eta\sigma^* + \frac{\lambda(\tau_0)}{4}\,\psi(\eta,\,\tau_0)\varphi(\zeta,\,\tau_0). \qquad (3.123)$$

Integration of equations (3.120)–(3.123) leads to the following expressions for the reflection and transmission coefficients in terms of the auxiliary functions $\varphi(\eta,\,\tau_0)$ and $\psi(\eta,\,\tau_0)$:

$$\varrho(\eta,\,\zeta,\,\tau_0) = \tfrac{1}{4}\int_0^{\tau_0}\lambda(\tau)\,\psi(\eta,\,\tau)\,\psi(\zeta,\,\tau)\,\frac{d\tau}{\eta\zeta}, \qquad (3.124)$$

$$\sigma(\eta,\,\zeta,\,\tau_0) = \tfrac{1}{4}\int_0^{\tau_0}\lambda(\tau)\varphi(\eta,\,\tau)\psi(\zeta,\,\tau)e^{-(\tau_0-\tau)/\eta}\,\frac{d\tau}{\eta\zeta}, \qquad (3.125)$$

$$\varrho^*(\eta,\,\zeta,\,\tau_0) = \tfrac{1}{4}\int_0^{\tau_0}\lambda(\tau)\varphi(\eta,\,\tau)\varphi(\zeta,\,\tau)e^{-(\tau_0-\tau)(1/\eta+1/\zeta)}\,\frac{d\tau}{\eta\zeta}, \qquad (3.126)$$

$$\sigma^*(\eta,\,\zeta,\,\tau_0) = \tfrac{1}{4}\int_0^{\tau_0}\lambda(\tau)\psi(\eta,\,\tau)\varphi(\zeta,\,\tau)e^{-(\tau_0-\tau)/\zeta}\,\frac{d\tau}{\eta\zeta}. \qquad (3.127)$$

The auxiliary functions are determined by the integral equations

$$\varphi(\eta,\,\tau_0) = 1+\tfrac{1}{2}\int_0^1\frac{d\zeta}{\zeta}\int_0^{\tau_0}\lambda(\tau)\varphi(\eta,\,\tau)\varphi(\zeta,\,\tau)e^{-(\tau_0-\tau)(1/\eta+1/\zeta)}\,d\tau, \qquad (3.128)$$

$$\psi(\eta,\,\tau_0) = e^{-\tau_0/\eta}+\tfrac{1}{2}\int_0^1\frac{d\zeta}{\zeta}\int_0^{\tau_0}\lambda(\tau)\psi(\eta,\,\tau)\varphi(\zeta,\,\tau)e^{-(\tau_0-\tau)/\zeta}\,d\tau, \qquad (3.129)$$

which may be obtained by substitution of equations (3.126) and (3.127) into equations (3.116) and (3.117).

Obviously, when λ is constant throughout the atmosphere, $\varrho^*(\eta,\,\zeta,\,\tau_0) = \varrho(\eta,\,\zeta,\,\tau_0)$ and $\sigma^*(\eta,\,\zeta,\,\tau_0) = \sigma(\eta,\,\zeta,\,\tau_0)$. In this case, equations (3.124)–(3.127) give two different expressions for both the reflection and transmission coefficients. One expression for each of these quantities was obtained in Section 3.2. However, equations for the auxiliary functions deduced from (3.128) and (3.129) for $\lambda = constant$ differ from equations (3.47) and (3.48) derived previously.

It is also possible to obtain an expression for the reflection coefficient of a semi-infinite inhomogeneous atmosphere in terms of an auxiliary function. The reflection coefficient takes the form

$$\varrho(\eta,\,\zeta) = \tfrac{1}{4}\int_0^{\infty}\lambda(\tau)\,\varphi(\eta,\,\tau)\varphi(\zeta,\,\tau)e^{-\tau(1/\eta+1/\zeta)}\,\frac{d\tau}{\eta\zeta}, \qquad (3.130)$$

where the auxiliary function $\varphi(\eta, \tau)$ is determined by

$$\varphi(\eta, \tau) = 1 + \frac{1}{2} \int_0^1 \frac{d\zeta}{\zeta} \int_\tau^\infty \lambda(\tau')\varphi(\eta, \tau')\varphi(\zeta, \tau')e^{-(\tau'-\tau)(1/\eta+1/\zeta)} \, d\tau'. \tag{3.131}$$

We consider two particular cases of equations (3.130) and (3.131).

1. Let $\lambda(\tau) = 0$ in the interval from 0 to τ_1 and maintain its previous values for other τ (i.e. we separate a layer of optical thickness τ_1 from a semi-infinite atmosphere). It is easy to see that in this case the reflection coefficient is again given by equation (3.130), in which τ is measured from the new atmospheric boundary and the function $\varphi(\eta, \tau)$ has the same value at τ as the previous auxiliary function did at optical depth $\tau_1+\tau$ from the old boundary. In other words, knowledge of the one auxiliary function determined by equation (3.131) allows us to find the reflection coefficient of any atmosphere obtained from the original one by separating off a layer of arbitrary optical thickness.

2. Let $\lambda(\tau) = 0$ in the interval from τ_0 to infinity. Then equations (3.130) and (3.131) (in which it is necessary to replace the upper limit ∞ by τ_0) determine the reflection coefficient of an atmosphere of finite optical thickness τ_0. If in this case we make the change of variable $\tau_0-\tau = t$ and $\tau_0-\tau' = t'$ in (3.130) and (3.131), we arrive at equations (3.126) and (3.128).

In the theory of inhomogeneous atmospheres, the case of a semi-infinite atmosphere was first considered, and equations (3.130) and (3.131) were obtained by V. V. Sobolev [9] and Bellman and Kalaba [10]. Subsequently, Ueno [11] obtained equations (3.124)–(3.127) and (3.128)–(3.129) for atmospheres of finite optical thickness τ_0. Busbridge [12] obtained these results by another method and established the relation between them.

The study of atmospheres with isotropic scattering and a single scattering albedo dependent on depth is of particular interest for the theory of stellar spectra. In planetary physics more significance is attached to the examination of atmospheres in which the phase function as well as the parameter λ change with depth. The reflection and transmission coefficients for this latter case may still be expressed in terms of auxiliary functions which satisfy equations of the kind considered above. This problem was solved in the work of E. G. Yanovitskii [13].

In real planetary atmospheres the dependence of the phase function on altitude is quite complicated. At those frequencies characteristic of spectral lines the single scattering albedo λ is also a complicated function of altitude. Accordingly, it is still difficult to say what method is most suitable for the determination of the radiation field in an inhomogeneous atmosphere. It may be that certain of the methods described in Section 1.7 will be useful for this purpose. If the functions $\lambda(\tau)$ and $x(\gamma, \tau)$ deviate only slightly from their mean values, the perturbation method proposed by Abhyankar and Fymat [14] may be used to find the reflection and transmission coefficients of the atmosphere.

It is important to point out that the asymptotic expressions for the reflection and transmission coefficients of a homogeneous atmosphere of large optical thickness ($\tau_0 \gg 1$) which were obtained in Section 3.3 may be relatively easily generalized to the case of inhomogeneous atmospheres. Quite simple formulas for the quantities $\varrho(\eta, \zeta, \tau_0)$ and $\sigma(\eta, \zeta, \tau_0)$ were obtained in this way by E. G. Yanovitskii [15]. We shall present his expressions for the case of pure scattering.

For homogeneous atmospheres with $\lambda = 1$, the reflection and transmission coefficients are given for arbitrary phase function by the asymptotic expressions (3.68) and (3.69). For an inhomogeneous atmosphere the corresponding expressions are

$$\varrho(\eta, \zeta, \tau_0) = \varrho(\eta, \zeta) - \frac{4u(\eta)u(\zeta)}{(3-\bar{x}_1)\tau_0+3\delta}, \tag{3.132}$$

$$\sigma(\eta, \zeta, \tau_0) = \frac{4\tilde{u}(\eta)u(\zeta)}{(3-\bar{x}_1)\tau_0+3\delta}, \tag{3.133}$$

which are very similar to equations (3.68) and (3.69). We have here signified the mean value of the quantity x_1 in the atmosphere by $\bar{x}_1$, so that

$$\bar{x}_1 = \frac{1}{\tau_0} \int_0^{\tau_0} x_1(\tau) \, d\tau. \tag{3.134}$$

We have also set

$$\delta = 2 \int_0^1 [u(\eta)+\tilde{u}(\eta)]\eta^2 \, d\eta. \tag{3.135}$$

The quantity $\varrho(\eta, \zeta)$ is the reflection coefficient of a semi-infinite atmosphere, while $u(\eta)$ is the relative intensity emerging from the atmosphere for the Milne problem (that is, when the energy sources are located at infinitely great optical depth). By $\tilde{u}(\eta)$ we mean the analogous quantity for a semi-infinite atmosphere extending upwards from the lower boundary of the given atmosphere. The functions $u(\eta)$ and $\tilde{u}(\eta)$ are normalized according to condition (2.112).

In the derivation of equations (3.132) and (3.133) Yanovitskii has used the same method as in Section 3.3. Thus, he has introduced an imaginary plane at a great depth within the semi-infinite atmosphere and recalled that in the case of pure scattering, the radiation is isotropic in these layers for arbitrary dependence of the phase function on depth.

References

1. V. A. AMBARTSUMYAN, On the problem of diffuse light reflection by a turbid medium, *Dokl. Akad. Nauk SSSR* **38**, 257 (1943).
2. S. CHANDRASEKHAR, On the radiative equilibrium of a stellar atmosphere XVII, *Astrophys. J.* **105**, 441 (1947).
3. H. C. VAN DE HULST, Radiative transfer in thick atmospheres with an arbitrary scattering function, *Bull. Astron. Inst. Netherlands* **20**, 77 (1968).
4. V. V. SOBOLEV, The diffusion of radiation in a medium of large optical thickness for anisotropic scattering, *Dokl. Akad. Nauk SSSR* **179**, 41 (1968) [*Sov. Phys. Doklady* **13**, 180 (1968)].
5. T. A. GERMOGENOVA, On the character of the solution of the transfer equation for a plane layer, *Zh. Vychisl. Matem. i Matem. Fiz.* **1**, 1001 (1961).
6. V. V. SOBOLEV, The diffusion of radiation in a medium of finite optical thickness, *Astron. Zh.* **34**, 336 (1957) [*Sov. Astron. A.J.* **1**, 332 (1957)].
7. V. V. IVANOV and V. V. LEONOV, Anisotropic scattering of light in an optically thick atmosphere, *Izv. Akad. Nauk SSSR, Fiz. Atmos. Okeana* **1**, 803 (1965) [*Atmospheric Oceanic Physics* **1**, 464 (1965)].
8. V. V. SOBOLEV, Anisotropic light scattering in an atmosphere of large optical thickness, *Astrofizika* **4**, 325 (1968) [*Astrophysics* **4**, 125 (1968)].

9. V. V. SOBOLEV, Radiative transfer in an inhomogeneous medium, *Dokl. Akad. Nauk SSSR*, **111**, 1000 (1956) [*Sov. Phys. Doklady* **1**, 747 (1956)].
10. R. E. BELLMAN and R. E. KALABA, On the principle of invariant imbedding and propagation through inhomogeneous media, *Proc. Nat. Acad. Sci. USA* **42**, 629 (1956).
11. S. UENO, The probabilistic method for problems of radiative transfer: X. Diffuse reflection and transmission in a finite inhomogeneous atmosphere, *Astrophys. J.* **132**, 729 (1960).
12. I. W. BUSBRIDGE, On inhomogeneous stellar atmospheres, *Astrophys. J.* **133**, 198 (1961).
13. E. G. YANOVITSKII, Diffuse reflection and transmission of light by a plane, inhomogeneous medium for anisotropic scattering, *Astron. Zh.* **38**, 912 (1961) [*Sov. Astron. A.J.* **5**, 697 (1962)].
14. K. D. ABHYANKAR and A. L. FYMAT, Theory of radiative transfer in inhomogeneous atmospheres, *Astrophys. J.* **158**, 325, 337 (1969); **159**, 1009, 1019 (1970).
15. E. G. YANOVITSKII, Anisotropic light scattering in an inhomogeneous atmosphere, *Astron. Zh.* **48**, 323 (1971) [*Sov. Astron. A.J.* **15**, 253 (1971)].

Chapter 4

ATMOSPHERES OVERLYING A REFLECTING SURFACE

Up to this point in our consideration of light scattering in an atmosphere we have not taken account of the presence of a planetary surface. Radiation falling on the surface will be partly absorbed and partly reflected. The reflected fraction may be quite large. The surface of the Earth, for example, reflects about 20% of the incident radiation in the summer, and approximately 80% when snow covered. Photons reflected by the surface may be scattered by the atmosphere and then reflected again by the surface, and this may be repeated many times. It is obvious that this process can strongly influence the radiation field in the atmosphere.

In order to take account of reflection by the surface, it is necessary to establish the appropriate boundary condition for $\tau = \tau_0$. This will replace the second of the boundary conditions (1.33) which we have used previously.

It might appear that the results obtained in the previous chapters for an atmosphere without an adjacent surface are of quite limited interest, but this is not the case. If we know the radiation field in such an atmosphere, it is relatively simple to determine the radiation field in an atmosphere bounded by a reflecting surface. This is particularly simple when the surface reflects radiation isotropically and the surface albedo a does not depend on the angle of incidence. The intensity for $a \neq 0$ may then be expressed explicitly in terms of the intensity for the case $a = 0$.

We begin this chapter by presenting an integral equation for the source function in an atmosphere bounded by a surface with an arbitrary law of reflection. We then derive the relation between the source function and the intensity emerging from the atmosphere both with and without an adjacent reflecting surface. The case of isotropic reflection by the surface ($a = $ constant) is examined in detail. Simple expressions for the atmospheric albedo and the illumination of the surface are presented for various values of a and for $\tau_0 \gg 1$. At the end of the chapter, we briefly consider the case of specular reflection by the surface, which corresponds to an atmosphere above the sea.

4.1. Basic Equations

We consider the problem of light scattering in a plane-parallel atmosphere of optical thickness τ_0 illuminated by parallel rays. As previously, we designate the cosine of the angle of incidence by ζ and the incident flux through an area normal to the solar radiation by πS.

We shall allow for the presence of a planetary surface which reflects radiation. The atmospheric layer will then be illuminated not only by parallel radiation from above, but also by diffuse radiation from below. The diffuse radiation from the planetary surface is, however, not known *a priori*. It must be determined as part of the solution of the given problem.

We shall designate the intensity of the diffuse radiation at an optical depth τ traveling in a direction characterized by a polar angle arccos η and an azimuthal angle φ by $\bar{I}(\tau, \eta, \zeta, \varphi, \tau_0)$ and the corresponding source function by $\bar{B}(\tau, \eta, \zeta, \varphi, \tau_0)$. In our previous discussion, which neglected light scattering by the planetary surface, these quantities were written as I and B, respectively. It is necessary to distinguish between these quantities because we shall obtain equations below which contain simultaneously $\bar{B}$ and I and B and $\bar{I}$. Other characteristics of the atmospheric radiation field in the presence of a reflecting surface will also be designated with a superior bar.

We wish to write an equation which will determine $\bar{B}$ and $\bar{I}$. It is clear that equation (1.36), which expresses B in terms of I, will remain valid in the present case. Consequently, we have

$$\bar{B}(\tau, \eta, \zeta, \varphi, \tau_0) = \frac{\lambda}{4\pi} \int_0^{2\pi} d\varphi' \int_{-1}^{1} \bar{I}(\tau, \eta', \zeta, \varphi', \tau_0)\, x(\gamma')\, d\eta' + \frac{\lambda}{4} Sx(\gamma)e^{-\tau/\zeta}, \qquad (4.1)$$

where $x(\gamma)$ is the phase function and the angles γ' and γ are determined by equations (1.37) and (1.38).

For radiation traveling downwards, the intensity $\bar{I}$ will be expressed in terms of the source function $\bar{B}$ as follows:

$$\bar{I}(\tau, \eta, \zeta, \varphi, \tau_0) = \int_0^{\tau} \bar{B}(\tau', \eta, \zeta, \varphi, \tau_0)e^{-(\tau-\tau')/\eta}\frac{d\tau'}{\eta}. \qquad (4.2)$$

For radiation moving upwards (taking $\eta > 0$), the intensity $\bar{I}$ is

$$\bar{I}(\tau, -\eta, \zeta, \varphi, \tau_0) = \int_{\tau}^{\tau_0} \bar{B}(\tau', -\eta, \zeta, \varphi, \tau_0)e^{-(\tau'-\tau)/\eta}\frac{d\tau'}{\eta} + I_R(\eta, \zeta, \varphi, \tau_0)e^{-(\tau_0-\tau)/\eta}. \qquad (4.3)$$

The quantity I_R which enters equation (4.3) is the intensity of radiation diffusely reflected by the planetary surface. In order to obtain an expression for I_R, we must give the law of light reflection for the surface.

Let $y(\eta, \eta_0, \varphi-\varphi_0)\, d\omega/2\pi$ be the probability that a photon incident at an angle arccos η_0 to the normal and with azimuth φ_0 is reflected by the surface at an angle arccos η to the normal with azimuth φ and within a solid angle $d\omega$. The incident intensity I_0 and the reflected intensity I are then related by the expression

$$I\eta = \frac{1}{2\pi} y(\eta, \eta_0, \varphi-\varphi_0) I_0\eta_0. \qquad (4.4)$$

The surface of the planet is illuminated by both unscattered but attenuated solar radiation and by diffuse radiation from the atmosphere. Bearing in mind equation (4.4), we thus obtain for I_R the equation

$$I_R(\eta, \zeta, \varphi, \tau_0)\eta = -\frac{1}{2\pi}\int_0^1 \eta'\,d\eta' \int_0^{2\pi} y(\eta, \eta', \varphi-\varphi')\bar{I}(\tau_0, \eta', \zeta, \varphi', \tau_0)\,d\varphi' + \tfrac{1}{2}Se^{-\tau_0/\zeta}y(\eta, \zeta, \varphi)\zeta.$$

(4.5)

We emphasize that this equation contains the quantity $\bar{I}$, and not I. In other words, we have included both the effects of the scattering of photons in the atmosphere and also their multiple reflection from the surface in determining the diffuse radiation incident on the surface.

According to (4.2), the intensity falling on the planetary surface is

$$\bar{I}(\tau_0, \eta, \zeta, \varphi, \tau_0) = \int_0^{\tau_0} \bar{B}(\tau, \eta, \zeta, \varphi, \tau_0)e^{-(\tau_0-\tau)/\eta}\frac{d\tau}{\eta}.$$

(4.6)

By substituting equations (4.2) and (4.3) into equation (4.1) and using (4.5) and (4.6), we obtain an integral equation which determines the source function $\bar{B}$ in an atmosphere overlying a reflecting surface.

We shall continue to assume that the phase function has the form (1.8). Then in equation (4.1) we have

$$x(\gamma') = p(\eta, \eta')+2\sum_{m=1}^{n} p^m(\eta, \eta')\cos m(\varphi-\varphi')$$

(4.7)

and an analogous expression for $x(\gamma)$. We shall assume that the quantity y is also expanded as a cosine series in the azimuth,

$$y(\eta, \eta', \varphi-\varphi') = y^0(\eta, \eta')+2\sum_{m=1}^{n_1} y^m(\eta, \eta')\cos m(\varphi-\varphi').$$

(4.8)

Substituting these last two expressions into (4.1)–(4.3) and (4.5), we deduce that the function $\bar{B}$ may always be written as

$$\bar{B}(\tau, \eta, \zeta, \varphi, \tau_0) = \bar{B}^0(\tau, \eta, \zeta, \tau_0)+2\sum_{m=1}^{n} \bar{B}^m(\tau, \eta, \zeta, \tau_0)\cos m\varphi,$$

(4.9)

while $\bar{I}$ and I_R take the form

$$\bar{I}(\tau, \eta, \zeta, \varphi, \tau_0) = \bar{I}^0(\tau, \eta, \zeta, \tau_0)+2\sum_{m=1}^{n} \bar{I}^m(\tau, \eta, \zeta, \tau_0)\cos m\varphi,$$

(4.10)

$$I_R(\eta, \zeta, \varphi, \tau_0) = I_R^0(\eta, \zeta, \tau_0)+2\sum_{m=1}^{n} I_R^m(\eta, \zeta, \tau_0)\cos m\varphi.$$

(4.11)

It is necessary, however, to note the following two points:

1. If $n_1 > n$, the summations for I_R and $\bar{I}$ must also contain terms with $m > n$, which correspond to direct solar radiation reflected by the surface. We have not included these terms in equations (4.10) and (4.11).

2. If $n > n_1$, then $y^m = 0$ and $I_R^m = 0$ for $m > n_1$. The presence of a reflecting surface thus does not affect the corresponding functions $\bar{B}^m$ and $\bar{I}^m$, so that they are equal to the

functions B^m and I^m. Nonetheless, in the interest of generality, we shall write n and not n_1 as the upper limit of the sum in (4.11), bearing in mind that $I_R^m = 0$ for $m > n_1$.

Using equations (4.1)–(4.3) and (4.9)–(4.11), we may obtain an integral equation for the determination of $\bar{B}^m(\tau, \eta, \zeta, \tau_0)$. This equation has the form

$$\bar{B}^m(\tau, \eta, \zeta, \tau_0) = \frac{\lambda}{2} \int_0^1 p^m(\eta, \eta')\, d\eta' \int_0^\tau \bar{B}^m(\tau', \eta', \zeta, \tau_0) e^{-(\tau-\tau')/\eta'}\, \frac{d\tau'}{\eta'}$$

$$+ \frac{\lambda}{2} \int_0^1 p^m(\eta, -\eta')\, d\eta' \int_\tau^{\tau_0} \bar{B}^m(\tau', -\eta', \zeta, \tau_0) e^{-(\tau'-\tau)/\eta'}\, \frac{d\tau'}{\eta'}$$

$$+ \frac{\lambda}{2} \int_0^1 p^m(\eta, -\eta') e^{-(\tau_0-\tau)/\eta'}\, I_R^m(\eta', \zeta, \tau_0)\, d\eta' + \frac{\lambda}{4} Sp^m(\eta, \zeta) e^{-\tau/\zeta} \qquad (m = 0, 1, \ldots, n). \quad (4.12)$$

In order to find an expression for the quantity I_R^m which enters (4.12), we must substitute (4.8) and (4.11) into equation (4.5). We obtain

$$I_R^m(\eta, \zeta, \tau_0)\, \eta = \int_0^1 y^m(\eta, \eta')\bar{I}^m(\tau_0, \eta', \zeta, \tau_0)\, \eta'\, d\eta' + \tfrac{1}{2} Se^{-\tau_0/\zeta} y^m(\eta, \zeta)\, \zeta. \qquad (4.13)$$

Equation (4.12) for $y^m = 0$ reduces to equation (3.1), which applies in the absence of a reflecting surface.

It is very important to point out that the quantities $\bar{B}^m$ and B^m which are determined respectively by equations (4.12) and (3.31) are related in such a manner that $\bar{B}^m$ may be found for an arbitrary reflection function $y^m(\eta, \eta')$ if B^m is known. To find this relation we compare equations (4.12), (3.1), and (3.31) (the last two being equivalent), and use the superposition principle for solutions of linear equations. This procedure yields

$$\bar{B}^m(\tau, \eta, \zeta, \tau_0) = B^m(\tau, \eta, \zeta, \tau_0) + \frac{2}{S} \int_0^1 B^m(\tau_0-\tau, -\eta, \eta', \tau_0) I_R^m(\eta', \zeta, \tau_0)\, d\eta'. \quad (4.14)$$

The quantity I_R^m which enters (4.14) is expressed in terms of the intensity $\bar{I}^m$ with the aid of equation (4.13). This intensity may in turn be expressed in terms of $\bar{B}^m$. Expression (4.14) may therefore be considered to be an integral equation for the desired function $\bar{B}^m$.

From equation (4.14) we may obtain a series of equations which directly determine the radiation intensity at the boundaries of the atmosphere, $\bar{I}^m(0, -\eta, \zeta, \tau_0)$ and $\bar{I}^m(\tau_0, \eta, \zeta, \tau_0)$. Instead of these intensities we shall use the reflection and transmission coefficients, which we shall designate by $\bar{\varrho}^m$ and $\bar{\sigma}^m$, respectively. The appropriate formulas are

$$\bar{I}^m(0, -\eta, \zeta, \tau_0) = S\bar{\varrho}^m(\eta, \zeta, \tau_0)\zeta = \int_0^{\tau_0} \bar{B}^m(\tau, -\eta, \zeta, \tau_0) e^{-\tau/\eta}\, \frac{d\tau}{\eta} + I_R^m(\eta, \zeta, \tau_0) e^{-\tau_0/\eta}, \quad (4.15)$$

$$\bar{I}^m(\tau_0, \eta, \zeta, \tau_0) = S\bar{\sigma}^m(\eta, \zeta, \tau_0)\zeta = \int_0^{\tau_0} \bar{B}^m(\tau_0-\tau, \eta, \zeta, \tau_0) e^{-\tau/\eta}\, \frac{d\tau}{\eta}. \qquad (4.16)$$

The last term in equation (4.15) represents that part of the intensity reflected by the surface which emerges directly, although attenuated, from the atmosphere.

We shall for simplicity introduce the notation

$$I_R^m(\eta, \zeta, \tau_0) = S\beta^m(\eta, \zeta, \tau_0)\zeta. \tag{4.17}$$

Equation (4.13) then takes the form

$$\beta^m(\eta, \zeta, \tau_0)\eta = \int_0^1 y^m(\eta, \eta')\bar{\sigma}^m(\eta', \zeta, \tau_0)\, \eta'\, d\eta' + \tfrac{1}{2}e^{-\tau_0/\zeta}y^m(\eta, \zeta). \tag{4.18}$$

In order to find equations determining $\bar{\varrho}^m$ and $\bar{\sigma}^m$, we multiply (4.14) first by $e^{-\tau/\eta}$ and then by $e^{-(\tau_0-\tau)/\eta}$ and integrate in each case over τ between 0 and τ_0. Using equations (4.15)–(4.17), and also the equations (3.4) and (3.5) relating ϱ^m and σ^m to the function B^m, we obtain

$$\bar{\varrho}^m(\eta, \zeta, \tau_0) = \varrho^m(\eta, \zeta, \tau_0) + 2\int_0^1 \sigma^m(\eta, \eta', \tau_0)\beta^m(\eta', \zeta, \tau_0)\, \eta'\, d\eta' + \beta^m(\eta, \zeta, \tau_0)e^{-\tau_0/\eta}, \tag{4.19}$$

$$\bar{\sigma}^m(\eta, \zeta, \tau_0) = \sigma^m(\eta, \zeta, \tau_0) + 2\int_0^1 \varrho^m(\eta, \eta', \tau_0)\beta^m(\eta', \zeta, \tau_0)\eta'\, d\eta'. \tag{4.20}$$

Equations (4.18), (4.19), and (4.20) constitute a system of three equations for three unknown functions: β^m, $\bar{\varrho}^m$, and $\bar{\sigma}^m$. In their solution we take as given the reflection coefficient ϱ^m and transmission coefficient σ^m of the atmosphere in the absence of the reflecting surface, and also the reflection function of the surface $y^m(\eta, \eta')$. Once the function β^m has been found, the source function $\bar{B}^m$ is determined by the equation

$$\bar{B}^m(\tau, \eta, \zeta, \tau_0) = B^m(\tau, \eta, \zeta, \tau_0) + 2\zeta\int_0^1 B^m(\tau_0-\tau, -\eta, \eta', \tau_0)\beta^m(\eta', \zeta, \tau_0)\, d\eta', \tag{4.21}$$

which is obtained by substitution of (4.18) into (4.14).

The equations presented in this section have not been studied in detail for many particular reflection functions $y^m(\eta, \eta')$. Application of these equations in practice is also limited by the fact that the form of the reflection function will vary within any large region on the surface of a planet. Average values of the reflection function are thus generally used.

In the following we shall apply the equations which we have just obtained to two particularly simple laws of light reflection by the surface. Initially, we shall consider the case of isotropic reflection, and subsequently, the case of specular reflection.

4.2. The Case of Isotropic Reflection

Let us assume that the planetary surface reflects radiation isotropically; that is, that the reflected intensity does not depend upon direction. In this case the reflection function will, according to equation (4.4), have the form

$$y(\eta, \eta_0) = 2a(\eta_0)\eta, \tag{4.22}$$

where $a(\eta_0)$ is the albedo of the surface. In general, the albedo will depend on the angle of incidence, arccos η_0. For simplicity in what follows, however, we shall assume that $a = $ constant.

It is obvious that the presence of an isotropically reflecting surface can affect only the radiation field averaged over azimuth (the quantities $\bar{I}^0$ and $\bar{B}^0$). We shall now determine these quantities, omitting the superscript zero as in previous chapters. For $m \geq 1$, we have simply $\bar{I}^m = I^m$, $\bar{B}^m = B^m$.

In accordance with the above discussion, we set $m = 0$ in the equations obtained above, define $\bar{I}^0 = \bar{I}$ and $\bar{B}^0 = \bar{B}$, and make use of expression (4.22) for the reflection function with $a = $ constant. In place of equations (4.18)–(4.20) we then obtain

$$\beta(\zeta, \tau_0) = a\left[2\int_0^1 \bar{\sigma}(\eta', \zeta, \tau_0)\, \eta'\, d\eta' + e^{-\tau_0/\zeta}\right], \tag{4.23}$$

$$\bar{\varrho}(\eta, \zeta, \tau_0) = \varrho(\eta, \zeta, \tau_0) + \beta(\zeta, \tau_0)\left[2\int_0^1 \sigma(\eta, \eta', \tau_0)\eta'\, d\eta' + e^{-\tau_0/\eta}\right], \tag{4.24}$$

$$\bar{\sigma}(\eta, \zeta, \tau_0) = \sigma(\eta, \zeta, \tau_0) + 2\beta(\zeta, \tau_0)\int_0^1 \varrho(\eta, \eta', \tau_0)\eta'\, d\eta'. \tag{4.25}$$

A significant feature of the present case is the fact that $\bar{\varrho}$ and $\bar{\sigma}$ may be expressed explicitly in terms of ϱ and σ. Substituting (4.25) into (4.23), we obtain

$$\beta(\zeta, \tau_0)\left[1 - 4a\int_0^1 \eta\, d\eta \int_0^1 \varrho(\eta, \eta', \tau_0)\eta'\, d\eta'\right] = a\left[2\int_0^1 \sigma(\eta, \zeta, \tau_0)\eta\, d\eta + e^{-\tau_0/\zeta}\right]. \tag{4.26}$$

If we introduce the notations

$$A(\zeta, \tau_0) = 2\int_0^1 \varrho(\eta, \zeta, \tau_0)\eta\, d\eta, \tag{4.27}$$

$$V(\zeta, \tau_0) = 2\int_0^1 \sigma(\eta, \zeta, \tau_0)\eta\, d\eta + e^{-\tau_0/\zeta}, \tag{4.28}$$

and

$$A_s(\tau_0) = 2\int_0^1 A(\eta, \tau_0)\eta\, d\eta, \tag{4.29}$$

we may rewrite equations (4.24)–(4.26) as

$$\beta(\zeta, \tau_0) = \frac{aV(\zeta, \tau_0)}{1 - aA_s(\tau_0)}, \tag{4.30}$$

$$\bar{\varrho}(\eta, \zeta, \tau_0) = \varrho(\eta, \zeta, \tau_0) + \frac{aV(\eta, \tau_0)\, V(\zeta, \tau_0)}{1 - aA_s(\tau_0)}, \tag{4.31}$$

$$\bar{\sigma}(\eta, \zeta, \tau_0) = \sigma(\eta, \zeta, \tau_0) + \frac{aA(\eta, \tau_0)\, V(\zeta, \tau_0)}{1 - aA_s(\tau)}. \tag{4.32}$$

Equations (4.31) and (4.32) express the reflection and transmission coefficients of an atmosphere bounded by a reflecting surface in terms of the reflection and transmission coefficients

7

for $a = 0$. We recall that the functions $\varrho(\eta, \zeta, \tau_0)$ and $\sigma(\eta, \zeta, \tau_0)$ are symmetric in η and ζ. The function $\bar{\varrho}(\eta, \zeta, \tau_0)$ is also symmetric, while $\bar{\sigma}(\eta, \zeta, \tau_0)$ is not.

Equations (4.31)–(4.32) were first obtained for isotropic scattering [1, 2]. It was subsequently noted that they remain valid for an arbitrary phase function (e.g. *TRT*, p. 237).

The quantities entering equations (4.31) and (4.32) have an obvious physical significance. The quantity $A(\zeta, \tau_0)$ is the plane albedo of the atmosphere, while $A_s(\tau_0)$ is its spherical albedo. By $V(\zeta, \tau_0)$ we denote the ratio of the illumination of the lower atmospheric boundary from above (which for $a = 0$ equals the illumination of a totally absorbing underlying surface) to the external illumination of the upper boundary.

A simple expression for the source function $\bar{B}(\tau, \eta, \zeta, \tau_0)$ may be obtained when the reflection law of the planetary surface has the form considered in this section (sometimes referred to as Lambert's law). Substituting (4.30) into (4.21), we have

$$\bar{B}(\tau, \eta, \zeta, \tau_0) = B(\tau, \eta, \zeta, \tau_0) + \frac{2a\zeta V(\zeta, \tau_0)}{1 - aA_s(\tau_0)} \int_0^1 B(\tau_0 - \tau, -\eta, \eta', \tau_0) \, d\eta'. \qquad (4.33)$$

This equation expresses the source function $\bar{B}$ for $a \neq 0$ in terms of the source function B for $a = 0$.

4.3. The Albedo of the Atmosphere and Illumination of the Surface

We are already acquainted with the problem of determining the atmospheric albedo and the illumination of the planetary surface (see Sections 1.5 and 3.1). We will now consider this problem in more detail, following the approach given in [3].

We initially consider an atmosphere in the absence of an adjacent reflecting surface, so that $a = 0$. Let E_0 be the illumination of the upper boundary of the atmosphere by solar radiation, so that $E_0 = \pi S \zeta$. We shall denote by $E(0, \zeta, \tau_0)$ the illumination of the upper boundary of the atmosphere by diffuse radiation from below, and by $E(\tau_0, \zeta, \tau_0)$ the illumination of the lower atmospheric boundary (from above) by both diffuse radiation and direct solar radiation attenuated by the atmosphere. Clearly,

$$E(0, \zeta, \tau_0) = A(\zeta, \tau_0) E_0, \qquad (4.34)$$

where $A(\zeta, \tau_0)$ is the atmospheric albedo, which is expressed in terms of the reflection coefficient ϱ by equation (4.27). Also,

$$E(\tau_0, \zeta, \tau_0) = V(\zeta, \tau_0) E_0, \qquad (4.35)$$

where $V(\zeta, \tau_0)$ is given by equation (4.28).

If the functions $A(\zeta, \tau_0)$ and $V(\zeta, \tau_0)$ are known, it is easy to find the energy U which experiences true absorption in the atmosphere (per column of unit cross-sectional area). This quantity equals

$$U(\zeta, \tau_0) = [1 - A(\zeta, \tau_0) - V(\zeta, \tau_0)] E_0. \qquad (4.36)$$

We now consider the corresponding quantities for the case $a \neq 0$, designating them with a superior bar. They will be related by the following expressions:

$$\bar{E}(0, \zeta, \tau_0) = \bar{A}(\zeta, \tau_0) E_0, \qquad (4.37)$$

where

$$\bar{A}(\zeta, \tau_0) = 2 \int_0^1 \bar{\varrho}(\eta, \zeta, \tau_0)\eta \, d\eta, \tag{4.38}$$

and

$$\bar{E}(\tau_0, \zeta, \tau_0) = \bar{V}(\zeta, \tau_0) E_0, \tag{4.39}$$

where

$$\bar{V}(\zeta, \tau_0) = 2 \int_0^1 \bar{\sigma}(\eta, \zeta, \tau_0)\eta \, d\eta + e^{-\tau_0/\zeta}. \tag{4.40}$$

In place of equation (4.36), we have

$$\bar{U}(\zeta, \tau_0) = [1 - \bar{A}(\zeta, \tau_0) - (1-a) \, \bar{V}(\zeta, \tau_0)] \, E_0, \tag{4.41}$$

where $(1-a)\bar{V}(\zeta, \tau_0) E_0$ is the energy absorbed by the planetary surface (per unit area per unit time).

These quantities for an atmosphere adjacent to a reflecting surface may be expressed in terms of the corresponding quantities for $a = 0$ with the aid of equations (4.31) and (4.32). Substituting (4.31) and (4.32) into (4.38) and (4.40) and making use of (4.27) and (4.28), we obtain

$$\bar{A}(\zeta, \tau_0) = A(\zeta, \tau_0) + \frac{aV_s(\tau_0)}{1 - aA_s(\tau_0)} \, V(\zeta, \tau_0) \tag{4.42}$$

and

$$\bar{V}(\zeta, \tau_0) = \frac{V(\zeta, \tau_0)}{1 - aA_s(\tau_0)}, \tag{4.43}$$

where

$$V_s(\tau_0) = 2 \int_0^1 V(\zeta, \tau_0)\zeta \, d\zeta. \tag{4.44}$$

Substitution of (4.42) and (4.43) into (4.41), and use of (4.36), yields

$$\bar{U}(\zeta, \tau_0) = U(\zeta, \tau_0) + a \, \frac{1 - A_s(\tau_0) - V_s(\tau_0)}{1 - aA_s(\tau_0)} \, V(\zeta, \tau_0)E_0. \tag{4.45}$$

These equations become significantly simpler for an atmosphere of large optical thickness $(\tau_0 \gg 1)$. In this case, the asymptotic expressions (3.55) and (3.57) may be used for the reflection and transmission coefficients. Introducing these expressions into equations (4.27) and (4.28) and neglecting the term $e^{-\tau_0/\zeta}$, we obtain

$$A(\zeta, \tau_0) = A(\zeta) - Cf(\tau_0)u(\zeta) \tag{4.46}$$

and

$$V(\zeta, \tau_0) = Cg(\tau_0)u(\zeta), \tag{4.47}$$

where

$$C = 2 \int_0^1 u(\eta)\eta \, d\eta, \tag{4.48}$$

7*

$A(\zeta)$ and $u(\zeta)$ are the albedo and transmission functions of a semi-infinite atmosphere with the same λ and $x(\gamma)$ as the given atmosphere, and $f(\tau_0)$ and $g(\tau_0)$ are determined by equations (3.56) and (3.58).

Equations (4.46) and (4.47) refer to an atmosphere for which $a = 0$. Substituting them into equations (4.42) and (4.43), we find the following expressions for $a \neq 0$:

$$\bar{A}(\zeta, \tau_0) = A(\zeta) - Cu(\zeta)\left[f(\tau_0) - \frac{aV_s(\tau_0)}{1 - aA_s(\tau_0)}\, g(\tau_0)\right] \tag{4.49}$$

and

$$\bar{V}(\zeta, \tau_0) = \frac{Cg(\tau_0)}{1 - aA_s(\tau_0)}\, u(\zeta), \tag{4.50}$$

where

$$A_s(\tau_0) = A_s - C^2 f(\tau_0), \quad V_s(\tau_0) = C^2 g(\tau_0). \tag{4.51}$$

The quantity A_s is the spherical albedo of the semi-infinite atmosphere.

If the role of true absorption is small ($k \ll 1$) in an atmosphere of large optical thickness ($\tau_0 \gg 1$), the formulas for the atmospheric albedo and illumination of the surface may be further simplified. In this case, the albedo of the semi-infinite atmosphere $A(\zeta)$ in equations (4.46) and (4.47) must be replaced by equation (2.107). In addition, the function $u(\zeta)$ becomes $u_0(\zeta)$ (the notation we have used for $k = 0$), and $C = 1$. As a result, we obtain

$$A(\zeta, \tau_0) = 1 - h(\tau_0)u_0(\zeta), \tag{4.52}$$

$$V(\zeta, \tau_0) = g(\tau_0)u_0(\zeta), \tag{4.53}$$

where $h(\tau_0)$ and $g(\tau_0)$ are given by equations (3.62)–(3.64). Equations (4.49) and (4.50) are then replaced by

$$\bar{A}(\zeta, \tau_0) = 1 - u_0(\zeta)\left[h(\tau_0) - \frac{ag^2(\tau_0)}{1 - a + ah(\tau_0)}\right] \tag{4.54}$$

and

$$\bar{V}(\zeta, \tau_0) = \frac{g(\tau_0)\, u_0(\zeta)}{1 - a + ah(\tau_0)}, \tag{4.55}$$

where we have made use of (2.107), (4.51), and (4.52).

Two specific cases of equations (4.54) and (4.55) are of particular interest. They correspond to different limiting values of the quantity $k\tau_0$, on which the functions $g(\tau_0)$ and $h(\tau_0)$ depend.

1. When $k\tau_0 \gg 1$, we find in place of (4.54) and (4.55)

$$\bar{A}(\zeta, \tau_0) = 1 - \frac{4k}{3 - x_1}\, u_0(\zeta), \tag{4.56}$$

$$\bar{V}(\zeta, \tau_0) = 8ke^{-k\tau_0}\, \frac{u_0(\zeta)}{4ka + (3 - x_1)(1 - a)}\,. \tag{4.57}$$

We see that equation (4.56) coincides with equation (2.107), so that the plane atmospheric albedo in this case is the same as that of a semi-infinite atmosphere. It follows from (4.57) that the illumination of the planetary surface is very small. In this case, the principle portion of the radiation incident on the planet is diffusely reflected by the atmosphere, a small fraction (of order k) undergoes true absorption in the atmosphere, and a negligible fraction (of order $e^{-k\tau_0}$ or smaller) is absorbed by the surface.

2. For $k\tau_0 \ll 1$ (in particular, in the case of pure scattering), we find from (4.54) and (4.55) that

$$\bar{A}(\zeta, \tau_0) = 1 - \frac{4(1-a)u_0(\zeta)}{4a + (1-a)\left[(3-x_1)\tau_0 + 3\delta\right]} \tag{4.58}$$

and

$$\bar{V}(\zeta, \tau_0) = \frac{4u_0(\zeta)}{4a + (1-a)\left[(3-x_1)\tau_0 + 3\delta\right]}, \tag{4.59}$$

where δ is given by equation (3.61). In this case, as in the previous one, the radiation incidens on the planet is principally diffusely reflected by the atmosphere. In contrast to the previous case, however, a certain fraction of the radiation (of order $1/\tau_0$ if a is not very close to 1) it absorbed by the surface. The fraction of radiation which undergoes true absorption in the atmosphere is very small. Equation (4.59), when evaluated for the phase function $x(\gamma) = 1 + x_1 \cos \gamma$ and for $a = 0$, reduces to an equation originally found by S. Piotrowski [4].

It should be noted that asymptotic expressions for $\bar{\varrho}(\eta, \zeta, \tau_0)$ and $\bar{\sigma}(\eta, \zeta, \tau_0)$ for $\tau_0 \gg 1$ may be obtained from equations (4.31) and (4.32). For this purpose one must use the asymptotic expressions for $\varrho(\eta, \zeta, \tau_0)$, $\sigma(\eta, \zeta, \tau_0)$, $A(\zeta, \tau_0)$ and $V(\zeta, \tau_0)$ derived in Section 3.3, and in the present section.

4.4. The Spherical Albedo of a Planet

The spherical albedo has already been defined in Section 1.5. It is the ratio of the total energy reflected by the planet to the solar energy incident upon it. The spherical albedo depends on the optical thickness of the atmosphere τ_0, the phase function $x(\gamma)$, the single scattering albedo λ, and the reflecting properties of the planetary surface. We designate the spherical albedo by $A_s(\tau_0)$ in the absence of reflection by the surface, and by $\bar{A}_s(\tau_0)$ when such reflection occurs.

Another quantity of interest is the ratio of the energy falling on the total surface of the planet to the solar energy incident upon the planet. We shall denote this quantity by $V_s(\tau_0)$ in the absence of surface reflection and by $\bar{V}_s(\tau_0)$ when such reflection occurs.

Equations (4.29) and (4.44), which were derived above, express $A_s(\tau_0)$ and $V_s(\tau_0)$ in terms of the quantities $A(\zeta, \tau_0)$ and $V(\zeta, \tau_0)$. These latter quantities represent, respectively, the fractions of radiation reflected and transmitted by the atmosphere at a location where the angle of incidence of solar radiation is arccos ζ. The analogous expressions when surface reflection occurs are

$$\bar{A}_s(\tau_0) = 2 \int_0^1 \bar{A}(\zeta, \tau_0)\zeta \, d\zeta, \quad \bar{V}_s(\tau_0) = 2 \int_0^1 \bar{V}(\zeta, \tau_0)\zeta \, d\zeta. \tag{4.60}$$

Let us find equations expressing $\bar{A}_s(\tau_0)$ and $\bar{V}_s(\tau_0)$ in terms of $A_s(\tau_0)$ and $V_s(\tau_0)$ for the case of isotropic reflection by the planetary surface with constant albedo a. This may be accomplished by substituting equations (4.42) and (4.43) into equation (4.60). We obtain

$$\bar{A}_s(\tau_0) = A_s(\tau_0) + \frac{aV_s^2(\tau_0)}{1 - aA_s(\tau_0)}, \tag{4.61}$$

$$\bar{V}_s(\tau_0) = \frac{V_s(\tau_0)}{1 - aA_s(\tau_0)}. \tag{4.62}$$

For the case of pure scattering in the atmosphere we must have

$$A_s(\tau_0) + V_s(\tau_0) = 1. \tag{4.63}$$

It is then possible to express $\bar{A}_s(\tau_0)$ in terms of $A_s(\tau_0)$ alone, and $\bar{V}_s(\tau_0)$ in terms of $\bar{V}_s(\tau_0)$ alone. In place of (4.61) and (4.62) we find

$$\bar{A}_s(\tau_0) = 1 - \frac{(1-a) \ [1 - A_s(\tau_0)]}{1 - aA_s(\tau_0)}, \tag{4.64}$$

$$\bar{V}_s(\tau_0) = \frac{V_s(\tau_0)}{1 - a + aV_s(\tau_0)}. \tag{4.65}$$

In a previous section we gave asymptotic expressions for $\bar{A}(\zeta, \tau_0)$ and $\bar{V}(\zeta, \tau_0)$ which are valid when $\tau_0 \gg 1$. By substituting these into (4.60) we may obtain asymptotic expressions for $\bar{A}_s(\tau_0)$ and $\bar{V}_s(\tau_0)$. We shall derive these expressions for the spherical albedo $\bar{A}_s(\tau_0)$.

Substituting (4.49) into the first of equations (4.60) and using (4.51), we find

$$\bar{A}_s(\tau_0) = A_s - C^2 \left[f(\tau_0) - \frac{aC^2 g^2(\tau_0)}{1 - aA_s - aC^2 f(\tau_0)} \right], \tag{4.66}$$

which is valid for $\tau_0 \gg 1$. The functions $f(\tau_0)$ and $g(\tau_0)$ are given by equations (3.56) and (3.58).

For $\tau_0 \gg 1$ and $k \ll 1$, we have in place of (4.66)

$$\bar{A}_s(\tau_0) = 1 - h(\tau_0) + \frac{ag^2(\tau_0)}{1 - a + ah(\tau_0)}, \tag{4.67}$$

where $g(\tau_0)$ and $h(\tau_0)$ are given by equations (3.62)–(3.64). We recall that for small true absorption, $k = \sqrt{(1-\lambda)(3-x_1)}$. From the relationships satisfied by $g(\tau_0)$ and $h(\tau_0)$ we then conclude that the quantity $\bar{A}_s(\tau_0)$ depends on four parameters: a, τ_0, λ and x_1 (the dependence on the phase function through δ is very weak).

Equation (4.67) has two limiting cases, depending upon the magnitude of the quantity $k\tau_0$. For $k\tau_0 \gg 1$ it follows that

$$\bar{A}_s(\tau_0) = 1 - \frac{4k}{3 - x_1}. \tag{4.68}$$

In this case the spherical albedo depends only on the parameters λ and x_1, and not on τ_0 and a. It is given by the same formula which holds when $\tau_0 = \infty$.

When $k\tau_0 \ll 1$ equation (4.67) becomes

$$\bar{A}_s(\tau_0) = 1 - \frac{4(1-a)}{4a + (1-a)\left[(3-x_1)\,\tau_0 + 3\delta\right]}. \tag{4.69}$$

In this case $\bar{A}_s(\tau_0)$ depends on a, τ_0, and x_1 but not on λ. It has the same value as in the case of pure scattering ($\lambda = 1$).

TABLE 4.1. SPHERICAL ALBEDO $\bar{A}_s(\tau_0)$ (IN %) FOR AN ATMOSPHERE OF LARGE OPTICAL THICKNESS τ_0, SMALL TRUE ABSORPTION k, A PHASE FUNCTION WITH $x_1 = 2$ AND AN UNDERLYING SURFACE ALBEDO a

τ_0	$a = 0$						$a = 0.2$					
	\multicolumn{6}{c}{k}							
	0	0.02	0.04	0.06	0.08	0.10	0	0.02	0.04	0.06	0.08	0.10
10	72	70	67	63	59	54	74	72	69	65	61	56
15	79	77	74	70	64	58	80	78	75	71	65	58
20	83	81	78	73	66	59	84	82	78	73	67	59
30	88	86	81	75	68	60	89	86	82	75	68	60
50	93	90	84	76	68	60	93	90	84	76	68	60
∞	100	92	84	76	68	60	100	92	84	76	68	60

τ_0	$a = 0.4$						$a = 0.6$					
	\multicolumn{6}{c}{k}							
	0	0.02	0.04	0.06	0.08	0.10	0	0.02	0.04	0.06	0.08	0.10
10	76	75	72	68	64	58	80	79	76	73	68	62
15	82	80	77	72	66	59	84	83	79	74	68	61
20	85	83	79	74	67	60	87	85	81	75	68	60
30	89	87	82	75	68	60	90	88	83	76	68	60
50	93	90	84	76	68	60	93	90	84	76	68	60
∞	100	92	84	76	68	60	100	92	84	76	68	60

τ_0	$a = 0.8$						$a = 1$					
	\multicolumn{6}{c}{k}							
	0	0.02	0.04	0.06	0.08	0.10	0	0.02	0.04	0.06	0.08	0.10
10	87	86	83	79	73	66	100	99	95	89	82	73
15	89	87	83	78	71	62	100	98	92	84	75	65
20	90	88	84	77	69	61	100	97	90	81	71	62
30	92	90	84	76	68	60	100	96	87	78	69	60
50	94	91	84	76	68	60	100	94	85	76	68	60
∞	100	92	84	76	68	60	100	92	84	76	68	60

Sample values of the spherical albedo computed from equation (4.67) for various values of the parameters a, τ_0 and k are presented in Table 4.1. The parameter x_1, which characterizes the degree of elongation of the phase function, was assigned the value $x_1 = 2$. Knowledge of the details of the phase function is not necessary for the present purpose. It is clear from the table that the calculated values of the spherical albedo are quite large. This is to be expected with the present assumption of large optical thickness τ_0 and small coefficient of true absorption relative to the coefficient of scattering (so that $(1 - \lambda) \ll 1$).

4.5. Specular Reflection of Light

A particularly important type of light reflection by a surface is specular reflection. In this case the reflected and incident rays are in the same vertical plane and the angle of reflection equals the angle of incidence. In general, the reflected intensity I is some fraction of the incident intensity I_0, the fraction depending upon the angle of incidence arccos ζ. We thus have

$$I = r(\zeta)\, I_0, \qquad (4.70)$$

where $0 \le r(\zeta) \le 1$.

We shall postulate that the atmosphere is bounded from below by a specularly reflecting surface. We shall expand the source function in the form (4.9) and obtain an integral equation determining each $\bar{B}^m$. As previously, we assume that parallel radiation is incident on the atmosphere from above at an angle arccos ζ to the normal, and that the incident flux at the upper boundary equals $\pi S \zeta$. This radiation is partially transmitted through the atmosphere and specularly reflected by the surface. As a result, parallel rays will be incident on the atmosphere from below. The angle of incidence of these latter rays again equals arccos ζ, and the corresponding illumination of the lower atmospheric boundary equals $\pi S \zeta e^{-\tau_0/\zeta} r(\zeta)$. In addition, the atmosphere is illuminated from below by diffuse radiation resulting from the specular reflection of the diffuse atmospheric radiation falling on the surface.

On the basis of the above discussion, we may write the following integral equation for the determination of $\bar{B}^m$:

$$\bar{B}^m(\tau, \eta, \zeta, \tau_0) = \frac{\lambda}{2} \int_0^1 p^m(\eta, \eta')\, d\eta' \int_0^\tau \bar{B}^m(\tau', \eta', \zeta, \tau_0)\, e^{-(\tau - \tau')/\eta'} \frac{d\tau'}{\eta'}$$

$$+ \frac{\lambda}{2} \int_0^1 p^m(\eta, -\eta')\, d\eta' \int_\tau^{\tau_0} \bar{B}^m(\tau', -\eta', \zeta, \tau_0)\, e^{-(\tau' - \tau)/\eta'} \frac{d\tau'}{\eta'}$$

$$+ \frac{\lambda}{2} \int_0^1 p^m(\eta, -\eta')\, e^{-(\tau_0 - \tau)/\eta'}\, I_R^m(\eta', \zeta, \tau_0)\, d\eta'$$

$$+ \frac{\lambda}{4} S e^{-\tau_0/\zeta} \zeta r(\zeta)\, p^m(\eta, -\zeta)\, e^{-(\tau_0 - \tau)/\zeta} + \frac{\lambda}{4} S p^m(\eta, \zeta)\, e^{-\tau/\zeta}, \qquad (4.71)$$

where I_R^m is the intensity of diffuse radiation reflected by the surface. Equation (4.71) may be obtained from equation (4.12) by dividing the next-to-the-last term in the latter into two parts (corresponding to reflection of direct solar radiation and of diffuse radiation by the surface).

For specular reflection the reflection function of the surface is

$$y^m(\eta, \eta') = \delta(\eta - \eta') \, r(\eta'),$$ (4.72)

where $\delta(\eta)$ is the Dirac δ-function. Equation (4.13) then gives

$$I_R^m(\eta, \zeta, \tau_0) = \bar{I}^m(\tau_0, \eta, \zeta, \tau_0) \, r(\eta),$$ (4.73)

where $\bar{I}^m(\tau_0, \eta, \zeta, \tau_0)$ is the intensity of diffuse radiation incident on the surface, and we have omitted the last term of (4.13) which has been included separately in equation (4.71) above. Equation (4.73) could, of course, also be written on the basis of (4.70).

We observe that the inhomogeneous terms in equation (4.71) are a superposition of the inhomogeneous terms of equations (3.1) and (3.31) which determine the function B^m. We thus obtain

$$\bar{B}^m(\tau, \eta, \zeta, \tau_0) = B^m(\tau, \eta, \zeta, \tau_0) + B^m(\tau_0 - \tau, -\eta, \zeta, \tau_0) \, e^{-\tau_0/\zeta} \, r(\zeta)$$

$$+ 2\zeta \int_0^1 B^m(\tau_0 - \tau, -\eta, \eta', \tau_0) \, \bar{\sigma}^m(\eta', \zeta, \tau_0) r(\eta') \, d\eta',$$ (4.74)

where we have used the relation

$$I_R^m(\eta, \zeta, \tau_0) = S\bar{\sigma}^m(\eta, \zeta, \tau_0) \, \zeta r(\eta),$$ (4.75)

which follows from (4.73) and (4.16). Equation (4.74) may be used to find $\bar{B}^m$ when $r \neq 0$ if B^m is known for $r = 0$. In order to do this, however, it is also necessary to know the transmission coefficient of the atmosphere $\bar{\sigma}^m$ for $r \neq 0$. Equations for the determination of this quantity and also $\bar{\varrho}^m$ may be easily obtained from equation (4.74).

Multiplying (4.74) by $e^{-\tau/\eta}$, integrating over τ between 0 and τ_0, and using equations (4.15), (4.16) and (4.75), we find

$$\bar{\varrho}^m(\eta, \zeta, \tau_0) = \varrho^m(\eta, \zeta, \tau_0) + \sigma^m(\eta, \zeta, \tau_0) \, e^{-\tau_0/\zeta} \, r(\zeta)$$

$$+ \bar{\sigma}^m(\eta, \zeta, \tau_0) \, e^{-\tau_0/\eta} r(\eta) + 2 \int_0^1 \sigma^m(\eta, \eta', \tau_0) \, \bar{\sigma}^m(\eta', \zeta, \tau_0) \, r(\eta') \, \eta' \, d\eta'.$$ (4.76)

After multiplying (4.74) by $e^{-(\tau_0 - \tau)/\eta}$, we may obtain in an analogous manner

$$\bar{\sigma}^m(\eta, \zeta, \tau_0) = \sigma^m(\eta, \zeta, \tau_0) + \varrho^m(\eta, \zeta, \tau_0) \, e^{-\tau_0/\zeta} \, r(\zeta)$$

$$+ 2 \int_0^1 \varrho^m(\eta, \eta', \tau_0) \, \bar{\sigma}^m(\eta', \zeta, \tau_0) \, r(\eta') \, \eta' \, d\eta'.$$ (4.77)

The problem of light scattering in an atmosphere bounded by a specularly reflecting surface is thus reduced to the determination of the transmission coefficient $\bar{\sigma}^m$ from equation (4.77). Once this coefficient has been found, the reflection coefficient $\bar{\varrho}^m$ is given by equation (4.76) and the source function $\bar{B}^m$ by equation (4.74).

The surface of the sea under quiet conditions is an example of a specularly reflecting surface. The function $r(\zeta)$ is given in this case by the well known Fresnel formula. The radiation field in the atmosphere above the sea may thus be determined with the aid of the equations obtained in the present section. This problem was considered in the book *TRT* (Sections 7.5 and 7.7) and in an article by S. D. Gutshabash [5].

Another problem whose solution requires consideration of specular reflection is the determination of the radiation field within the sea. A fraction $1 - r(\zeta)$ of those photons which are incident on the surface penetrate into the sea and experience scattering and true absorption. In the course of such scattering these photons may be reflected from the surface of the sea (when they are incident at sufficiently large angles they experience total internal reflection). This process of reflection of photons by the surface greatly influences the radiation field within the sea.

This problem of diffuse radiation within the sea differs from the previous problem in that a reflecting surface bounds the medium under consideration from above (that is, from the side on which radiation is incident on the medium), rather than from below. In the case of isotropic scattering, this problem was examined in the articles of V. A. Ambartsumyan [6] and V. V Sobolev [7]. In the first of these references, the intensity of radiation emerging from the sea was determined, while in the second reference the source function was found.

We should note, however, that in a rigorous treatment the two problems described above cannot be considered independently of one another, since photons may travel back and forth between the atmosphere and the sea many times. Consequently, there arises a rather complicated problem of the simultaneous determination of the radiation field in the atmosphere and in the sea. In solving this problem it is necessary to take account of the refraction of radiation and the change in its intensity during passage from one medium into the other (see *TRT*, §8, Chapter 7).

The problem of diffuse reflection and transmission of light in an atmosphere bounded by a specularly reflecting surface has recently been considered by Casti, Kalaba and Ueno [8]. These authors assumed that the atmospheric phase function was isotropic. The problem was reduced to the solution of equations which determine the reflection and transmission coefficients directly, with the optical thickness as independent variable. Numerical results were given for the case when $r(\zeta) = 1$.

References

1. H. C. van de Hulst, Scattering in a planetary atmosphere, *Ap. J.* **107**, 220 (1948).
2. V. V. Sobolev, On the brightness coefficients of a plane layer of a turbid medium, *Dokl. Akad. Nauk SSSR* **61**, 803 (1948).
3. V. V. Sobolev, The albedo and surface illumination for a planet possessing an atmosphere, *Astron. Zh.* **46**, 419 (1969) [*Sov. Astron.—AJ* **13**, 330 (1969)].
4. S. Piotrowski, Asymptotic case of the diffusion of light through an optically thick scattering layer, *Acta Astronomica* **6**, 61 (1956).
5. S. D. Gutshabash, Scattering of light in an atmosphere adjoining a specularly reflecting surface, *Izv. Akad. Nauk SSSR, Ser. geofiz.* No.12, 1812 (1960) [*Bull. (Izv.) Acad. Sci. USSR, Geophys. Ser.*, No.12, 1213 (1960)].
6. V. A. Ambartsumyan, On the problem of multiple scattering of light in a plane parallel medium with internal reflection at the bounding surface, *Trudy Astron. Obs. Leningrad. Gos. Univ.* **20** (1964).
7. V. V. Sobolev, The diffusion of radiation in a medium with a specularly reflecting boundary, *Dokl. Akad. Nauk SSSR* **136**, 571 (1961) [*Sov. Phys. Doklady* **6**, 21 (1961)].
8. J. L. Casti, R. Kalaba and S. Ueno, Reflection and transmission functions for finite isotropically scattering atmospheres with specular reflectors, *J. Quant. Spectrosc. Radiat. Transfer* **9**, 537 (1969).

Chapter 5

GENERAL THEORY

IN THE preceding chapters we have examined the problem of diffuse reflection and transmission of light by an atmosphere. We shall now consider the determination of the diffuse radiation intensity at any optical depth within the atmosphere. As a specific example of the resulting equations, we shall again obtain the expressions for the intensity emerging from the atmosphere.

We shall assume, as previously, that the source function B and the radiation intensity I are expanded as cosine series in azimuth, and we shall seek the coefficients B^m and I^m of the expansions. For B^m we have the integral equation (1.53), while the quantity I^m may be found from equations (1.51) and (1.52).

The function B^m may, however, be determined by another method. We shall now show that the function of three variables $B^m(\tau, \eta, \zeta)$ may be expressed in terms of a function $\Phi^m(\tau)$, which depends on only one variable. For the determination of the function $\Phi^m(\tau)$, we shall give a linear integral equation with a kernel which depends on the absolute value of the difference of its arguments. Equations of this type have been thoroughly studied. When $\tau_0 = \infty$, they may be solved in explicit form.

The functions $\Phi^m(\tau)$ $(m = 0, 1, \ldots, n)$ are determined independently of each other. If they are all known, then the problem of finding the radiation field in the atmosphere is completely solved.

It is important that the resolvent of the integral equation (1.53) may also be expressed in terms of $\Phi^m(\tau)$. Consequently, knowledge of $\Phi^m(\tau)$ allows us to determine the radiation field in the atmosphere for any distribution of radiation sources (not just for the case of illumination by parallel radiation). The function $\Phi^m(\tau)$ thus plays a fundamental role in the theory of radiative transfer.

In the present chapter we restrict ourselves to the examination of the radiation field in a semi-infinite atmosphere. Generalization to atmospheres of finite optical thickness will be carried out in the following chapter.

5.1. Transformation of the Basic Integral Equation

As has been previously stated, we are seeking the function $B^m(\tau, \eta, \zeta)$. It is determined by equation (1.53) for arbitrary τ_0, and for $\tau_0 = \infty$ by equation (2.31). In these equations the integration is carried out over both variables, τ and η. It is possible, however, to obtain an integral equation for the function $B^m(\tau, \eta, \zeta)$ in which the integration is only over optical depth τ. Clearly, the existence of such an equation is of great interest for the present problem.

Equations of this kind were first found by Kuščer [1, 2] in connection with the solution of two specific problems: the Milne problem and the problem of diffuse reflection by a semi-infinite atmosphere for $m = 0$. Subsequently, Sobolev [3] obtained the more general results which are presented below.

In order to find the desired equation for the function $B^m(\tau, \eta, \zeta)$, we return to equation (1.46). Since the quantity $p^m(\eta, \eta')$ entering that equation is given by equation (1.41), we have

$$B^m(\tau, \eta, \zeta) = \sum_{i=m}^{n} c_i^m B_i^m(\tau, \zeta) \, P_i^m(\eta), \tag{5.1}$$

where

$$B_i^m(\tau, \zeta) = \frac{\lambda}{2} \int_{-1}^{1} P_i^m(\eta) \, I^m(\tau, \eta, \zeta) \, d\eta + \frac{\lambda}{4} S P_i^m(\zeta) \, e^{-\tau/\zeta}. \tag{5.2}$$

For the function $I^m(\tau, \eta, \zeta)$ we find upon substitution of (5.1) into (1.51) and (1.52) that

$$I^m(\tau, \eta, \zeta) = \sum_{i=m}^{n} c_i^m I_i^m(\tau, \eta, \zeta) \, P_i^m(\eta), \tag{5.3}$$

where

$$I_i^m(\tau, \eta, \zeta) = \int_{0}^{\tau} B_i^m(\tau', \zeta) e^{-(\tau-\tau')/\eta} \frac{d\tau'}{\eta} \quad (\eta > 0), \tag{5.4}$$

$$I_i^m(\tau, \eta, \zeta) = -\int_{\tau}^{\tau_0} B_i^m(\tau', \zeta) \, e^{-(\tau-\tau')/\eta} \frac{d\tau'}{\eta} \quad (\eta < 0). \tag{5.5}$$

Using the recurrence relation for the Legendre functions,

$$(i-m+1) \, P_{i+1}^m(\eta) + (i+m) \, P_{i-1}^m(\eta) = (2i+1) \, \eta \, P_i^m(\eta), \tag{5.6}$$

we may obtain a recurrence relation for the functions $B_i^m(\tau, \zeta)$. In fact, from (5.2) with the aid of (5.6), we find

$$(i-m+1) \frac{dB_{i+1}^m(\tau, \zeta)}{d\tau} + (i+m) \frac{dB_{i-1}^m(\tau, \zeta)}{d\tau}$$

$$= (2i+1) \frac{\lambda}{2} \int_{-1}^{1} P_i^m(\eta) \eta \frac{dI^m(\tau, \eta, \zeta)}{d\tau} \, d\eta - (2i+1) P_i^m(\zeta) \frac{\lambda}{4} S e^{-\tau/\zeta}. \tag{5.7}$$

Use of equations (1.45), (5.1), and (5.2) then gives

$$(i-m+1) \frac{dB_{i+1}^m(\tau, \zeta)}{d\tau} + (i+m) \frac{dB_{i-1}^m(\tau, \zeta)}{d\tau} = -z_i B_i^m(\tau, \zeta), \tag{5.8}$$

where

$$z_i = 2i+1-\lambda x_i. \tag{5.9}$$

The following recurrence relations for the quantities $I_i^m(\tau, \eta, \zeta)$ follow from equations (5.4), (5.5), and (5.8)

$$(i-m+1)\, I_{i+1}^m(\tau, \eta, \zeta) + (i+m)\, I_{i-1}^m(\tau, \eta, \zeta)$$
$$= z_i\eta I_i^m(\tau, \eta, \zeta) + (i-m+1)\, [N_{i+1}^m(\tau, \zeta) - N_{i-1}^m(0, \zeta)\, e^{-\tau/\zeta}] \quad (\eta > 0), \qquad (5.10)$$

$$(i-m+1)\, I_i^m\ (\tau, \eta, \zeta) + (i+m)\, I_{i-1}^m(\tau, \eta, \zeta)$$
$$= z_i\eta I_i^m(\tau, \eta, \zeta) + (i-m+1)\, N_{i+1}^m(\tau, \zeta) \quad (\eta < 0), \qquad (5.11)$$

where we have introduced the notation

$$N_{i+1}^m(\tau, \zeta) = B_{i+1}^m(\tau, \zeta) + \frac{i+m}{i-m+1}\, B_{i-1}^m(\tau, \zeta). \qquad (5.12)$$

With the aid of these recurrence relations for the quantities $B_i^m(\tau, \zeta)$ and $I_i^m(\tau, \eta, \zeta)$, we may find separate integral equations for each of the functions $B_i^m(\tau, \zeta)$. Let us first consider the function $B_m^m(\tau, \zeta)$.

Substituting equation (5.3) into equation (5.2) and setting $i = m$, we have

$$B_m^m(\tau, \zeta) = \frac{\lambda}{2} \int_{-1}^{1} P_m^m(\eta)\, d\eta \sum_{i=m}^{n} c_i^m I_i^m(\tau, \eta, \zeta)\, P_i^m(\eta) + \frac{\lambda}{4}\, SP_m^m(\zeta)\, e^{-\tau/\zeta}. \qquad (5.13)$$

Expressing the function $I_i^m(\tau, \eta, \zeta)$ in terms of $I_m^m(\tau, \eta, \zeta)$ with the aid of equation (5.11), we obtain

$$I_i^m(\tau, \eta, \zeta) = R_{im}^m(\eta)\, I_m^m(\tau, \eta, \zeta) + \sum_{k=m+1}^{n} R_{ik}^m(\eta)\, N_k^m(\tau, \zeta), \qquad (5.14)$$

where the $R_{ik}^m(\eta)$ are determined by the recurrence relation

$$(i-m+1)\, R_{i+1,k}^m(\eta) + (i+m)\, R_{i-1,k}^m(\eta) = z_i\eta R_{ik}^m(\eta) \qquad (5.15)$$

with $R_{kk}^m(\eta) = 1$, $R_{ik}^m(\eta) = 0$ for $i < k$. Substitution of equation (5.14) and the analogous expression for $\eta > 0$ into equation (5.13) gives

$$B_m^m(\tau, \zeta) = \int_{-1}^{1} \Psi^m(\eta)\, I_m^m(\tau, \eta, \zeta)\, d\eta + f_m^m(\tau, \zeta), \qquad (5.16)$$

where

$$f_m^m(\tau, \zeta) = -\sum_{k=m+1}^{n} N_k^m(0, \zeta) \int_{0}^{1} \Psi_k^m(\eta)\, e^{-\tau/\eta}\, d\eta + \frac{\lambda}{4}\, SP_m^m(\zeta)\, e^{-\tau/\zeta}, \qquad (5.17)$$

$$\Psi_k^m(\eta) = \frac{\lambda}{2}\, P_m^m(\eta) \sum_{i=k}^{n} c_i^m R_{ik}^m(\eta)\, P_i^m(\eta), \qquad (5.18)$$

and for simplicity, $\Psi_m^m(\eta)$ is written as $\Psi^m(\eta)$.

We then obtain the desired integral equation for $B_m^m(\tau, \zeta)$ by substituting equations (5.4) and (5.5) (for $i = m$) into (5.16):

$$B_m^m(\tau, \zeta) = \int_{0}^{\infty} K^m(|\tau-t|)\, B_m^m(t, \zeta)\, dt + f_m^m(\tau, \zeta), \qquad (5.19)$$

where

$$K^m(\tau) = \int_0^1 \Psi^m(\eta) \, e^{-\tau/\eta} \frac{d\eta}{\eta} . \tag{5.20}$$

Equations similar to (5.19) may also be obtained for the other functions $B_i^m(\tau, \zeta)$. We have for any i

$$B_i^m(\tau, \zeta) = \int_0^\infty K^m(|\tau - t|) \, B_i^m(t, \zeta) \, dt + f_i^m(\tau, \zeta), \tag{5.21}$$

where the functions $f_i^m(\tau, \zeta)$ are related by the recurrence relation

$$f_{i+1}^m(\tau, \zeta) = N_{i+1}^m(0, \zeta) \int_0^1 \Psi^m(\eta) \, e^{-\tau/\eta} \, d\eta + \frac{1}{i-m+1} \left[z_i \int_\tau^\infty f_i^m(t, \zeta) \, dt - (i+m) f_{i-1}^m(\tau, \zeta) \right]. \tag{5.22}$$

These last results may be easily proven with the aid of the equation

$$\frac{dB_i^m(\tau, \zeta)}{d\tau} = \int_0^\infty K^m(|\tau - t|) \frac{dB_i^m(t, \zeta)}{dt} \, dt + B_i^m(0, \zeta) \, K^m(\tau) + \frac{df_i^m(\tau, \zeta)}{d\tau}, \tag{5.23}$$

which follows from (5.21) and the recurrence relation (5.8).

The functions $f_{m+1}^m(\tau, \zeta), f_{m+2}^m(\tau, \zeta)$, etc., may be found from the function $f_m^m(\tau, \zeta)$ with the aid of equations (5.17) and (5.22). We obtain

$$f_i^m(\tau, \zeta) = R_{im}^m(\zeta) \, P_m^m(\zeta) \frac{\lambda}{4} \, S e^{-\tau/\zeta} + \sum_{k=m+1}^n N_k^m(0, \zeta) \int_0^1 g_{ik}^m(\eta) \, e^{-\tau/\eta} \, d\eta, \tag{5.24}$$

where

$$g_{ik}^m(\eta) = \Psi_m^m(\eta) \, R_{ik}^m(\eta) - \Psi_k^m(\eta) \, R_{im}^m(\eta). \tag{5.25}$$

Thus, $B_i^m(\tau, \zeta)$ is determined by the integral equation (5.21), in which the kernel is given by equation (5.20) and the inhomogeneous term by equation (5.24).

Substitution of (5.21) into (5.1) then yields an equation for the determination of the function $B^m(\tau, \eta, \zeta)$ which has the form

$$B^m(\tau, \eta, \zeta) = \int_0^\infty K^m(|\tau - t|) \, B^m(t, \eta, \zeta) \, dt + f^m(\tau, \eta, \zeta), \tag{5.26}$$

where

$$f^m(\tau, \eta, \zeta) = A_m^m(\zeta, \eta) \, P_m^m(\zeta) \frac{\lambda}{4} \, S e^{-\tau/\zeta} + \sum_{k=m+1}^n N_k^m(0, \zeta) \int_0^1 G_k^m(\eta', \eta) \, e^{-\tau/\eta'} \, d\eta', \tag{5.27}$$

$$G_k^m(\eta', \eta) = \Psi_m^m(\eta') \, A_k^m(\eta', \eta) - \Psi_k^m(\eta') \, A_m^m(\eta', \eta), \tag{5.28}$$

$$A_k^m(\eta', \eta) = \sum_{i=m}^n c_i^m \, R_{ik}^m(\eta') \, P_i^m(\eta). \tag{5.29}$$

The integral equations (5.21) and (5.26) are the expressions we have been seeking. The kernel $K^m(\tau)$ of these equations depends on the function $\Psi^m(\eta)$, which we will call the *characteristic function*. On the basis of equations (5.18) and (5.29), it may be written in the form

$$\Psi^m(\eta) = \frac{\lambda}{2} P_m^m(\eta) A_m^m(\eta, \eta). \tag{5.30}$$

In order to calculate the characteristic function from this formula, it is necessary to know the phase function $x(\gamma)$ and the single scattering albedo λ.

The functions $R_{ik}^m(\eta), g_{ik}^m(\eta), A_k^m(\eta', \eta)$ and $G_k^m(\eta', \eta)$ which enter the inhomogeneous terms of equations (5.21) and (5.26) may also be easily determined if $x(\gamma)$ and λ are given. As is evident from equation (5.12), however, in order to find the quantity $N_k^m(0, \zeta)$, it is necessary to know the boundary values of the function $B_k^m(\tau, \zeta)$; that is, the quantities $B_k^m(0, \zeta)$. The determination of these quantities will be considered in the next section.

5.2. The Auxiliary Equation

The inhomogeneous terms of equations (5.21) and (5.26) are superpositions of exponentials. Thus, it is useful to introduce the auxiliary equation

$$D^m(\tau, \zeta) = \int_0^\infty K^m(|\tau - t|) D^m(t, \zeta) \, dt + \frac{\lambda}{4} S e^{-\tau/\zeta}, \tag{5.31}$$

which determines the auxiliary function $D^m(\tau, \zeta)$. Once equation (5.31) has been solved, the functions $B_i^m(\tau, \zeta)$ and $B^m(\tau, \eta, \zeta)$ which we are seeking may be found from the equations

$$B_i^m(\tau, \zeta) = R_{im}^m(\zeta) P_m^m(\zeta) D^m(\tau, \zeta) + \frac{4}{\lambda S} \sum_{k=m+1}^n N_k^m(0, \zeta) \int_0^1 g_{ik}^m(\eta') D^m(\tau, \eta') \, d\eta', \tag{5.32}$$

$$B^m(\tau, \eta, \zeta) = A_m^m(\zeta, \eta) P_m^m(\zeta) D^m(\tau, \zeta)$$

$$+ \frac{4}{\lambda S} \sum_{k=m+1}^n N_k^m(0, \zeta) \int_0^1 G_k^m(\eta', \eta) D^m(\tau, \eta') \, d\eta'. \tag{5.33}$$

We shall return to the solution of equation (5.31) in Section 5.4.

We shall now concentrate on the determination of the quantities $N_i^m(0, \zeta)$. On the basis of (5.12) we have

$$N_{i+1}^m(0, \zeta) = B_{i+1}^m(0, \zeta) + \frac{i+m}{i-m+1} B_{i-1}^m(0, \zeta). \tag{5.34}$$

Setting $\tau = 0$ in equation (5.32) and using (5.34), we obtain

$$B_i^m(0, \zeta) = R_{im}^m(\zeta) P_m^m(\zeta) D^m(0, \zeta)$$

$$+ \frac{4}{\lambda S} \sum_{k=m+1}^n \left[B_k^m(0, \zeta) + \frac{k+m-1}{k-m} B_{k-2}^m(0, \zeta) \right] \int_0^1 g_{ik}^m(\eta') D^m(0, \eta') \, d\eta'. \tag{5.35}$$

In this manner, we arrive at a system of linear algebraic equations for the determination of $B_i^m(0, \zeta)$. The solution of this system expresses $B_i^m(0, \zeta)$ (and hence the quantities $N_i^m(0, \zeta)$ which we are seeking) in terms of the function $D^m(0, \zeta)$.

It is convenient to introduce the following notation:

$$B_i^m(0, \zeta) = \frac{\lambda}{4} S\varphi_i^m(\zeta), \tag{5.36}$$

$$D^m(0, \zeta) = \frac{\lambda}{4} SH^m(\zeta). \tag{5.37}$$

Then we obtain in place of equation (5.35)

$$\varphi_i^m(\zeta) = R_{im}^m(\zeta) \, P_m^m(\zeta) \, H^m(\zeta)$$

$$+ \sum_{k=m+1}^{n} \left[\varphi_k^m(\zeta) + \frac{k+m-1}{k-m} \varphi_{k-2}^m(\zeta) \right] \int_0^1 g_{ik}^m(\eta) \, H^m(\eta) \, d\eta. \tag{5.38}$$

The system of linear algebraic equations (5.38) allows us to express the functions $\varphi_i^m(\zeta)$ in terms of the function $H^m(\zeta)$. It follows from these equations that $\varphi_i^m(\zeta)$ has the form

$$\varphi_i^m(\zeta) = q_i^m(\zeta) \, P_m^m(\zeta) \, H^m(\zeta), \tag{5.39}$$

where $q_i^m(\zeta)$ is a polynomial in ζ of order $(n-m)$.

We recall that the functions $\varphi_i^m(\zeta)$ were introduced in Chapter 2 in connection with the problem of diffuse reflection of light by a semi-infinite atmosphere. They were specified by a system of nonlinear integral equations (2.42). We now see that it is sufficient for the determination of all of the functions $\varphi_i^m(\zeta)$ ($i=m, m+1, \ldots, n$) to know the one function $H^m(\zeta)$ and the polynomials $q_i^m(\zeta)$. In the following section we will obtain an equation for the determination of $H^m(\zeta)$. After it has been solved, the polynomials $q_i^m(\zeta)$ may be found from the system of linear algebraic equations

$$q_i^m(\zeta) = R_{im}^m(\zeta) + \sum_{k=m+1}^{n} \left[q_k^m(\zeta) + \frac{k+m-1}{k-m} q_{k-2}^m(\zeta) \right] \int_0^1 g_{ik}^m(\eta) \, H^m(\eta) \, d\eta, \tag{5.40}$$

obtained by substitution of (5.39) into (5.38).

A different method for determining these polynomials will be given in Chapter 7.

5.3. The Function $H^m(\eta)$

If the auxiliary equation (5.31) has been solved, then $H^m(\eta)$ may be found from equation (5.37). However, an equation determining this function directly may be obtained from (5.31). We shall proceed in the same manner that was used in the examination of diffuse light reflection by an atmosphere (see Section 1.2).

We may rewrite equation (5.31) in the form

$$D^m(\tau, \zeta) = \int_0^\tau K^m(\alpha) \, D^m(\tau-\alpha, \zeta) \, d\alpha + \int_0^\infty K^m(\alpha) \, D^m(\tau+\alpha, \zeta) \, d\alpha + \frac{\lambda}{4} S e^{-\tau/\zeta}. \tag{5.41}$$

Differentiating with respect to τ, we obtain

$$\frac{dD^m(\tau, \zeta)}{d\tau} = \int_0^\infty K^m(|\tau - t|) \frac{dD^m(t, \zeta)}{dt} dt + K^m(\tau) D^m(0, \zeta) - \frac{\lambda}{4\zeta} Se^{-\tau/\zeta}. \tag{5.42}$$

Comparison of (5.42) with (5.31) and use of equations (5.20) and (5.37) gives

$$\frac{dD^m(\tau, \zeta)}{d\tau} = -\frac{1}{\zeta} D^m(\tau, \zeta) + H^m(\zeta) \int_0^1 \Psi^m(\eta') D^m(\tau, \eta') \frac{d\eta'}{\eta'}. \tag{5.43}$$

Multiplying (5.43) by $e^{-\tau/\eta}$, integrating with respect to τ between zero and infinity, and introducing the notation

$$S\beta^m(\eta, \zeta) = \int_0^\infty D^m(\tau, \zeta) e^{-\tau/\eta} \frac{d\tau}{\eta\zeta}, \tag{5.44}$$

we find

$$(\eta + \zeta)\beta^m(\eta, \zeta) = \frac{\lambda}{4} H^m(\zeta) \left[1 + \frac{4}{\lambda} \eta \int_0^1 \Psi^m(\eta')\beta^m(\eta, \eta') d\eta' \right]. \tag{5.45}$$

On the other hand, setting $\tau = 0$ in equation (5.31) and using equations (5.20), (5.37), and (5.44), we have

$$H^m(\zeta) = 1 + \frac{4}{\lambda} \zeta \int_0^1 \Psi^m(\eta')\beta^m(\eta', \zeta) d\eta'. \tag{5.46}$$

Since the function $\beta^m(\eta, \zeta)$ is symmetric, a property of the solution of equations with symmetric kernels, it follows from (5.45) and (5.46) that

$$\beta^m(\eta, \zeta) = \frac{\lambda}{4} \frac{H^m(\eta)H^m(\zeta)}{\eta + \zeta}. \tag{5.47}$$

Substituting (5.47) into (5.46), we are led to the desired equation for the determination of $H^m(\eta)$:

$$H^m(\zeta) = 1 + \zeta H^m(\zeta) \int_0^1 \Psi^m(\eta) \frac{H^m(\eta)}{\eta + \zeta} d\eta. \tag{5.48}$$

In the case of isotropic scattering, equation (5.48) reduces to equation (2.44) for the function $\varphi(\zeta)$ obtained by V. A. Ambartsumyan in solving the problem of diffuse light reflection. Subsequently, Chandrasekhar examined the same problem for other simple phase functions and obtained equation (5.48). Equation (5.48) is therefore generally known as the "Ambartsumyan–Chandrasekhar equation".

Since the function $H^m(\eta)$ plays a significant role in the theory of anisotropic light scattering, let us examine some of its basic properties.

We begin by obtaining integral relations which this function must satisfy. Multiplying (5.48) by $\Psi^m(\zeta)$ and integrating over ζ between 0 and 1, we obtain after a short transformation

$$\int_0^1 H^m(\zeta)\Psi^m(\zeta)\,d\zeta = 1 - \left[1 - 2\int_0^1 \Psi^m(\zeta)\,d\zeta\right]^{1/2}. \tag{5.49}$$

Using this result, we may rewrite equation (5.48) in the form

$$\frac{1}{H^m(\zeta)} = \left[1 - 2\int_0^1 \Psi^m(\eta)\,d\eta\right]^{1/2} + \int_0^1 \frac{\eta H^m(\eta)}{\eta+\zeta}\Psi^m(\eta)\,d\eta. \tag{5.50}$$

In order to obtain another integral relation, we multiply equation (5.48) by $\Psi^m(\zeta)\zeta^2$ and integrate over ζ between 0 and 1. Bearing in mind the equality

$$\frac{\zeta^3}{\eta+\zeta} = \zeta^2 - \eta\zeta + \eta^2 - \frac{\eta^3}{\eta+\zeta}$$

and using equation (5.49), we obtain

$$\left[1 - 2\int_0^1 \Psi^m(\zeta)\,d\zeta\right]^{1/2}\int_0^1 H^m(\zeta)\Psi^m(\zeta)\zeta^2\,d\zeta + \tfrac{1}{2}\left[\int_0^1 H^m(\zeta)\Psi^m(\zeta)\zeta\,d\zeta\right]^2 = \int_0^1 \Psi^m(\zeta)\zeta^2\,d\zeta. \tag{5.51}$$

For the calculation of the radical which enters equations (5.49)–(5.51), we may obtain a simple formula which does not require knowledge of the function $\Psi^m(\eta)$. We note that this function is even, that is, that

$$\Psi^m(-\eta) = \Psi^m(\eta), \tag{5.52}$$

which follows from the defining equations (5.30) and (5.29) and the relations $R_{ik}^m(-\eta) = (-1)^{i+k}R_{ik}^m(\eta)$ and $P_i^m(-\eta) = (-1)^{i+m}P_i^m(\eta)$. We may then write

$$2\int_0^1 \Psi^m(\eta)\,d\eta = \frac{\lambda}{2}\sum_{i=m}^n c_i^m \int_{-1}^1 P_m^m(\eta)R_{im}^m(\eta)P_i^m(\eta)\,d\eta. \tag{5.53}$$

From the recurrence relation (5.15) we have

$$R_{im}^m(\eta) = \frac{z_m z_{m+1}\cdots z_{i-1}}{(i-m)!}\,\eta^{i-m} + \ldots, \tag{5.54}$$

where we have retained only the term containing the highest power of η. In an analogous manner, we have

$$P_i^m(\eta) = \frac{(2i)!}{2^i i!\,(i-m)!}\,(1-\eta^2)^{m/2}\eta^{i-m} + \ldots \tag{5.55}$$

It follows from equations (5.54) and (5.55) that

$$P_m^m(\eta)R_{im}^m(\eta) = \frac{z_m z_{m+1}\cdots z_{i-1}}{(2m+1)(2m+3)\ldots(2i-1)}\,P_i^m(\eta) + \ldots, \tag{5.56}$$

where we have not written the terms containing the functions $P_k^m(\eta)$ for $k < i$. Substitution of (5.56) into (5.53) and use of the orthogonality properties of the Legendre functions yields

$$2 \int_0^1 \Psi^m(\eta) \, d\eta = \frac{\lambda}{2} \sum_{i=m}^n c_i^m \frac{z_m z_{m+1} \dots z_{i-1}}{(2m+1)(2m+3)\dots(2i-1)} \frac{2}{2i+1} \frac{(i+m)!}{(i-m)!}. \qquad (5.57)$$

Making use of equations (5.9) and (2.42), we find

$$2 \int_0^1 \Psi^m(\eta) \, d\eta = \lambda \sum_{i=m}^n \left(1 - \lambda \frac{x_m}{2m+1}\right) \dots \left(1 - \lambda \frac{x_{i-1}}{2i-1}\right) \frac{x_i}{2i+1}. \qquad (5.58)$$

From (5.58) it is easy to obtain the desired expression:

$$1 - 2 \int_0^1 \Psi^m(\eta) \, d\eta = \prod_{i=m}^n \left(1 - \lambda \frac{x_i}{2i+1}\right). \qquad (5.59)$$

The quantity determined by equation (5.59) enters equation (5.50) under the radical sign. Since the function $H^m(\eta)$ is real, the relation

$$2 \int_0^1 \Psi^m(\eta) \, d\eta \le 1 \qquad (5.60)$$

is always satisfied. This inequality also follows from (5.59), since for any phase function $x_0 = 1$ and $x_i < (2i+1)$ for $i \ge 1$. The equality sign applies in (5.60) only when $\lambda = 1$ and $m = 0$. It follows that in the case of pure scattering we have for $m = 0$

$$2 \int_0^1 \Psi^0(\eta) \, d\eta = 1. \qquad (5.61)$$

In this case the integral relations (5.49) and (5.51) have the form

$$\int_0^1 H^0(\eta) \Psi^0(\eta) \, d\eta = 1, \qquad (5.62)$$

$$\int_0^1 H^0(\eta) \Psi^0(\eta) \eta \, d\eta = \left[2 \int_0^1 \Psi^0(\eta) \eta^2 \, d\eta\right]^{1/2}. \qquad (5.63)$$

In addition to the nonlinear integral equation (5.48), we may obtain a linear integral equation for the determination of $H^m(\eta)$. For this purpose we multiply (5.48) by $\Psi^m(\zeta)/(\zeta - \zeta_1)$ and integrate over ζ between 0 and 1. Separating the last term into two parts, we find

$$\int_0^1 \Psi^m(\zeta) \frac{H^m(\zeta)}{\zeta - \zeta_1} \, d\zeta = \int_0^1 \frac{\Psi^m(\zeta)}{\zeta - \zeta_1} \, d\zeta + \zeta_1 \int_0^1 \frac{\Psi^m(\zeta) H^m(\zeta)}{\zeta - \zeta_1} \, d\zeta \int_0^1 \frac{\Psi^m(\eta) H^m(\eta)}{\zeta_1 + \eta} \, d\eta$$

$$+ \int_0^1 \frac{\Psi^m(\eta) H^m(\eta)}{\zeta_1 + \eta} \eta \, d\eta \int_0^1 \frac{\Psi^m(\zeta) H^m(\zeta)}{\zeta + \eta} \, d\zeta. \qquad (5.64)$$

8*

Expressing the internal integral in the last two terms of (5.64) with the aid of (5.48), we have

$$-\frac{1}{H^m(\zeta_1)} \int\limits_0^1 \frac{\Psi^m(\zeta) H^m(\zeta)}{\zeta - \zeta_1} \, d\zeta = \int\limits_0^1 \frac{\Psi^m(\eta) H^m(\eta)}{\zeta_1 + \eta} \, d\eta + \int\limits_{-1}^1 \frac{\Psi^m(\zeta)}{\zeta - \zeta_1} \, d\zeta. \qquad (5.65)$$

We multiply (5.65) by $\zeta_1 H^m(\zeta_1)$ and again make use of equation (5.48). Then, writing η in place of ζ_1 for simplicity, we obtain

$$H^m(\eta) T^m(\eta) = 1 + \eta \int\limits_0^1 \frac{\Psi^m(\zeta) H^m(\zeta)}{\zeta - \eta} \, d\zeta, \qquad (5.66)$$

where we have introduced the notation

$$T^m(\eta) = 1 + \eta \int\limits_{-1}^1 \frac{\Psi^m(\zeta)}{\zeta - \eta} \, d\zeta. \qquad (5.67)$$

Equation (5.66) is a singular integral equation with a Cauchy kernel and its solution may be found in explicit form. The nature of the roots of the equation

$$T^m\left(\frac{1}{k}\right) = 0, \qquad (5.68)$$

plays a central role in the solution of equation (5.66). It may be shown that for $m = 0$, equation (5.68) has only real roots which lie in the interval from -1 to $+1$, while for $m \geq 1$ there are usually no roots.

When equation (5.68) does not possess roots, Mullikin [4] has obtained the following expression for $H^m(\eta)$:

$$H^m(\eta) = \exp\left[\eta \int\limits_0^1 \frac{\theta^m(\zeta)}{\zeta(\zeta + \eta)} \, d\zeta\right], \qquad (5.69)$$

where

$$\theta^m(\zeta) = \frac{1}{\pi} \arctan \frac{\pi \zeta \Psi^m(\zeta)}{T^m(\zeta)}. \qquad (5.70)$$

If for $m = 0$ equation (5.68) has a single root, then Mullikin has found that

$$H^0(\eta) = \frac{1 + \eta}{1 + k\eta} \exp\left[\eta \int\limits_0^1 \frac{\theta^0(\zeta)}{\zeta(\zeta + \eta)} \, d\zeta\right], \qquad (5.71)$$

where $\theta^0(\zeta)$ is determined by equation (5.70) for $m = 0$.

Other formulas for the function $H^m(\eta)$ have also been given. The relation between them has been established in the work of D. I. Nagirner [5].

In practice, values of $H^m(\eta)$ are more often computed by numerical solution of the non-linear integral equation (5.48), rather than with the equations mentioned above.

5.4. The Fundamental Function $\Phi^m(\tau)$

We shall now consider the solution of the auxiliary equation (5.31). In the interest of generality, however, we shall examine the following equation:

$$D^m(\tau) = \int_0^\infty K^m(|\tau-t|)\, D^m(t)\, dt + D_*^m(\tau),\qquad (5.72)$$

where $D_*^m(\tau)$ is an arbitrary function. Equation (5.31) is a particular case of (5.72). We shall make use of a general method of solution which was developed for equations with kernels depending on the absolute value of the difference of two arguments [6, 7].

Let $\Gamma^m(\tau, \tau')$ be the resolvent of the integral equation (5.72). The solution of this equation may then be written in the form

$$D^m(\tau) = D_*^m(\tau) + \int_0^\infty \Gamma^m(\tau, \tau')\, D_*^m(\tau')\, d\tau'.\qquad (5.73)$$

It is well known that the resolvent satisfies the equation

$$\Gamma^m(\tau, \tau') = \int_0^\infty K^m(|\tau-\tau''|)\, \Gamma^m(\tau'', \tau')\, d\tau'' + K^m(|\tau-\tau'|)\qquad (5.74)$$

and, because of the symmetry of the kernel, is a symmetric function of τ and τ'; that is, $\Gamma^m(\tau, \tau') = \Gamma^m(\tau', \tau)$.

We may rewrite equation (5.74) in the form

$$\Gamma^m(\tau, \tau') = \int_0^\tau K^m(\alpha)\Gamma^m(\tau-\alpha, \tau')\, d\alpha + \int_0^\infty K^m(\alpha)\Gamma^m(\tau+\alpha, \tau')\, d\alpha + K^m(|\tau-\tau'|). \quad (5.75)$$

Differentiating (5.75) first with respect to τ, and then with respect to τ', and adding the resulting equations, we find

$$\frac{\partial \Gamma^m}{\partial \tau} + \frac{\partial \Gamma^m}{\partial \tau'} = K^m(\tau)\Gamma^m(0,\tau') + \int_0^\infty K^m(|\tau-\tau''|)\left(\frac{\partial \Gamma^m}{\partial \tau} + \frac{\partial \Gamma^m}{\partial \tau'}\right)d\tau''. \qquad (5.76)$$

Comparing equations (5.76) and (5.74) for $\tau' = 0$, we have

$$\frac{\partial \Gamma^m}{\partial \tau} + \frac{\partial \Gamma^m}{\partial \tau'} = \Phi^m(\tau)\Phi^m(\tau'),\qquad (5.77)$$

where we have set

$$\Gamma^m(\tau, 0) = \Phi^m(\tau).\qquad (5.78)$$

From equation (5.77) it follows (for $\tau' > \tau$) that

$$\Gamma^m(\tau, \tau') = \Phi^m(\tau'-\tau) + \int_0^\tau \Phi^m(\alpha)\Phi^m(\alpha+\tau'-\tau)\, d\alpha.\qquad (5.79)$$

In this manner, the resolvent $\Gamma^m(\tau, \tau')$ is expressed in terms of the function $\Phi^m(\tau)$ which depends on only a single argument.

From equation (5.70) for $\tau' = 0$ we may deduce the equation

$$\Phi^m(\tau) = \int_0^\infty K^m(|\tau - \tau'|)\Phi^m(\tau')\,d\tau' + K^m(\tau),\tag{5.80}$$

for the determination of $\Phi^m(\tau)$. If the function $\Phi^m(\tau)$ is known, then the solution of equation (5.72) for an arbitrary inhomogeneous term is easily found with the aid of equations (5.73) and (5.79).

In particular, we may obtain in this manner the solution to the auxiliary equation (5.31), in which $D_*^m(\tau) = \lambda S e^{-\tau/\zeta}/4$. In this case equation (5.73) becomes

$$D^m(\tau, \zeta) = \frac{\lambda}{4} S\left[e^{-\tau/\zeta} + \int_0^\infty \Gamma^m(\tau, \tau')e^{-\tau'/\zeta}\,d\tau'\right].\tag{5.81}$$

Integrating by parts and using equation (5.77), we find

$$D^m(\tau, \zeta) = D^m(0, \zeta)\left[e^{-\tau/\zeta} + \int_0^\tau \Phi^m(\tau')e^{-(\tau-\tau')/\zeta}\,d\tau'\right],\tag{5.82}$$

where

$$D^m(0, \zeta) = \frac{\lambda}{4} S\left[1 + \int_0^\infty \Phi^m(\tau)e^{-\tau/\zeta}\,d\tau\right].\tag{5.83}$$

From (5.83) and (5.37) we may obtain an important relation expressing $H^m(\zeta)$ in terms of $\Phi^m(\tau)$:

$$H^m(\zeta) = 1 + \int_0^\infty \Phi^m(\tau)e^{-\tau/\zeta}\,d\tau.\tag{5.84}$$

Knowledge of the functions $\Phi^m(\tau)$ for all m thus provides the complete solution to the problem of radiative transfer in the atmosphere. Once they have been determined, we may find the auxiliary functions $D^m(\tau, \zeta)$ with the aid of equations (5.82) and (5.83). The quantities $B_i^m(0, \zeta)$ are then determined from the algebraic equation (5.35), and finally the functions $B^m(\tau, \eta, \zeta)$ which we are seeking from equation (5.33).

We now turn to the determination of the fundamental function $\Phi^m(\tau)$. For this purpose we compare the integral equations (5.80) and (5.31). Since they have identical kernels and the inhomogeneous term of equation (5.80) is, according to (5.20), a superposition of inhomogeneous terms of the kind contained in (5.31), we obtain

$$\Phi^m(\tau) = \frac{4}{\lambda S}\int_0^1 D^m(\tau, \zeta)\,\Psi^m(\zeta)\,\frac{d\zeta}{\zeta}.\tag{5.85}$$

Using this relation and the definition (5.37), we find from equation (5.82)

$$\Phi^m(\tau) = L^m(\tau) + \int_0^\tau L^m(\tau - \tau')\Phi^m(\tau')\,d\tau',\tag{5.86}$$

where

$$L^m(\tau) = \int_0^1 H^m(\zeta)\,\Psi^m(\zeta)e^{-\tau/\zeta}\,\frac{d\zeta}{\zeta}. \tag{5.87}$$

We see that $\Phi^m(\tau)$ satisfies not only equation (5.80), but also the Volterra-type equation (5.86). In order to determine $\Phi^m(\tau)$ from equation (5.86), we need to know $H^m(\zeta)$, which has been discussed in detail in the previous section.

Let us apply a Laplace transform to equation (5.86). Setting

$$\overline{\Phi}^m(s) = \int_0^\infty \Phi^m(\tau)e^{-\tau s}\,d\tau, \tag{5.88}$$

we obtain

$$\overline{\Phi}^m(s) = \frac{1}{1 - \displaystyle\int_0^1 H^m(\zeta)\,\Psi^m(\zeta)\,\frac{d\zeta}{1+s\zeta}} - 1. \tag{5.89}$$

This equation may be rewritten with the aid of (5.66) in the form

$$\overline{\Phi}^m(s) = \frac{1}{H^m\left(-\dfrac{1}{s}\right)T^m\left(-\dfrac{1}{s}\right)} - 1. \tag{5.90}$$

Thus, determination of the function $\Phi^m(\tau)$ reduces to inverting a Laplace transform. This may be done by the method of contour integration, for which it is necessary to know the singularities of the function $\overline{\Phi}^m(s)$ in the complex plane (for more details see D. I. Nagirner [8]).

Let us examine the case $m = 0$. We assume that the function $\overline{\Phi}^0(s)$ has only one simple pole, $s = -k$, determined by equation (5.68) for $m = 0$, and a branch point at $s = -1$. Inversion of the Laplace transform then gives

$$\overline{\Phi}^0(\tau) = Ce^{-k\tau} + \int_0^1 e^{-\tau/\eta}\,\frac{\Psi^0(\eta)}{[T^0(\eta)]^2 + [\pi\eta\Psi^0(\eta)]^2}\,\frac{d\eta}{\eta H^0(\eta)}, \tag{5.91}$$

where

$$C = \frac{1}{\displaystyle\int_0^1 \frac{\Psi^0(\eta)}{(1-k\eta)^2}\,H^0(\eta)\eta\,d\eta}. \tag{5.92}$$

At large optical depths $(\tau \gg 1)$, $\Phi^0(\tau)$ assumes the asymptotic form

$$\Phi^0(\tau) = Ce^{-k\tau}. \tag{5.93}$$

The quantity k in this expression has the same value as in the asymptotic formulas obtained earlier (k was initially introduced in Section 2.1).

The functions $\Phi^m(\tau)$ for $m \geq 1$ which determine the dependence of the radiation field on azimuth may be found in an analogous manner. If equation (5.68) has no roots, the expression for $\Phi^m(\tau)$ will not contain an exponential term similar to (5.93). In this case we have

$$\Phi^m(\tau) = \int_0^1 e^{-\tau/\eta} \frac{\Psi^m(\eta)}{[T^m(\eta)]^2 + [\pi\eta\Psi^m(\eta)]^2} \frac{d\eta}{\eta H^m(\eta)}. \tag{5.94}$$

It is not difficult to show from (5.94) that the function $\Phi^m(\tau)$ for large optical depths generally falls off as

$$\Phi^m(\tau) \sim \frac{e^{-\tau}}{\tau^{m+1}} \tag{5.95}$$

and always decreases much more rapidly than the function $\Phi^0(\tau)$. Thus, for very large values of τ, the radiation field may be considered independent of azimuth.

The asymptotic behavior of the function $\Phi^m(\tau)$ for $\tau \gg 1$ has been examined in detail by A. S. Anikonov [9].

5.5. Particular Cases

We have seen above that it is sufficient to know the functions $\Phi^m(\tau)$ in order to completely determine the functions $B(\tau, \eta, \zeta, \varphi)$ and $I(\tau, \eta, \zeta, \varphi)$ and hence to determine the radiation field in the atmosphere. The $\Phi^m(\tau)$ depend only on optical depth τ. The number of these functions equals the number of terms in the expansion of the phase function in Legendre polynomials. More precisely, each of the functions $B^m(\tau, \eta, \zeta)$, which are the coefficients of the expansion of the source function as a cosine series in azimuth, is expressed in terms of a corresponding function $\Phi^m(\tau)$.

For a semi-infinite atmosphere, $\Phi^m(\tau)$ is in turn expressed explicitly in terms of the function $H^m(\eta)$. It is thus possible to say that the determination of the radiation field in a semi-infinite atmosphere reduces to the determination of $H^m(\eta)$.

As examples of the use of the results which we have obtained, we shall now consider the determination of the source function $B(\tau, \eta, \zeta, \varphi)$ in a semi-infinite atmosphere for two simple phase functions.

Two-term phase function

Let us consider the case where the phase function has the form $x(\gamma) = 1 + x_1 \cos \gamma$. The source function is then given by the equation

$$B(\tau, \eta, \zeta, \varphi) = B^0(\tau, \eta, \zeta) + 2B^1(\tau, \eta, \zeta) \cos \varphi, \tag{5.96}$$

where, according to equation (5.1),

$$B^0(\tau, \eta, \zeta) = B_0^0(\tau, \zeta) + x_1 B_1^0(\tau, \zeta)\eta, \tag{5.97}$$

$$B^1(\tau, \eta, \zeta) = \frac{x_1}{2} B_1^1(\tau, \zeta) \sqrt{1-\eta^2}. \tag{5.98}$$

According to equation (5.32), the functions $B_0^0(\tau, \zeta)$ and $B_1^0(\tau, \zeta)$ may be expressed in terms of the single auxiliary function $D^0(\tau, \zeta)$ through the equations

$$B_0^0(\tau, \zeta) = D^0(\tau, \zeta) - \frac{2}{S} x_1 B_1^0(0, \zeta) \int_0^1 D^0(\tau, \eta)\eta \, d\eta \qquad (5.99)$$

and

$$B_1^0(\tau, \zeta) = (1-\lambda)\zeta \, D^0(\tau, \zeta) + \frac{2}{S} B_1^0(0, \zeta) \int_0^1 D^0(\tau, \eta) \, d\eta, \qquad (5.100)$$

while the function $B_1^1(\tau, \zeta)$ has the form

$$B_1^1(\tau, \zeta) = D^1(\tau, \zeta) \sqrt{1-\zeta^2}. \qquad (5.101)$$

The quantity $B_1^0(0, \zeta)$ which enters equations (5.99) and (5.100) is easily found from equation (5.100) by setting $\tau = 0$. The auxiliary functions $D^0(\tau, \zeta)$ and $D^1(\tau, \zeta)$ are expressed in terms of $\Phi^0(\tau)$ and $\Phi^1(\tau)$ by equation (5.82) for $m = 0$ and $m = 1$, respectively.

Equation (5.91) may be used for the determination of $\Phi^0(\tau)$. For the given phase function we have

$$\Psi^0(\eta) = \frac{\lambda}{2} [1 + x_1(1-\lambda)\eta^2], \qquad (5.102)$$

$$T^0(\eta) = 1 - \lambda + \Psi^0(\eta)\left(2 - \eta \ln \frac{1+\eta}{1-\eta}\right), \qquad (5.103)$$

and the constant k may be found from equation (2.18).

The function $\Phi^1(\tau)$ may be found in a similar manner. The equation $T^1(1/k) = 0$ has no solution, however, and as a result the exponential term will be absent in the expression for $\Phi^1(\tau)$. We thus obtain

$$\Phi^1(\tau) = \int_0^1 e^{-\tau/\eta} \frac{\Psi^1(\eta)}{[T^1(\eta)]^2 + [\pi\eta\Psi^1(\eta)]^2} \frac{d\eta}{\eta H^1(\eta)}, \qquad (5.104)$$

where

$$\Psi^1(\eta) = \frac{\lambda}{2} x_1(1-\eta^2), \qquad (5.105)$$

and

$$T^1(\eta) = 1 - \frac{\lambda}{2} x_1 + \Psi^1(\eta)\left(2 - \eta \ln \frac{1+\eta}{1-\eta}\right). \qquad (5.106)$$

The asymptotic expressions for the functions $\Phi^0(\tau)$ and $\Phi^1(\tau)$ for large optical depths are of considerable interest. We have already shown that $\Phi^0(\tau)$ is determined by equation (5.93) for $\tau \gg 1$. For $\Phi^1(\tau)$ we deduce from (5.104) the following asymptotic expression:

$$\Phi^1(\tau) = \frac{2\lambda x_1}{(2-\lambda x_1)^2 H^1(1)} \frac{e^{-\tau}}{\tau^2}, \qquad (5.107)$$

which is in agreement with equation (5.95).

We shall now derive equations expressing the functions $\varphi_0^0(\zeta)$ and $\varphi_1^0(\zeta)$ in terms of $H^0(\zeta)$, and the function $\varphi_1^1(\zeta)$ in terms of $H^1(\zeta)$. These may be obtained from equations (5.99)–(5.101) if we set in them $\tau = 0$ and make use of the definitions (5.36) and (5.37). We find

$$\varphi_0^0(\zeta) = H^0(\zeta)\left[1 - \frac{\lambda}{2}x_1(1-\lambda)\frac{h_1^0\zeta}{1-\frac{\lambda}{2}h_0^0}\right], \qquad (5.108)$$

$$\varphi_1^0(\zeta) = H^0(\zeta)\frac{(1-\lambda)\zeta}{1-\frac{\lambda}{2}h_0^0}, \qquad (5.109)$$

$$\varphi_1^1(\zeta) = H^1(\zeta)\sqrt{1-\zeta^2}. \qquad (5.110)$$

We have denoted the zeroth and first moments of $H^0(\zeta)$ by h_0^0 and h_1^0, respectively.

Three-term phase function

For the phase function $x(\gamma) = 1 + x_1 \cos\gamma + x_2 P_2(\cos\gamma)$, the source function has the form

$$B(\tau, \eta, \zeta, \varphi) = B^0(\tau, \eta, \zeta) + 2B^1(\tau, \eta, \zeta)\cos\varphi + 2B^2(\tau, \eta, \zeta)\cos 2\varphi. \qquad (5.111)$$

We shall first obtain $B^0(\tau, \eta, \zeta)$. From equation (5.1) we have

$$B^0(\tau, \eta, \zeta) = B_0^0(\tau, \zeta) + x_1 B_1^0(\tau, \zeta)\,\eta + x_2 B_2^0(\tau, \zeta)\,P_2(\eta). \qquad (5.112)$$

All of the functions $B_i^0(\tau, \zeta)$ may be expressed in terms of a single auxiliary function $D^0(\tau, \zeta)$. For this purpose we may use the following equation which is obtained from (5.32) with the aid of (5.34):

$$B_i^0(\tau, \zeta) = R_{i0}^0(\zeta)\,D^0(\tau, \zeta) + \frac{4}{\lambda S}B_1^0(0,\zeta)\int_0^1 g_{i0}^0(\eta)\,D^0(\tau, \eta)\,d\eta$$

$$+ \frac{4}{\lambda S}\left[B_2^0(0,\zeta) + \frac{1}{2}B_0^0(0,\zeta)\right]\int_0^1 g_{i2}^0(\eta)\,D^0(\tau, \eta)\,d\eta. \qquad (5.113)$$

The polynomials $R_{i0}^0(\zeta)$ which enter (5.113) may be found from the recurrence relation (5.15). In the present case they are just

$$\left.\begin{array}{l} R_{00}^0(\zeta) = 1, \quad R_{10}^0(\zeta) = (1-\lambda)\,\zeta, \\ R_{20}^0(\zeta) = \frac{1}{2}[(1-\lambda)(3-\lambda x_1)\,\zeta^2 - 1]. \end{array}\right\} \qquad (5.114)$$

The quantities $g_{i1}^0(\eta)$ and $g_{i2}^0(\eta)$ are determined from equation (5.25). We obtain

$$g_{01}^0(\eta) = -\frac{\lambda}{2}\eta\left[x_1 + \frac{x_2}{4}(3-\lambda x_1)(3\eta^2 - 1)\right], \qquad (5.115)$$

$$g_{02}^0(\eta) = -\frac{\lambda}{4}x_2(3\eta^2 - 1), \qquad (5.116)$$

$$g_{11}^0(\eta) = \frac{\lambda}{2}\left[1 - \frac{x_2}{4}(3\eta^2 - 1)\right], \tag{5.117}$$

$$g_{12}^0(\eta) = -\frac{\lambda}{4}x_2(1-\lambda)\eta(3\eta^2 - 1), \tag{5.118}$$

$$g_{21}^0(\eta) = \frac{\lambda}{4}\eta[3 + x_1(1-\lambda)], \tag{5.119}$$

$$g_{22}^0(\eta) = \frac{\lambda}{2}[1 + x_1(1-\lambda)\,\eta^2]. \tag{5.120}$$

The auxiliary function $D^0(\tau, \zeta)$ is expressed in terms of $\Phi^0(\tau)$, which may in turn be expressed in terms of $H^0(\eta)$. Determination of $\Phi^0(\tau)$ and $H^0(\eta)$ requires utilization of the characteristic function

$$\Psi^0(\eta) = \frac{\lambda}{2}\left\{1 + x_1(1-\lambda)\eta^2 + \frac{x_2}{4}(2\eta^2 - 1)[(1-\lambda)(3-\lambda x_1)\eta^2 - 1]\right\}, \tag{5.121}$$

which is obtained from equation (5.18).

The function $B^1(\tau, \eta, \zeta)$ may be found in a similar manner. According to (5.1), it takes the form

$$B^1(\tau, \eta, \zeta) = \frac{x_1}{2}\,B_1^1(\tau, \zeta)\,\sqrt{1-\eta^2} + \frac{x_2}{2}\,B_2^1(\tau, \zeta)\eta\,\sqrt{1-\eta^2}. \tag{5.122}$$

The quantities $B_1^1(\tau, \zeta)$ and $B_2^1(\tau, \zeta)$ may be expressed in terms of the auxiliary function $D^1(\tau, \zeta)$ as follows:

$$B_i^1(\tau, \zeta) = R_{i1}^1(\zeta)\,\sqrt{1-\zeta^2}\,D^1(\tau, \zeta)$$
$$+ \frac{4}{\lambda S}\,B_2^1(0, \zeta)\int_0^1 g_{i2}^1(\eta)\,D^1(\tau, \eta)\,d\eta \qquad (i = 1, 2), \tag{5.123}$$

where

$$R_{11}^1(\zeta) = 1, \qquad R_{21}^1(\zeta) = (3 - \lambda x_1)\,\zeta, \tag{5.124}$$

$$g_{12}^1(\eta) = -\frac{\lambda}{4}x_2\eta(1-\eta^2), \tag{5.125}$$

$$g_{22}^1(\eta) = \frac{\lambda}{4}x_1(1-\eta^2). \tag{5.126}$$

The functions $\Phi^1(\tau)$ and $H^1(\tau)$ are determined with the aid of the characteristic function

$$\Psi^1(\eta) = \frac{\lambda}{4}(1-\eta^2)[x_1 + (3 - \lambda x_1)x_2\eta^2]. \tag{5.127}$$

For the determination of the function $B^2(\tau, \eta, \zeta)$ we obtain, as above, the equation

$$B^2(\tau, \eta, \zeta) = \frac{x_2}{8}\,B_2^2(\tau, \zeta)(1-\eta^2), \tag{5.128}$$

where

$$B_2^2(\tau, \zeta) = 3(1-\zeta^2)\, D^2(\tau, \zeta). \tag{5.129}$$

The characteristic function required for determining $\Phi^2(\tau)$ and $H^2(\eta)$ is

$$\Psi^2(\eta) = \frac{3}{16}\, \lambda x_2 (1-\eta^2)^2. \tag{5.130}$$

The quantities $B_i^0(0, \zeta)$ entering equation (5.113) may be found from that equation by setting $\tau = 0$. Proceeding in this manner and using the definitions (5.36) and (5.37), we obtain the following equation which expresses the function $\varphi_i^0(\zeta)$ in terms of $H^0(\zeta)$:

$$\varphi_i^0(\zeta) = R_{i0}^0(\zeta)\, H^0(\zeta) + \varphi_1^0(\zeta) \int_0^1 g_{i1}^0(\eta)\, H^0(\eta)\, d\eta$$

$$+ \left[\varphi_2^0(\zeta) + \frac{1}{2}\varphi_0^0(\zeta) \right] \int_0^1 g_{i2}^0(\eta)\, H^0(\eta)\, d\eta \qquad (i = 0, 1, 2). \tag{5.131}$$

In a similar manner we may find from (5.123) a relation expressing $\varphi_i^1(\zeta)$ in terms of $H^1(\zeta)$:

$$\varphi_i^1(\zeta) = R_{i1}^1(\zeta)\, \sqrt{1-\zeta^2}\, H^1(\zeta) + \varphi_2^1(\zeta) \int_0^1 g_{i2}^1(\eta)\, H^1(\eta)\, d\eta \qquad (i = 1, 2). \tag{5.132}$$

From equation (5.129) we have

$$\varphi_2^2(\zeta) = 3(1-\zeta^2)\, H^2(\zeta). \tag{5.133}$$

We shall subsequently obtain by other methods expressions for the functions $\varphi_i^m(\zeta)$ in terms of $H^m(\zeta)$ for a three-term phase function (see Section 7.4).

References

1. I. KUŠČER, Milne's problem for anisotropic scattering, *J. Math. Phys.* **34**, 256 (1955).
2. I. KUŠČER, Diffuse reflection of light from a semi-infinite scattering medium, *J. Math. Phys.* **37**, 52 (1958).
3. V. V. SOBOLEV, Anisotropic scattering of light in a semi-infinite atmosphere, III, *Astron. Zh.* **46**, 512 (1969) [*Sov. Astron. A.J.* **13**, 403 (1969)].
4. T. W. MULLIKIN, Chandrasekhar's X and Y equations, *Trans. Amer. Math.* **113**, 316 (1964).
5. D. I. NAGIRNER, Multiple scattering of light in a semi-infinite medium, *Uchon. Zap. Leningrad. Gos. Univ.* No. 337 [*Trudy Astron. Obs. Leningrad. Gos. Univ.* **25** (1968)].
6. V. V. SOBOLEV, Diffusion of radiation in a semi-infinite medium, *Dokl. Akad. Nauk SSSR* **116**, 45 (1957) [*Sov. Phys. Doklady* **2**, 426 (1957)].
7. V. V. SOBOLEV, On the theory of radiation diffusion in stellar atmospheres, *Astron. Zh.* **36**, 573 (1959) [*Sov. Astron. A.J.* **3**, 563 (1960)].
8. D. I. NAGIRNER, On the solution of the integral equations of the theory of light scattering, *Astron. Zh.* **41**, 669 (1964) [*Sov. Astron A.J.* **8**, 533 (1965)].
9. A. S. ANIKONOV, The light regime in the deep layers of planetary atmospheres, *Astron. Zh.* **50**, 137 (1973) [*Sov. Astron. A.J.* **17**, 88 (1973)].

Chapter 6

GENERAL THEORY (CONTINUED)

IN THE preceding chapter we set forth the general theory of radiative transfer in a semi-infinite atmosphere. In this chapter we shall generalize those results to atmospheres of finite optical thickness τ_0.

As before, the problem may be reduced to the determination of certain fundamental functions $\Phi^m(\tau, \tau_0)$. Once these functions are known, the radiation field in the atmosphere is completely determined. In particular, with the aid of $\Phi^m(\tau, \tau_0)$ we may express the functions $X^m(\eta, \tau_0)$ and $Y^m(\eta, \tau_0)$, knowledge of which is sufficient for determining the intensity of radiation emerging from the atmosphere.

The radiation field within an atmosphere of finite optical thickness may also be determined through the use of equations containing the optical thickness τ_0 as independent variable. Equations of this type are presented in Section 6.5.

At the end of the chapter we shall examine atmospheres of large optical thickness ($\tau_0 \gg 1$). We shall obtain asymptotic expressions for the functions $\Phi^0(\tau, \tau_0)$, $X^0(\eta, \tau_0)$, and $Y^0(\eta, \tau_0)$ in terms of the corresponding functions for a semi-infinite atmosphere, $\Phi^0(\tau)$ and $H^0(\eta)$, which will be assumed known.

6.1. Expression of the Source Function in Terms of Auxiliary Functions

As previously, we shall assume that the source function has been expanded in a cosine series in the azimuth, so that

$$B(\tau, \eta, \zeta, \varphi, \tau_0) = B^0(\tau, \eta, \zeta, \tau_0) + 2 \sum_{m=1}^{n} B^m(\tau, \eta, \zeta, \tau_0) \cos m\varphi. \tag{6.1}$$

The quantities $B^m(\tau, \eta, \zeta, \tau_0)$ then take the form

$$B^m(\tau, \eta, \zeta, \tau_0) = \sum_{i=m}^{n} c_i^m P_i^m(\eta) B_i^m(\tau, \zeta, \tau_0), \tag{6.2}$$

where c_i^m is given by equation (1.42), and $P_i^m(\eta)$ is an associated Legendre polynomial.

In Chapter 1 we obtained the system of equations (1.55) for the determination of the $B_i^m(\tau, \zeta, \tau_0)$. It is possible, however, to find for each of these quantities a separate integral equation with a kernel which depends on the modulus of the difference of its arguments. This was shown in the previous chapter for the case $\tau_0 = \infty$. We shall now present the

107

analogous equations for an atmosphere of finite optical thickness (see also the derivation in [1]).

Proceeding as in Section 5.1, we may obtain the following equation determining the function $B_i^m(\tau, \zeta, \tau_0)$:

$$B_i^m(\tau, \zeta, \tau_0) = \int_0^{\tau_0} K^m(|\tau-t|)\, B_i^m(t, \zeta, \tau_0)\, dt + f_i^m(\tau, \zeta, \tau_0), \tag{6.3}$$

where

$$
\begin{aligned}
f_i^m(\tau, \zeta, \tau_0) = R_{im}^m(\zeta)\, P_m^m(\zeta)\, e^{-\tau/\zeta} \\
+ \sum_{k=m+1}^{n} \left[N_k^m(0, \zeta, \tau_0) \int_0^1 g_{ik}^m(\eta)\, e^{-\tau/\eta} d\eta + N_k^m(\tau_0, \zeta, \tau_0) \int_0^1 g_{ik}^m(-\eta)\, e^{-(\tau_0-\tau)/\eta}\, d\eta \right]
\end{aligned}
\tag{6.4}
$$

and

$$N_{i+1}^m(\tau, \zeta, \tau_0) = B_{i+1}^m(\tau, \zeta, \tau_0) + \frac{i+m}{i-m+1}\, B_{i-1}^m(\tau, \zeta, \tau_0). \tag{6.5}$$

The kernel of the integral equation (6.3) is given by the expression

$$K^m(\tau) = \int_0^1 \Psi^m(\eta)\, e^{-\tau/\eta}\, \frac{d\eta}{\eta}, \tag{6.6}$$

and the quantities $\Psi^m(\eta)$ and $g_{ik}^m(\eta)$ which enter (6.6) and (6.4) are defined by (5.18) anp (5.25).

Because the inhomogeneous term of equation (6.3) is a superposition of exponentials, it is expedient to introduce the auxiliary equation

$$D^m(\tau, \zeta, \tau_0) = \int_0^{\tau_0} K^m(|\tau-t|)\, D^m(t, \zeta, \tau_0)\, dt + \frac{\lambda}{4}\, S e^{-\tau/\zeta}. \tag{6.7}$$

Then all the functions $B_i^m(\tau, \zeta, \tau_0)$ $(i = m+1, m+2, \ldots, n)$ may be expressed in terms of the single auxiliary function $D^m(\tau, \zeta, \tau_0)$ by means of the equation

$$
\begin{aligned}
B_i^m(\tau, \zeta, \tau_0) = R_{im}^m(\zeta)\, P_m^m(\zeta)\, D^m(\tau, \zeta, \tau_0) \\
+ \frac{4}{\lambda S} \sum_{k=m+1}^{n} \left[N_k^m(0, \zeta, \tau_0) \int_0^1 g_{ik}^m(\eta)\, D^m(\tau, \eta, \tau_0)\, d\eta \right. \\
\left. + N_k^m(\tau_0, \zeta, \tau_0) \int_0^1 g_{ik}^m(-\eta) D^m(\tau_0-\tau, \eta, \tau_0)\, d\eta \right].
\end{aligned}
\tag{6.8}
$$

In order to find the quantities $N_k^m(0, \zeta, \tau_0)$ and $N_k^m(\tau_0, \zeta, \tau_0)$ which enter equation (6.8), we set $\tau = 0$ in that equation and make use of equation (6.5). This leads to a system of algebraic equations for the quantities $B_i^m(0, \zeta, \tau_0)$ and $B_i^m(\tau_0, \zeta, \tau_0)$ which determine $N_k^m(0, \zeta, \tau_0)$ and $N_k^m(\tau_0, \zeta, \tau_0)$ through equation (6.5).

The function $B^m(\tau, \eta, \zeta, \tau_0)$ may also be expressed in terms of $D^m(\tau, \zeta, \tau_0)$ as

$$B^m(\tau, \eta, \zeta, \tau_0) = A_m^m(\zeta, \eta) P_m^m(\zeta) D^m(\tau, \zeta, \tau_0)$$

$$+\frac{4}{\lambda S} \sum_{k=m+1}^{n} \left[N_k^m(0, \zeta, \tau_0) \int_0^1 G_k^m(\eta', \eta) D^m(\tau, \eta', \tau_0) d\eta' \right.$$

$$\left. +N_k^m(\tau_0, \zeta, \tau_0) \int_0^1 G_k^m(-\eta', \eta) D^m(\tau_0-\tau, \eta', \tau_0) d\eta' \right], \tag{6.9}$$

where the quantities $A_m^m(\zeta, \eta)$ and $G_k^m(\eta', \eta)$ are given by equations (5.28) and (5.29).

In this way the determination of the source function is reduced to finding the auxiliary functions $D^m(\tau, \zeta, \tau_0)$. Each of these functions may be expressed, as we shall show below, in terms of a function $\Phi^m(\tau, \tau_0)$. This process successively lowers the number of arguments of the functions we are seeking.

6.2. The Fundamental Function $\Phi^m(\tau, \tau_0)$

The auxiliary equation (6.7) is a particular case of the following equation:

$$D^m(\tau, \tau_0) = \int_0^{\tau_0} K^m(|\tau-t|) D^m(t, \tau_0) dt + D_*^m(\tau), \tag{6.10}$$

where $D_*^m(\tau)$ is an arbitrary function. We shall first consider equation (6.10), and then return to equation (6.7). In doing so, we shall use the same method as in Section 5.4 (see [2], and also [7] in the References to Chapter 5).

Denoting the resolvent of equation (6.10) by $\Gamma^m(\tau, \tau', \tau_0)$, we have

$$D^m(\tau, \tau_0) = D_*^m(\tau)+ \int_0^{\tau_0} \Gamma^m(\tau, \tau', \tau_0) D_*^m(\tau') d\tau'. \tag{6.11}$$

The resolvent is obviously a symmetric function of τ and τ', so that $\Gamma^m(\tau, \tau', \tau_0) = \Gamma^m(\tau', \tau, \tau_0)$. It is determined in the well-known fashion by

$$\Gamma^m(\tau, \tau', \tau_0) = \int_0^{\tau_0} K^m(|\tau-\tau''|) \Gamma^m(\tau'', \tau', \tau_0) d\tau''+K^m(|\tau-\tau''|). \tag{6.12}$$

Let us differentiate (6.12) first with respect to τ and then with respect to τ'. Adding the resulting expressions term by term, we find

$$\frac{\partial \Gamma^m}{\partial \tau}+\frac{\partial \Gamma^m}{\partial \tau'} = K^m(\tau) \Gamma^m(0, \tau', \tau_0)-K^m(\tau_0-\tau)$$

$$\times \Gamma^m(\tau_0, \tau', \tau_0)+ \int_0^{\tau_0} K^m(|\tau-\tau''|) \left(\frac{\partial \Gamma^m}{\partial \tau''}+\frac{\partial \Gamma^m}{\partial \tau'} \right) d\tau''. \tag{6.13}$$

On the other hand, setting $\tau' = 0$ in (6.12) and introducing the notation

$$\Phi^m(\tau \ \tau_0) = \Gamma^m(\tau, 0, \tau_0),\tag{6.14}$$

we have

$$\Phi^m(\tau, \tau_0) = \int_0^{\tau_0} K^m(|\tau - t|)\,\Phi^m(t, \tau_0)\,dt + K^m(\tau).\tag{6.15}$$

Comparison of (6.13) and (6.15) yields

$$\frac{\partial \Gamma^m}{\partial \tau} + \frac{\partial \Gamma^m}{\partial \tau'} = \Phi^m(\tau, \tau_0)\,\Phi^m(\tau', \tau_0) - \Phi^m(\tau_0 - \tau, \tau_0)\,\Phi^m(\tau_0 - \tau', \tau_0),\tag{6.16}$$

from which it follows that (for $\tau' > \tau$)

$$\Gamma^m(\tau, \tau', \tau_0) = \Phi^m(\tau' - \tau, \tau_0) + \int_0^\tau [\Phi^m(\alpha + \tau' - \tau, \tau_0)\,\Phi^m(\alpha, \tau_0)$$

$$- \Phi^m(\tau_0 - \alpha - \tau' + \tau, \tau_0)\,\Phi^m(\tau_0 - \alpha, \tau_0)]\,d\alpha.\tag{6.17}$$

In this manner, the function $\Gamma^m(\tau, \tau', \tau_0)$ of two variables may be expressed in terms of the function $\Phi^m(\tau, \tau_0)$ of one variable (τ_0 is a parameter). The solution of equation (6.10) for any inhomogeneous term may be expressed in terms of $\Phi^m(\tau, \tau_0)$ with the aid of equations (6.17) and (6.11).

For the auxiliary equation (6.7), $D_*^m(\tau) = \lambda S e^{-\tau/\zeta}/4$. On the basis of equation (6.11) we thus obtain

$$D^m(\tau, \zeta, \tau_0) = \frac{\lambda}{4}\, S \left[e^{-\tau/\zeta} + \int_0^{\tau_0} \Gamma^m(\tau, \tau', \tau_0)\, e^{-\tau'/\zeta}\, d\tau' \right].\tag{6.18}$$

By using the relation (6.16) and integrating by parts, we find after a short computation

$$D^m(\tau, \zeta, \tau_0) = \frac{\lambda}{4}\, S \left\{ X^m(\zeta, \tau_0) \left[e^{-\tau/\zeta} + \int_0^\tau \Phi^m(t, \tau_0)\, e^{-(\tau - t)/\zeta}\, dt \right] \right.$$

$$\left. - Y^m(\zeta, \tau_0) \int_0^\tau \Phi^m(\tau_0 - t, \tau_0)\, e^{-(\tau - t)/\zeta}\, dt \right\},\tag{6.19}$$

where we have introduced the notation

$$D^m(0, \zeta, \tau_0) = \frac{\lambda}{4}\, S X^m(\zeta, \tau_0),$$

$$D^m(\tau_0, \zeta, \tau_0) = \frac{\lambda}{4}\, S Y^m(\zeta, \tau_0). \tag{6.20}$$

From (6.18) with the aid of (6.14) we obtain the following equations which express the functions $X^m(\zeta, \tau_0)$ and $Y^m(\zeta, \tau_0)$ in terms of $\Phi^m(\tau, \tau_0)$:

$$X^m(\zeta, \tau_0) = 1 + \int_0^{\tau_0} \Phi^m(\tau, \tau_0)\, e^{-\tau/\zeta}\, d\tau,\tag{6.21}$$

$$Y^m(\zeta, \tau_0) = e^{-\tau_0/\zeta} + \int_0^{\tau_0} \Phi^m(\tau_0 - \tau, \tau_0)\, e^{-\tau/\zeta}\, d\tau.\tag{6.22}$$

It should be noted that equation (6.19) may be found by a different procedure which does not require introducing the resolvent. Differentiating (6.7) with respect to τ, we have

$$\frac{dD^m(\tau, \zeta, \tau_0)}{d\tau} = \int_0^{\tau_0} K^m(|\tau-t|) \frac{dD^m(t, \zeta, \tau_0)}{dt} dt$$

$$+ K^m(\tau) D^m(0, \zeta, \tau_0) - K^m(\tau_0-\tau) D^m(\tau_0, \zeta, \tau_0) - \frac{\lambda}{4\zeta} S e^{-\tau/\zeta}. \tag{6.23}$$

Comparison of equations (6.23) and (6.7) and use of (6.6) yields

$$\frac{dD^m(\tau, \zeta, \tau_0)}{d\tau} = -\frac{1}{\zeta} D^m(\tau, \zeta, \tau_0) + D^m(0, \zeta, \tau_0) \Phi^m(\tau, \tau_0)$$

$$- D^m(\tau_0, \zeta, \tau_0) \Phi^m(\tau_0-\tau, \tau_0), \tag{6.24}$$

with the notation

$$\Phi^m(\tau, \tau_0) = \frac{4}{\lambda S} \int_0^1 D^m(\tau, \zeta, \tau_0) \Psi^m(\zeta) \frac{d\zeta}{\zeta}. \tag{6.25}$$

By integrating (6.24) and using the definitions (6.20), we are led to equation (6.19). Equation (6.15) for $\Phi^m(\tau, \tau_0)$ then follows from (6.7) and (6.25). By setting $\tau = 0$ and $\tau = \tau_0$ in equation (6.7), we may obtain expressions for the functions $X^m(\zeta, \tau_0)$ and $Y^m(\zeta, \tau_0)$.

For example, when $\tau = 0$ equation (6.7) gives

$$X^m(\zeta, \tau_0) = \frac{4}{\lambda S} \int_0^{\tau_0} K^m(\tau) D^m(\tau, \zeta, \tau_0) d\tau + 1. \tag{6.26}$$

But from (6.7) and (6.15) it follows that

$$\int_0^{\tau_0} K^m(\tau) D^m(\tau, \zeta, \tau_0) d\tau = \frac{\lambda}{4} S \int_0^{\tau_0} \Phi^m(\tau, \tau_0) e^{-\tau/\zeta} d\tau, \tag{6.27}$$

where we have made use of the properties of equations which differ only in their inhomogeneous terms. By substituting (6.27) into (6.26), we then find equation (6.21). Equation (6.22) may be found in an analogous manner.

We see that determination of the source function $B(\tau, \eta, \zeta, \varphi, \tau_0)$ reduces to determination of the fundamental functions $\Phi^m(\tau, \tau_0)$ from equation (6.15). The number of these functions equals the number of terms in the expansion (6.1). Once (6.15) has been solved, each auxiliary function $D^m(\tau, \zeta, \tau_0)$ may be found from equation (6.19), and subsequently the $B^m(\tau, \eta, \zeta, \tau_0)$ from equation (6.9).

It is important that the functions $\Phi^m(\tau, \tau_0)$ are determined independently of each other. If, for example, we are only interested in the radiation field averaged over azimuth, then it is sufficient to find the single function $\Phi^0(\tau, \tau_0)$.

9

6.3. The $X^m(\zeta, \tau_0)$ and $Y^m(\zeta, \tau_0)$ Functions

Let us consider the functions $X^m(\zeta, \tau_0)$ and $Y^m(\zeta, \tau_0)$ which were introduced in equation (6.20). These functions are of particular interest because with their aid the intensity of radiation emerging from the atmosphere may be expressed.

Probably the simplest method for determining $X^m(\zeta, \tau_0)$ and $Y^m(\zeta, \tau_0)$ consists of finding the fundamental function $\Phi^m(\tau, \tau_0)$ from the integral equation (6.15) and subsequently using equations (6.21) and (6.22). We may, however, obtain equations which determine these functions directly.

In order to find these latter equations, we return to equation (6.24). Multiplying it by $e^{-\tau/\zeta}$, integrating over τ between zero and τ_0, and using equations (6.21) and (6.22), we obtain

$$\left(\frac{1}{\eta}+\frac{1}{\zeta}\right)\int_0^{\tau_0} D^m(\tau, \eta, \tau_0)e^{-\tau/\eta}\,d\tau = \frac{\lambda}{4} S[X^m(\eta, \tau_0)X^m(\zeta, \tau_0)-Y^m(\eta, \tau_0)Y^m(\zeta, \tau_0)]. \quad (6.28)$$

On the other hand, setting $\tau = 0$ in equation (6.7), using (6.6), and interchanging the order of integration, we have

$$D^m(0, \zeta, \tau_0) = \int_0^1 \Psi^m(\eta)\frac{d\eta}{\eta}\int_0^{\tau_0} D^m(\tau, \zeta, \tau_0)e^{-\tau/\eta}\,d\tau+\frac{\lambda}{4} S. \quad (6.29)$$

Substitution of the integral from (6.28) into (6.29) gives the first equation relating $X^m(\zeta, \tau_0)$ and $Y^m(\zeta, \tau_0)$:

$$X^m(\zeta, \tau_0) = 1+\zeta\int_0^1 \Psi^m(\eta)\frac{X^m(\zeta, \tau_0)\,X^m(\eta, \tau_0)-Y^m(\zeta, \tau_0)\,Y^m(\eta, \tau_0)}{\eta+\zeta}\,d\eta. \quad (6.30)$$

The second equation may be obtained in an analogous manner, after multiplying (6.24) by $e^{-(\tau_0-\tau)/\eta}$, integrating over τ, and then using (6.7) with $\tau = \tau_0$. The resulting equation has the form

$$Y^m(\zeta, \tau_0) = e^{-\tau_0/\zeta}+\zeta\int_0^1 \Psi^m(\eta)\frac{Y^m(\zeta, \tau_0)\,X^m(\eta, \tau_0)-Y^m(\eta, \tau_0)\,X^m(\zeta, \tau_0)}{\zeta-\eta}\,d\eta. \quad (6.31)$$

The nonlinear system of equations (6.30) and (6.31) may serve for the determination of $X^m(\zeta, \tau_0)$ and $Y^m(\zeta, \tau_0)$. It is the generalization of equations (3.22) and (3.23), which were obtained by V. A. Ambartsumyan in solving the problem of diffuse reflection and transmission of light for isotropic scattering. Chandrasekhar [3] obtained the system of equations (6.30) and (6.31) while examining the same problem for the simplest anisotropic phase functions.

We may also obtain a system of linear integral equations for the determination of $X^m(\zeta, \tau_0)$ and $Y^m(\zeta, \tau_0)$. The derivation from equations (6.30) and (6.31) proceeds in the same manner as in the derivation of equation (5.66), and yields

$$X^m(\zeta, \tau_0)T^m(\zeta) = 1 + \zeta \int_0^1 \frac{\Psi^m(\eta)}{\eta-\zeta} X^m(\eta, \tau_0) \, d\eta - e^{-\tau_0/\zeta} \zeta \int_0^1 \frac{\Psi^m(\eta)}{\eta+\zeta} Y^m(\eta, \tau_0) \, d\eta, \qquad (6.32)$$

$$Y^m(\zeta, \tau_0)T^m(\zeta) = e^{-\tau_0/\zeta} + \zeta \int_0^1 \frac{\Psi^m(\eta)}{\eta-\zeta} Y^m(\eta, \tau_0) \, d\eta - e^{-\tau_0/\zeta} \zeta \int_0^1 \frac{\Psi^m(\eta)}{\eta+\zeta} X^m(\eta, \tau_0) \, d\eta, \qquad (6.33)$$

where $T^m(\eta)$ is given by equation (5.67).

The functions $X^m(\zeta, \tau_0)$ and $Y^m(\zeta, \tau_0)$ were examined in the books by Chandrasekhar and Busbridge to which we have already referred (see the References to Chapter 1). Subsequently these functions were studied by Mullikin, who examined the question of the uniqueness of the solutions to equations (6.30) and (6.31) and to (6.32) and (6.33). His results were presented in the article by Carlstedt and Mullikin [4], which also contains detailed tables of $X(\zeta, \tau_0)$ and $Y(\zeta, \tau_0)$ for the case of isotropic scattering.

We note that if the functions $X^m(\zeta, \tau_0)$ and $Y^m(\zeta, \tau_0)$ are known, the function $\Phi^m(\tau, \tau_0)$ may be found from the equation

$$\Phi^m(\tau, \tau_0) = L^m(\tau, \tau_0) + \int_0^\tau \Phi^m(t, \tau_0) L^m(\tau-t, \tau_0) \, dt - \int_0^\tau \Phi^m(\tau_0-t, \tau_0) M^m(\tau-t, \tau_0) \, dt, \qquad (6.34)$$

where we have set

$$L^m(\tau, \tau_0) = \int_0^1 e^{-\tau/\eta} X^m(\eta, \tau_0) \frac{d\eta}{\eta}, \qquad (6.35)$$

$$M^m(\tau, \tau_0) = \int_0^1 e^{-\tau/\eta} Y^m(\eta, \tau_0) \frac{d\eta}{\eta}. \qquad (6.36)$$

Equation (6.34) follows from (6.19) and (6.25).

As we have already mentioned, the intensity of radiation diffusely reflected and diffusely transmitted by the atmosphere can be expressed in terms of $X^m(\zeta, \tau_0)$ and $Y^m(\zeta, \tau_0)$. This results from the fact that knowledge of the functions $X^m(\zeta, \tau_0)$ and $Y^m(\zeta, \tau_0)$ allows us to determine the auxiliary functions $\varphi_i^m(\zeta, \tau_0)$ and $\psi_i^m(\zeta, \tau_0)$ which were introduced in Chapter 3. We recall that the reflection and transmission coefficients are expressed in terms of these latter functions by equations (3.16) and (3.17).

In order to obtain expressions for the functions $\varphi_i^m(\zeta, \tau_0)$ and $\psi_i^m(\zeta, \tau_0)$, we first note the relationships

$$B_i^m(0, \zeta, \tau_0) = \frac{\lambda}{4} S\varphi_i^m(\zeta, \tau_0),$$

$$B_i^m(\tau_0, \zeta, \tau_0) = \frac{\lambda}{4} S\psi_i^m(\zeta, \tau_0), \qquad (6.37)$$

9*

which may be established by comparing equations (3.12) and (3.13) with the equations which follow from (6.2) when $\tau = 0$ and when $\tau = \tau_0$. Let us next set $\tau = 0$ and $\tau = \tau_0$ in the relations (6.8). Using the equations (6.5), (6.20) and (6.37), we find

$$\varphi_i^m(\zeta, \tau_0) = R_{im}^m(\zeta) P_m^m(\zeta) X^m(\zeta, \tau_0)$$

$$+ \sum_{k=m+1}^n \left[\varphi_k^m(\zeta, \tau_0) + \frac{k+m-1}{k-m} \varphi_{k-2}^m(\zeta, \tau_0) \right] \int_0^1 g_{ik}^m(\eta) X^m(\eta, \tau_0)\, d\eta$$

$$+ \sum_{k=m+1}^n \left[\psi_k^m(\zeta, \tau_0) + \frac{k+m-1}{k-m} \psi_{k-2}^m(\zeta, \tau_0) \right] \int_0^1 g_{ik}^m(-\eta) Y^m(\eta, \tau_0)\, d\eta \qquad (6.38)$$

and

$$\psi_i^m(\zeta, \tau_0) = R_{im}^m(\zeta) P_m^m(\zeta) Y^m(\zeta, \tau_0) + \sum_{k=m+1}^n \left[\varphi_k^m(\zeta, \tau_0) \right.$$

$$\left. + \frac{k+m-1}{k-m} \varphi_{k-2}^m(\zeta, \tau_0) \right] \int_0^1 g_{ik}^m(\eta) Y^m(\eta, \tau_0)\, d\eta$$

$$+ \sum_{k=m+1}^n \left[\psi_k^m(\zeta, \tau_0) + \frac{k+m-1}{k-m} \psi_{k-2}^m(\zeta, \tau_0) \right] \int_0^1 g_{ik}^m(-\eta) X^m(\eta, \tau_0)\, d\eta. \qquad (6.39)$$

The functions $\varphi_i^m(\zeta, \tau_0)$ and $\psi_i^m(\zeta, \tau_0)$ may thus be found from the system of algebraic equations (6.38) and (6.39) if $X^m(\zeta, \tau_0)$ and $Y^m(\zeta, \tau_0)$ are known. The number of equations equals $2(n-m+1)$ for given m.

After taking account of the relations

$$R_{ik}^m(-\eta) = (-1)^{i+k} R_{ik}^m(\eta),$$
$$g_{ik}^m(-\eta) = (-1)^{i+k} g_{ik}^m(\eta), \qquad (6.40)$$

we find from equations (6.38) and (6.39) that the functions $\varphi_i^m(\zeta, \tau_0)$ and $\psi_i^m(\zeta, \tau_0)$ have the form

$$\varphi_i^m(\zeta, \tau_0) = P_m^m(\zeta)\,[X^m(\zeta, \tau_0) q_i^m(\zeta, \tau_0) + (-1)^{i+m} Y^m(\zeta, \tau_0) s_i^m(-\zeta, \tau_0)], \qquad (6.41)$$

$$\psi_i^m(\zeta, \tau_0) = P_m^m(\zeta)\,[X^m(\zeta, \tau_0) s_i^m(\zeta, \tau_0) + (-1)^{i+m} Y^m(\zeta, \tau_0) q_i^m(-\zeta, \tau_0)], \qquad (6.42)$$

where $q_i^m(\zeta, \tau_0)$ and $s_i^m(\zeta, \tau_0)$ are polynomials in ζ of order $(n-m)$. These polynomials may be determined from the equations

$$q_i^m(\zeta, \tau_0) = R_{im}^m(\zeta) + \sum_{k=m+1}^n \left[q_k^m(\zeta, \tau_0) + \frac{k+m-1}{k-m} q_{k-2}^m(\zeta, \tau_0) \right] \int_0^1 g_{ik}^m(\eta) X^m(\eta, \tau_0)\, d\eta$$

$$+ \sum_{k=m+1}^n \left[s_k^m(\zeta, \tau_0) + \frac{k+m-1}{k-m} s_{k-2}^m(\zeta, \tau_0) \right] \int_0^1 g_{ik}^m(-\eta) Y^m(\eta, \tau_0)\, d\eta, \qquad (6.43)$$

and

$$s_i^m(\zeta, \tau_0) = \sum_{k=m+1}^{n} \left[q_k^m(\zeta, \tau_0) + \frac{k+m-1}{k-m} q_{k-2}^m(\zeta, \tau_0) \right]$$

$$\times \int_0^1 g_{ik}^m(\eta) Y^m(\eta, \tau_0) \, d\eta + \sum_{k=m+1}^{n} \left[s_k^m(\zeta, \tau_0) \right.$$

$$\left. + \frac{k+m-1}{k-m} s_{k-2}^m(\zeta, \tau_0) \right] \int_0^1 g_{ik}^m(-\eta) X^m(\eta, \tau_0) \, d\eta. \tag{6.44}$$

In the next chapter we shall present an alternative method for finding the polynomials $q_i^m(\zeta, \tau_0)$ and $s_i^m(\zeta, \tau_0)$.

6.4. Particular Cases

We have assumed heretofore that the source function has been expanded in a cosine series in azimuth, and we have sought the coefficients of this expansion, $B^m(\tau, \eta, \zeta, \tau_0)$. Let us recount all the stages of the computation which are necessary for the determination of each of these coefficients:

1. Solution of the integral equation (6.15) for the function $\Phi^m(\tau, \tau_0)$.
2. Computation of the functions $X^m(\zeta, \tau_0)$ and $Y^m(\zeta, \tau_0)$ from equations (6.21) and (6.22).
3. Determination of the functions $\varphi_i^m(\zeta, \tau_0)$ and $\psi_i^m(\zeta, \tau_0)$ from the system of algebraic equations (6.38) and (6.39).
4. Computation of the function $D^m(\tau, \zeta, \tau_0)$ from equation (6.19).
5. Determination of the quantity $B^m(\tau, \eta, \zeta, \tau_0)$ from equation (6.9).

If we are only interested in the radiation emerging from the atmosphere, then we may stop at the third step of the above procedure, since the intensity of this radiation is given in terms of $\varphi_i^m(\zeta, \tau_0)$ and $\psi_i^m(\zeta, \tau_0)$.

We shall now use the equations which we have obtained to compute the radiation field in an atmosphere for several particular phase functions.

Isotropic scattering

In this case the source function is determined by the integral equation (1.57) and does not depend on the azimuth or on the variable η. It may be expressed in terms of a single fundamental function $\Phi^0(\tau, \tau_0)$. In order to determine the intensity of radiation emerging from the atmosphere, it is necessary to know only the two functions $X^0(\zeta, \tau_0)$ and $Y^0(\zeta, \tau_0)$, which are identical to the functions $\varphi_0^0(\zeta, \tau_0)$ and $\psi_0^0(\zeta, \tau_0)$. For simplicity we shall omit the super- and sub-zero in the designation of these functions.

From equation (6.15) it follows that the function $\Phi(\tau, \tau_0)$ is determined by

$$\Phi(\tau, \tau_0) = \frac{\lambda}{2} \int_0^{\tau_0} E_1(|\tau-t|) \, \Phi(t, \tau_0) \, dt + \frac{\lambda}{2} E_1(\tau), \tag{6.45}$$

while equations (6.21) and (6.22) take the form

$$\varphi(\zeta, \tau_0) = 1 + \int_0^{\tau_0} \Phi(\tau, \tau_0) e^{-\tau/\zeta}\, d\tau, \tag{6.46}$$

$$\psi(\zeta, \tau_0) = e^{-\tau_0/\zeta} + \int_0^{\tau_0} \Phi(\tau, \tau_0) e^{-(\tau_0-\tau)/\zeta}\, d\tau. \tag{6.47}$$

The source function coincides with the auxiliary function $D^0(\tau, \zeta, \tau_0)$, so that on the basis of (6.19) we have

$$B(\tau, \zeta, \tau_0) = \frac{\lambda}{4} S \left\{ \varphi(\zeta, \tau_0) \left[e^{-\tau/\zeta} + \int_0^{\tau} \Phi(t, \tau_0)\, e^{-(\tau-t)/\zeta}\, dt \right] \right.$$

$$\left. - \psi(\zeta, \tau_0) \int_0^{\tau} \Phi(\tau_0 - t, \tau_0) e^{-(\tau-t)/\zeta}\, dt \right\}. \tag{6.48}$$

The relations (6.45)–(6.48), originally found in [5], were subsequently used for calculations [6]. We shall now illustrate some of the results obtained.

We shall seek the function $\Phi(\tau, \tau_0)$ in the form of a powers series in λ:

$$\Phi(\tau, \tau_0) = \frac{\lambda}{2} E_1(\tau) + \lambda^2 \Phi_2(\tau, \tau_0) + \lambda^3 \Phi_3(\tau, \tau_0) + \ldots, \tag{6.49}$$

where

$$\Phi_n(\tau, \tau_0) = \frac{1}{2} \int_0^{\tau_0} E_1(|\tau - \tau'|) \Phi_{n-1}(\tau', \tau_0)\, d\tau'. \tag{6.50}$$

It is clear that the nth term of the series (6.49) corresponds to scattering of the nth order. This series is useful for small values of τ_0 or for values of λ that are not very close to 1. As an example, Table 6.1 presents values of the first six functions $\Phi_n(\tau, \tau_0)$ for $\tau_0 = 0.3$ (accurate to 1 or 2 units in the last significant figure). With the aid of this table and equation (6.49) we may find the value of the function $\Phi(\tau, \tau_0)$ for any λ.

TABLE 6.1. FUNCTION $\Phi_n(\tau, \tau_0)$ FOR $\tau_0 = 0.3$ (ISOTROPIC SCATTERING)

n	τ						
	0	0.05	0.10	0.15	0.20	0.25	0.30
1	∞	1.234	0.911	0.732	0.611	0.522	0.453
2	0.300	0.336	0.323	0.285	0.274	0.243	0.196
3	0.079	0.099	0.105	0.104	0.098	0.089	0.071
4	0.025	0.032	0.034	0.035	0.034	0.031	0.024
5	0.008	0.010	0.012	0.012	0.011	0.010	0.008
6	0.003	0.003	0.004	0.004	0.004	0.004	0.003

We shall likewise expand $\varphi(\zeta, \tau_0)$ and $\psi(\zeta, \tau_0)$ in the series

$$\varphi(\zeta, \tau_0) = 1 + \lambda\varphi_1(\zeta, \tau_0) + \lambda^2\varphi_2(\zeta, \tau_0) + \dots, \tag{6.51}$$

$$\psi(\zeta, \tau_0) = e^{-\tau_0/\zeta} + \lambda\psi_1(\zeta, \tau_0) + \lambda^2\psi_2(\zeta, \tau_0) + \dots, \tag{6.52}$$

where

$$\varphi_n(\zeta, \tau_0) = \int_0^{\tau_0} \Phi_n(\tau, \tau_0)e^{-\tau/\zeta} \, d\tau, \quad \psi_n(\zeta, \tau_0) = \int_0^{\tau_0} \Phi_n(\tau, \tau_0)e^{-(\tau_0-\tau)/\zeta} \, d\tau. \tag{6.53}$$

Values of the functions $\varphi_n(\zeta, \tau_0)$ and $\psi_n(\zeta, \tau_0)$ are given in Tables 6.2 and 6.3 for $\tau_0 = 0.3$. These tables together with equations (6.51) and (6.52) allow us to find $\varphi(\zeta, \tau_0)$ and $\psi(\zeta, \tau_0)$ for arbitrary λ. We recall that the reflection and transmission coefficients of the atmosphere are determined by these functions in accordance with equations (3.20) and (3.21).

TABLE 6.2. FUNCTION $\varphi_n(\zeta, \tau_0)$ FOR $\tau_0 = 0.3$
(ISOTROPIC SCATTERING)

n	ζ					
	0	0.2	0.4	0.6	0.8	1.0
1	0	0.165	0.205	0.223	0.232	0.238
2	0	0.047	0.061	0.069	0.072	0.075
3	0	0.015	0.020	0.023	0.024	0.025
4	0	0.005	0.007	0.008	0.008	0.008
5	0	0.002	0.002	0.003	0.003	0.003
6	0	0.001	0.001	0.001	0.001	0.001

TABLE 6.3. FUNCTION $\psi_n(\zeta, \tau_0)$ FOR $\tau_0 = 0.3$
(ISOTROPIC SCATTERING)

n	ζ					
	0	0.2	0.4	0.6	0.8	1.0
1	0	0.114	0.170	0.196	0.211	0.221
2	0	0.042	0.058	0.066	0.071	0.073
3	0	0.014	0.020	0.022	0.024	0.025
4	0	0.005	0.007	0.008	0.008	0.008
5	0	0.002	0.002	0.002	0.003	0.003
6	0	0.001	0.001	0.001	0.001	0.001

If the functions $\Phi(\tau, \tau_0)$, $\varphi(\zeta, \tau_0)$ and $\psi(\zeta, \tau_0)$ are known, then it is easy to find the source function from equation (6.48). Values of this function are given in Table 6.4, for $\lambda = 1$ and $\tau_0 = 0.3$.

We see that the source function may be found quite easily after we have determined the function $\Phi(\tau, \tau_0)$ from equation (6.45). Knowledge of the source function is sufficient for the determination of the radiation intensity at any optical depth from the equations which have been given.

TABLE 6.4. SOURCE FUNCTION $B(\tau, \zeta, \tau_0)/S$ FOR $\tau_0 = 0.3$,
$\lambda = 1$ AND ISOTROPIC SCATTERING

τ	ζ					
	0.0	0.2	0.4	0.6	0.8	1.0
0	0.250	0.308	0.324	0.331	0.335	0.338
0.1	0.000	0.223	0.290	0.318	0.333	0.342
0.2	0.000	0.154	0.241	0.280	0.302	0.317
0.3	0.000	0.100	0.182	0.226	0.251	0.268

It should also be noted that the resolvent $\Gamma(\tau, \tau', \tau_0)$ of the integral equation with kernel $E_1(|\tau - \tau'|)\lambda/2$ may be expressed in terms of the function $\Phi(\tau, \tau_0)$. Thus, the source function for any distribution of primary radiation sources in the atmosphere may be expressed in terms of the function $\Phi(\tau, \tau_0)$.

Let us suppose, for example, that the primary sources are distributed uniformly and radiate isotropically. Then the source function satisfies the equation

$$B(\tau, \tau_0) = \frac{\lambda}{2} \int_0^{\tau_0} E_1(|\tau - t|)B(t, \tau_0)\, dt + 1. \tag{6.54}$$

It is easy to show that $B(\tau, \tau_0)$ is expressed in terms of $\Phi(\tau, \tau_0)$ by

$$B(\tau, \tau_0) = B(0, \tau_0)\left[1 + \int_0^{\tau} \Phi(t, \tau_0)\, dt - \int_{\tau_0 - \tau}^{\tau} \Phi(t, \tau_0)\, dt\right], \tag{6.55}$$

where

$$B(0, \tau_0) = B(\tau_0, \tau_0) = 1 + \int_0^{\tau_0} \Phi(\tau, \tau_0)\, d\tau. \tag{6.56}$$

The boundary value for the source function may also be determined from the equation

$$B(0, \tau_0) = \frac{1}{1 - \dfrac{\lambda}{2}(\alpha_0 - \beta_0)}, \tag{6.57}$$

where α_0 and β_0 are the zeroth moments of the functions $\varphi(\zeta, \tau_0)$ and $\psi(\zeta, \tau_0)$.

Three-term phase function

When the phase function is $x(\gamma) = 1 + x_1 \cos \gamma + x_2 P_2 (\cos \gamma)$, the source function takes the form

$$B(\tau, \eta, \zeta, \varphi, \tau_0) = B^0(\tau, \eta, \zeta, \tau_0) + 2B^1(\tau, \eta, \zeta, \tau_0)\cos \varphi + 2B^2(\tau, \eta, \zeta, \tau_0)\cos 2\varphi. \tag{6.58}$$

In order to determine each of the quantities $B^m(\tau, \eta, \zeta, \tau_0)$, we must find the corresponding function $\Phi^m(\tau, \tau_0)$. We have already examined the similar problem for a semi-infinite atmosphere (in Section 5.5), in which case the function $\Phi^m(\tau)$ was expressed explicitly in terms of the function $H^m(\zeta)$. For an atmosphere of finite optical thickness it is first necessary to determine the $\Phi^m(\tau, \tau_0)$ from equation (6.15), and then compute the functions $X^m(\zeta, \tau_0)$ and $Y^m(\zeta, \tau_0)$ from equations (6.21) and (6.22). The functions $K^m(\tau)$ which enter equation (6.15) are determined by equation (6.6). In the present case we must substitute into (6.6) the expressions for the functions $\Psi^m(\eta)$ which are given by equations (5.121), (5.127), and (5.130). Proceeding in this manner, we obtain

$$K^0(\tau) = \frac{\lambda}{2}\left(1-\frac{x_2}{4}\right)E_1(\tau)+\frac{\lambda}{2}\left[(1-\lambda)x_1-\frac{x_2}{4}(1-\lambda)(3-\lambda x_1)-\frac{3}{4}x_2\right]E_3(\tau)$$

$$+\frac{3}{8}\lambda x_2(1-\lambda)(3-\lambda x_2)E_5(\tau),\tag{6.58}$$

$$K^1(\tau) = \frac{\lambda}{2}x_1E_1(\tau)-\frac{\lambda}{2}[x_1-(3-\lambda x_1)x_2]E_3(\tau)-\frac{\lambda}{2}(3-\lambda x_1)x_2E_5(\tau),\tag{6.59}$$

$$K^2(\tau) = \frac{3}{16}\lambda x_2[E_1(\tau)-2E_3(\tau)+E_5(\tau)].\tag{6.60}$$

The solution of equation (6.15) has been carried out numerically by an iterative method for the two phase functions: (A) $x_1 = 1$, $x_2 = 1$ and (B) $x_1 = \frac{3}{2}$, $x_2 = 1$. Sample values of the function $\Phi^m(\tau, \tau_0)$ are presented in Table 6.5 for the case $\lambda = 1$ and $\tau_0 = 1$.

TABLE 6.5. THE FUNCTION $\Phi^m(\tau, \tau_0)$ FOR $\lambda = 1$, $\tau_0 = 1$
AND TWO SAMPLE THREE-TERM PHASE FUNCTIONS

τ	$m = 0$	$m = 1$		$m = 2$	τ	$m = 0$	$m = 1$		$m = 2$
	A, B	A	B	A, B		A, B	A	B	A, B
0	∞	∞	∞	∞	0.10	1.947	0.569	0.839	0.254
0.01	3.193	1.120	1.644	0.648	0.20	1.599	0.410	0.611	0.156
0.02	2.791	0.951	1.396	0.524	0.30	1.387	0.320	0.481	0.108
0.03	2.575	0.854	1.253	0.452	0.40	1.225	0.258	0.392	0.079
0.04	2.401	0.785	1.152	0.403	0.50	1.086	0.212	0.325	0.059
0.05	2.303	0.732	1.075	0.365	0.60	0.961	0.177	0.272	0.045
0.06	2.208	0.689	1.013	0.335	0.70	0.842	0.148	0.228	0.035
0.07	2.128	0.652	0.960	0.310	0.80	0.725	0.123	0.190	0.028
0.08	2.040	0.621	0.915	0.289	0.90	0.606	0.102	0.157	0.022
0.09	2.000	0.594	0.880	0.270	1.00	0.461	0.082	0.123	0.017

In order to compute the functions $\varphi_i^m(\zeta, \tau_0)$, $\psi_i^m(\zeta, \tau_0)$, $D^m(\tau, \zeta, \tau_0)$ and $B^m(\tau, \eta, \zeta, \tau_0)$ by the method described above, it is necessary to use the expressions for the functions $R_{im}^m(\zeta)$ and $g_{ik}^m(\zeta)$ given in Section 5.5. We shall not consider the results of these computations here.

6.5. Equations Containing Derivatives with Respect to τ_0

Heretofore in this chapter we have taken the optical thickness of the atmosphere τ_0 as given. It is possible, however, to obtain equations for the various quantities characterizing the radition field which describe how these quantities change with changing τ_0. These equations may be used to determine the desired quantities. Equations of this type (i.e. containing derivatives with respect to τ_0) have in fact already been obtained in Section 3.2 for the reflection and transmission coefficients of the atmosphere and also for the auxiliary functions $\varphi_i^m(\zeta, \tau_0)$ and $\psi_i^m(\zeta, \tau_0)$. We shall now find similar equations for the functions $\Phi^m(\tau, \tau_0)$, $X^m(\zeta, \tau_0)$ and $Y^m(\zeta, \tau_0)$.

We shall begin with the integral equation (6.12) for the resolvent $\Gamma^m(\tau, \tau', \tau_0)$. Differentiating with respect to τ_0, we obtain

$$\frac{\partial \Gamma^m(\tau, \tau', \tau_0)}{\partial \tau_0} = \int_0^{\tau_0} K^m(|\tau - \tau''|) \frac{\partial \Gamma^m(\tau'', \tau', \tau_0)}{\partial \tau_0} \, d\tau'' + K^m(\tau_0 - \tau) \Gamma^m(\tau_0, \tau', \tau_0). \quad (6.61)$$

Comparison of (6.61) and (6.15) gives

$$\frac{\partial \Gamma^m(\tau, \tau', \tau_0)}{\partial \tau_0} = \Phi^m(\tau_0 - \tau, \tau_0) \Phi^m(\tau_0 - \tau', \tau_0), \quad (6.62)$$

where we recall that $\Gamma^m(\tau_0, \tau', \tau_0) = \Phi^m(\tau_0 - \tau', \tau_0)$. Setting $\tau' = 0$, we obtain

$$\frac{\partial \Phi^m(\tau, \tau_0)}{\partial \tau_0} = \Phi^m(\tau_0 - \tau, \tau_0) \Phi^m(\tau_0, \tau_0). \quad (6.63)$$

We have thus arrived at an important equation which must be satisfied by the function $\Phi^m(\tau, \tau_0)$.

Equation (6.63), however, is insufficient for the determination of this function, since the characteristic function $\Psi^m(\eta)$ does not enter it. In order to find $\Phi^m(\tau, \tau_0)$ with the aid of (6.63), we must add a constraint condition. The relation

$$\Phi^m(\tau_0, \tau_0) = \int_0^{\tau_0} K^m(\tau_0 - \tau) \Phi^m(\tau, \tau_0) \, d\tau + K^m(\tau_0), \quad (6.64)$$

which follows from (6.15) for $\tau = \tau_0$, may serve as such a condition.

We have already pointed out that the function $\Phi^m(\tau, \tau_0)$ for small τ_0 may usefully be determined iteratively from equation (6.15). As τ_0 increases, it is clearly necessary to switch to equations (6.63) and (6.64) for the determination of $\Phi^m(\tau, \tau_0)$.

It is also easy to obtain equations for $X^m(\zeta, \tau_0)$ and $Y^m(\zeta, \tau_0)$ which involve differentiation with respect to τ_0. We may use equations (6.21) and (6.22) for this purpose. Differentiating them with respect to τ_0 and using equation (6.63), we find

$$\frac{\partial X^m(\zeta, \tau_0)}{\partial \tau_0} = Y^m(\zeta, \tau_0) \Phi^m(\tau_0, \tau_0), \quad (6.65)$$

$$\frac{\partial Y^m(\zeta, \tau_0)}{\partial \tau_0} = -\frac{1}{\zeta} Y^m(\zeta, \tau_0) + X^m(\zeta, \tau_0) \Phi^m(\tau_0, \tau_0). \quad (6.66)$$

The quantity $\Phi^m(\tau_0, \tau_0)$ which enters here may be expressed in terms of $Y^m(\zeta, \tau_0)$. For this purpose we set $\tau = 0$ in equation (6.25) and substitute into that equation the second of equations (6.20). We obtain as a result

$$\Phi^m(\tau_0, \tau_0) = \int_0^1 Y^m(\zeta, \tau_0)\,\Psi^m(\zeta)\,\frac{d\zeta}{\zeta}.$$ (6.67)

Equations (6.65)–(6.67) must be solved subject to the boundary conditions

$$X^m(\zeta, 0) = Y^m(\zeta, 0) = 1,$$ (6.68)

which follow from (6.21) and (6.22).

The equations which we have obtained for $X^m(\zeta, \tau_0)$ and $Y^m(\zeta, \tau_0)$ may be replaced by the following system of integral equations:

$$X^m(\zeta, \tau_0) = 1 + \int_0^{\tau_0} Y^m(\zeta, \tau)\,d\tau \int_0^1 Y^m(\eta, \tau)\,\Psi^m(\eta)\,\frac{d\eta}{\eta},$$ (6.69)

$$Y^m(\zeta, \tau_0) = e^{-\tau_0/\zeta} + \int_0^{\tau_0} e^{-(\tau_0-\tau)/\zeta}X^m(\zeta, \tau)\,d\tau \int_0^1 Y^m(\eta, \tau)\,\Psi^m(\eta)\,\frac{d\eta}{\eta}.$$ (6.70)

These equations represent a generalization of the integral equations (3.47) and (3.48) which apply to the case of isotropic scattering.

6.6. Atmospheres of Large Optical Thickness

If the optical thickness of the atmosphere is large ($\tau_0 \gg 1$), then asymptotic expressions may be obtained for the various quantities which characterize the radiation field. These expressions become more exact as τ_0 increases. Such expressions have already been obtained in Chapter 3 for the reflection and transmission coefficients averaged over azimuth. The basis for the derivation of these expressions is our knowledge of the radiation field in the deep layers of a semi-infinite medium. By introducing an imaginary horizontal plane at these depths and considering the conditions at the boundary so created, we arrive at the formulas which we are seeking. We emphasize that this procedure may be used only for quantities which are averaged over azimuth, since the radiation field in the deep layers does not depend upon azimuth.

We shall now find an asymptotic expression for the fundamental function $\Phi^0(\tau, \tau_0)$, which we shall express in terms of the corresponding function $\Phi^0(\tau)$ for a semi-infinite atmosphere. For simplicity in what follows we shall omit the index "0" on all quantities averaged over azimuth (e.g. we shall write $\Phi(\tau, \tau_0)$ instead of $\Phi^0(\tau, \tau_0)$, $\Phi(\tau)$ instead of $\Phi^0(\tau)$, etc.).

The function $\Phi(\tau)$ is determined by the integral equation (5.80) with $m = 0$. Splitting the integral into two parts and using equation (6.6) for $K(\tau)$, we may rewrite this equation in the form

$$\Phi(\tau) = \int_0^{\tau_0} K(|\tau - t|)\Phi(t)\,dt + \int_0^1 \Psi(\eta)\,d\eta \int_{\tau_0}^\infty \Phi(t)e^{-(t-\tau)/\eta}\,\frac{dt}{\eta} + K(\tau).$$ (6.71)

When $\tau_0 \gg 1$ we may use in the second integral the asymptotic expression for $\Phi(t)$ valid in deep layers,

$$\Phi(t) = Ce^{-kt}, \tag{6.72}$$

where C is given by equation (5.92) and k by equation (2.22). Substituting equation (6.72) into (6.71), we find

$$\Phi(\tau) = \int_0^{\tau_0} K(|\tau - t|)\Phi(t)\,dt + \Phi(\tau_0) \int_0^1 \Psi(\eta)\,e^{-(\tau_0-\tau)/\eta}\,\frac{d\eta}{1+k\eta} + K(\tau). \tag{6.73}$$

By comparing equation (6.73) with (6.7) and (6.15) we see that

$$\Phi(\tau) = \Phi(\tau, \tau_0) + \frac{4}{\lambda S}\Phi(\tau_0) \int_0^1 D(\tau_0 - \tau,\, \eta,\, \tau_0)\, \Psi(\eta)\, \frac{d\eta}{1+k\eta}. \tag{6.74}$$

An analogous expression may be obtained for the auxiliary function $D(\tau, \zeta, \tau_0)$:

$$D(\tau, \zeta) = D(\tau, \zeta, \tau_0) + \frac{4}{\lambda S}\, D(\tau_0, \zeta) \int_0^1 D(\tau_0 - \tau,\, \eta,\, \tau_0)\, \Psi(\eta)\frac{d\eta}{1+k\eta}. \tag{6.75}$$

Introducing the notations

$$\left.\begin{array}{l} \Delta(\tau) = \dfrac{4}{\lambda S} \displaystyle\int_0^1 D(\tau, \eta)\, \Psi(\eta)\, \dfrac{d\eta}{1+k\eta}, \\[4mm] \Delta(\tau, \tau_0) = \dfrac{4}{\lambda S} \displaystyle\int_0^1 D(\tau, \eta, \tau_0)\, \Psi(\eta)\, \dfrac{d\eta}{1+k\eta}, \end{array}\right\} \tag{6.76}$$

we have from (6.75) that

$$\Delta(\tau_0 - \tau, \tau_0) = \frac{\Delta(\tau_0 - \tau) - \Delta(\tau_0)\, \Delta(\tau)}{1 - \Delta^2(\tau_0)}. \tag{6.77}$$

Substituting (6.77) into (6.74) we find the following asymptotic expression for the function $\Phi(\tau, \tau_0)$:

$$\Phi(\tau, \tau_0) = \Phi(\tau) - \Phi(\tau_0)\, \frac{\Delta(\tau_0 - \tau) - \Delta(\tau_0)\, \Delta(\tau)}{1 - \Delta^2(\tau_0)}. \tag{6.78}$$

The quantity $\Delta(\tau)$ which enters this last equation may easily be expressed in terms of $\Phi(\tau)$. In order to obtain this expression, we use the relation

$$\frac{dD(t, \eta)}{dt} = -\frac{1}{\eta}\, D(t, \eta) + D(0, \eta)\, \Phi(t), \tag{6.79}$$

which follows from (5.82). Multiplying (6.79) by e^{-kt} and integrating over t between τ and ∞, we find

$$\int_\tau^\infty D(t, \eta)\, e^{-kt}\, \frac{dt}{\eta} = \frac{D(\tau, \eta)}{1+k\eta}\, e^{-k\tau} + \frac{D(0, \eta)}{1+k\eta} \int_\tau^\infty \Phi(t)\, e^{-kt}\, dt. \tag{6.80}$$

After multiplication of (6.80) by $\Psi(\eta)$ and integration over η between 0 and 1, we obtain

$$\Delta(\tau) = \left[1 - \int_0^1 \frac{H(\eta)}{1+k\eta} \Psi(\eta) \, d\eta \right] \int_\tau^\infty \Phi(t) \, e^{k(\tau-t)} dt, \tag{6.81}$$

where we have utilized equations (5.85) and (6.76).

In this way, the function $\Phi(\tau, \tau_0)$ for $\tau_0 \gg 1$ is determined by the asymptotic expression (6.78), in which the quantity $\Delta(\tau)$ is given by equation (6.81). We emphasize that this asymptotic expression for $\Phi(\tau, \tau_0)$ is valid for any τ in the interval from 0 to τ_0.

We may obtain a very simple expression for the quantity $\Delta(\tau_0)$ from equation (6.81). Setting $\tau = \tau_0$ in this equation and substituting equation (6.72) into it, we find that

$$\Delta(\tau_0) = \left[1 - \int_0^1 \frac{H(\eta)}{1+k\eta} \Psi(\eta) \, d\eta \right] \frac{C}{2k} e^{-k\tau_0}. \tag{6.82}$$

We shall need the relation

$$\int_0^1 \frac{H(\eta)}{1-k\eta} \Psi(\eta) \, d\eta = 1, \tag{6.83}$$

which follows from (5.66) and (5.68). From (6.83) we may obtain

$$1 - \int_0^1 \frac{H(\eta)}{1+k\eta} \Psi(\eta) \, d\eta = 2k \int_0^1 \frac{H(\eta)}{1-k^2\eta^2} \Psi(\eta)\eta \, d\eta. \tag{6.84}$$

Substituting (6.84) into (6.82) and using equation (5.92) for the constant C, we find

$$\Delta(\tau_0) = be^{-k\tau_0}, \tag{6.85}$$

where

$$b = \frac{\displaystyle\int_0^1 \frac{H(\eta)}{1-k^2\eta^2} \Psi(\eta)\eta \, d\eta}{\displaystyle\int_0^1 \frac{H(\eta)}{(1-k\eta)^2} \Psi(\eta)\eta \, d\eta}. \tag{6.86}$$

Equation (6.81) now takes the form

$$\Delta(\tau) = \frac{2kb}{C} \int_\tau^\infty \Phi(t) \, e^{k(\tau-t)} \, dt. \tag{6.87}$$

Asymptotic expressions for the functions $X(\zeta, \tau_0)$ and $Y(\zeta, \tau_0)$ may be obtained with the aid of the asymptotic formula (6.78). To do this we must substitute (6.78) into equations (6.21) and (6.22). It is clear from equation (6.21) that values of $\Phi(\tau, \tau_0)$ for small τ play the predominant role in determining $X(\zeta, \tau_0)$. In this case equation (6.78) may be rewritten in the form

$$\Phi(\tau, \tau_0) = \Phi(\tau) - \Phi(\tau_0) \frac{\Delta(\tau_0) [e^{k\tau} - \Delta(\tau)]}{1 - \Delta^2(\tau_0)} . \tag{6.88}$$

Introducing (6.88) into (6.21) and taking account of (5.84) and (5.87), we find

$$X(\zeta, \tau_0) = H(\zeta) - \frac{\Phi(\tau_0) \Delta(\tau_0)}{1 - \Delta^2(\tau_0)} \frac{\zeta}{1 - k\zeta} \left\{ 1 - \frac{2kb}{C} \left[1 + \int_0^\infty \Phi(t) e^{-kt} dt - H(\zeta) \right] \right\}. \tag{6.89}$$

In order to find the integral entering the last equation, we may use the relation (6.80) for $\tau = 0$. It follows that

$$\int_0^\infty D(t, \eta) e^{-kt} \frac{dt}{\eta} = \frac{D(0, \eta)}{1 + k\eta} \left[1 + \int_0^\infty \Phi(t) e^{-kt} dt \right]. \tag{6.90}$$

Multiplying (6.90) by $\Psi(\eta)$, integrating between 0 and 1, and using (5.37) and (5.85), we obtain

$$\left[1 + \int_0^\infty \Phi(t) e^{-kt} dt \right] \left[1 - \int_0^1 \frac{H(\eta)}{1 + k\eta} \Psi(\eta) d\eta \right] = 1 \tag{6.91}$$

where

$$1 + \int_0^\infty \Phi(t) e^{-kt} dt = \frac{C}{2kb} . \tag{6.92}$$

Substitution of equations (6.72) (for $t = \tau_0$), (6.85), and (6.92) into (6.89) gives

$$X(\zeta, \tau_0) = H(\zeta) - \frac{2kb e^{-2k\tau_0}}{1 - b^2 e^{-2k\tau_0}} \frac{\zeta}{1 - k\zeta} H(\zeta). \tag{6.93}$$

In an analogous manner, we may obtain an asymptotic expression for the function $Y(\zeta, \tau_0)$:

$$Y(\zeta, \tau_0) = \frac{2kb e^{-k\tau_0}}{1 - b^2 e^{-2k\tau_0}} \frac{\zeta}{1 - k\zeta} H(\zeta). \tag{6.94}$$

The asymptotic expressions (6.93) and (6.94) are valid for arbitrary values of λ; that is, for any ratio of the scattering coefficient to the true absorption coefficient in the atmosphere. In the case of pure scattering ($\lambda = 1, k = 0$), however, an indeterminacy arises in these equations. It may be resolved by finding an expression for the quantity b for small k. It is not difficult to show (see [7]) that the function $H(\eta)$ is given for small k by

$$H(\eta) = H_0(\eta) (1 - k\eta), \tag{6.95}$$

where $H_0(\eta)$ denotes the function $H(\eta)$ for $\lambda = 1$. The function $\Psi(\eta)$ may be expanded in powers of $(1-\lambda)$; that is, in its expansion there are no terms of order k (which is of order $\sqrt{1-\lambda}$). It then follows from equation (1.86) for small k that

$$b = 1 - k\delta, \tag{6.96}$$

where

$$\delta = 2 \frac{\int\limits_0^1 H_0(\eta)\,\Psi_0(\eta)\eta^2\,d\eta}{\int\limits_0^1 H_0(\eta)\,\Psi_0(\eta)\eta\,d\eta}, \tag{6.97}$$

and $\Psi_0(\eta)$ is the function $\Psi(\eta)$ for $\lambda = 1$. Substituting (6.96) into (6.93) and (6.94), we find the following asymptotic expressions for $X(\zeta, \tau_0)$ and $Y(\zeta, \tau_0)$ for pure scattering:

$$X(\zeta, \tau_0) = H(\zeta) - \frac{\zeta H(\zeta)}{\tau_0 + \delta}, \tag{6.98}$$

$$Y(\zeta, \tau_0) = \frac{\zeta H(\zeta)}{\tau_0 + \delta}. \tag{6.99}$$

The asymptotic expressions for $X(\zeta, \tau_0)$ and $Y(\zeta, \tau_0)$ are generalizations of those introduced in Section 3.4 for the functions $\varphi(\zeta, \tau_0)$ and $\psi(\zeta, \tau_0)$ in the case of isotropic scattering. Asymptotic expressions for $X(\zeta, \tau_0)$ and $Y(\zeta, \tau_0)$ were examined by a different method and in more detail in the article by Carlstedt and Mullikin [4]. The asymptotic expression for the fundamental function $\Phi(\tau, \tau_0)$ was first obtained for the case of isotropic scattering by Sobolev [8]. Equation (6.78), which generalizes the previous result, is found here by the same procedure.

References

1. V. V. SOBOLEV, Anisotropic scattering of light in an atmosphere of finite optical thickness, *Astrofizika* **5**, 343 (1969) [*Astrophysics* **5**, 161 (1969)].
2. V. V. SOBOLEV, On the theory of the diffusion of radiation, *Izv. Akad. Nauk Arm. SSR, Ser. Fiz.-Matem. Nauk* **11**, 38 (1958).
3. S. CHANDRASEKHAR, On radiative equilibrium of a stellar atmosphere. XXII, *Astrophys. J.* **107**, 48 (1948).
4. J. L. CARLSTEDT and T. W. MULLIKIN, Chandrasekhar's X- and Y-functions, *Astrophys. J. Suppl.*, **12**, 449 (1966).
5. V. V. SOBOLEV, Diffusion of radiation in a plane layer, *Dokl. Akad. Nauk SSSR* **120**, 69 (1958) [*Sov. Phys. Doklady* **3**, 541 (1958)].
6. V. V. SOBOLEV and I. N. MININ, Isotropic scattering of light in an atmosphere of finite optical thickness, *Astron. Zh.* **38**, 1025 (1961) [*Sov. Astron. A.J.* **5**, 785 (1962)].
7. A. K. KOLESOV and V. V. SOBOLEV, Some asymptotic expressions in the theory of anisotropic light scattering, *Astrofizika* **5,**175 (1969) [*Astrophysics* **5**, 87 (1969)].
8. V. V. SOBOLEV, Diffusion of radiation in a plane layer of large optical thickness, *Dokl. Akad. Nauk SSSR* **155**, 416 (1964) [*Sov. Phys. Doklady* **9**, 222 (1964)].

Chapter 7

LINEAR INTEGRAL EQUATIONS FOR THE
REFLECTION AND TRANSMISSION COEFFICIENTS

THE intensity of radiation diffusely reflected and transmitted by an atmosphere is of paramount importance in practical applications of the theory. The problem of determining these intensities has already been discussed: in Chapter 2 for a semi-infinite atmosphere and in Chapter 3 for atmospheres of finite optical thickness. In those discussions the reflection and transmission coefficients were expressed in terms of the auxiliary functions $\varphi_i^m(\eta)$ for $\tau_0 = \infty$, and $\varphi_i^m(\eta, \tau_0)$ and $\psi_i^m(\eta, \tau_0)$ for finite τ_0. A system of nonlinear integral equations was obtained for the determination of these functions.

We shall now present linear integral equations for the determination of the reflection and transmission coefficients, and with their aid we shall also obtain linear integral equations for the auxiliary functions. For the case of a semi-infinite atmosphere these equations have kernels of Cauchy type, and their solution may be obtained explicitly.

As we have shown earlier in Chapters 5 and 6, all the auxiliary functions $\varphi_i^m(\eta)$ for given m may be expressed in terms of only one function $H^m(\eta)$, and all the auxiliary functions $\varphi_i^m(\eta, \tau_0)$ and $\psi_i^m(\eta, \tau_0)$ in terms of the two functions $X^m(\eta, \tau_0)$ and $Y^m(\eta, \tau_0)$. We shall give an alternative method for obtaining these expressions.

7.1. Semi-infinite Atmosphere

The linear integral equations for the reflection and transmission coefficients have been known for a long time [1, 2]. We shall now obtain these equations for the case of a semi-infinite atmosphere. They will be generalized in Section 7.4 to atmospheres of finite optical thickness.

As previously, we shall assume that the source function B and the intensity I are expanded in cosine series in azimuth, so that they are represented by equations (1.43) and (1.44). The coefficients of these expansions are determined by the equations (see Section 1.3)

$$\eta \frac{dI^m(\tau, \eta, \zeta)}{d\tau} = -I^m(\tau, \eta, \zeta) + B^m(\tau, \eta, \zeta), \tag{7.1}$$

$$B^m(\tau, \eta, \zeta) = \frac{\lambda}{2} \int_{-1}^{1} p^m(\eta, \eta') I^m(\tau, \eta', \zeta) \, d\eta' + \frac{\lambda}{4} Sp^m(\eta, \zeta) e^{-\tau/\zeta}, \tag{7.2}$$

with the boundary condition

$$I^m(0, \eta, \zeta) = 0, \qquad \eta > 0. \tag{7.3}$$

Our problem consists of finding equations which will directly determine the quantities $\varrho^m(\eta, \zeta)$ through which the reflection coefficient $\varrho(\eta, \zeta, \varphi)$ is expressed by means of equation (2.29). The quantity $\varrho^m(\eta, \zeta)$ is related to the functions I^m and B^m by

$$S\varrho^m(\eta, \zeta)\zeta = I^m(0, -\eta, \zeta) = \int_0^\infty B^m(\tau, -\eta, \zeta) \, e^{-\tau/\eta} \frac{d\tau}{\eta}. \tag{7.4}$$

With this goal in mind we introduce the new unknown functions $F^m(\eta'', \eta, \zeta)$ according to the equation

$$SF^m(\eta'', \eta, \zeta)\zeta = \int_0^\infty B^m(\tau, -\eta, \zeta) \, e^{-\tau/\eta''} \frac{d\tau}{\eta''}. \tag{7.5}$$

It is clear that

$$\varrho^m(\eta, \zeta) = F^m(\eta, \eta, \zeta). \tag{7.6}$$

Let us replace η by $-\eta$ in equation (7.2), multiply the resulting equation by $e^{-\tau/\eta''} \, d\tau/\eta''$, and integrate from 0 to ∞. Upon use of (7.5) we then obtain

$$SF^m(\eta'', \eta, \zeta)\zeta$$
$$= \frac{\lambda}{2} \int_{-1}^1 p^m(\eta, \eta') \, d\eta' \int_0^\infty I^m(\tau, -\eta', \zeta) e^{-\tau/\eta''} \frac{d\tau}{\eta''} + \frac{\lambda}{4} S \frac{p^m(-\eta, \zeta)}{\eta''+\zeta} \zeta, \tag{7.7}$$

where we have also replaced η' by $-\eta'$ and used the fact that $p^m(-\eta, -\eta') = p^m(\eta, \eta')$. In order to evaluate the integral with respect to τ in (7.7), we return to equation (7.1), multiply it by $e^{-\tau/\eta''}$, and integrate over τ between 0 and ∞. This procedure yields

$$(\eta+\eta'') \int_0^\infty I^m(\tau, \eta, \zeta) e^{-\tau/\eta''} \frac{d\tau}{\eta''} = \eta I^m(0, \eta, \zeta) + \eta'' \int_0^\infty B^m(\tau, \eta, \zeta) e^{-\tau/\eta''} \frac{d\tau}{\eta''}. \tag{7.8}$$

Replacing η by $-\eta'$ and using equations (7.4) and (7.5), we then find

$$\int_0^\infty I^m(\tau, -\eta', \zeta) e^{-\tau/\eta''} \frac{d\tau}{\eta''} = \frac{\eta''F^m(\eta'', \eta', \zeta) - \eta'F^m(\eta', \eta', \zeta)}{\eta''-\eta'} S. \tag{7.9}$$

Substitution of (7.9) into (7.7) leads to the following equation for the determination of the function $F^m(\eta'', \eta, \zeta)$:

$$F^m(\eta'', \eta, \zeta)$$
$$= \frac{\lambda}{2} \int_{-1}^1 p^m(\eta, \eta') \frac{\eta''F^m(\eta'', \eta', \zeta) - \eta'F^m(\eta', \eta', \zeta)}{\eta''-\eta'} \, d\eta' + \frac{\lambda}{4} \frac{p^m(-\eta, \zeta)}{\eta''+\zeta}. \tag{7.10}$$

The quantity $p^m(\eta, \eta')$ which enters (7.10) is given by equation (1.41). The function $F^m(\eta'', \eta, \zeta)$ may thus be written in the form

$$F^m(\eta'', \eta, \zeta) = \sum_{i=m}^{n} u_i^m(\eta'', \zeta) \, P_i^m(\eta). \tag{7.11}$$

By substituting (1.41) and (7.11) into (7.10) we obtain

$$u_i^m(\eta, \zeta)$$

$$= \frac{\lambda}{2} c_i^m \sum_{j=m}^{n} \int_{-1}^{1} \frac{\eta u_j^m(\eta, \zeta) - \eta' u_j^m(\eta', \zeta)}{\eta - \eta'} \, P_i^m(\eta') P_j^m(\eta') \, d\eta' + \frac{\lambda}{4} c_i^m \frac{P_i^m(-\zeta)}{\eta + \zeta}. \tag{7.12}$$

We note that it follows from (7.3) that $u_i^m(\eta', \zeta) = 0$ for $\eta' < 0$. We thus have a system of linear integral equations (7.12) for the determination of the functions $u_i^m(\eta, \zeta)$. After this system has been solved, the quantity $\varrho^m(\eta, \zeta)$ is given by the equation

$$\varrho^m(\eta, \zeta) = \sum_{i=m}^{n} u_i^m(\eta, \zeta) \, P_i^m(\eta), \tag{7.13}$$

which follows from (7.6) and (7.11).

It is not difficult, however, to obtain a single equation for the determination of the quantities $\varrho^m(\eta, \zeta)$ which we are seeking. To do this, we rewrite equation (7.12) in the form

$$u_i^m(\eta, \zeta) = \frac{\lambda}{2} c_i^m \left[\eta \sum_{j=m}^{n} u_j^m(\eta, \zeta) \int_{-1}^{1} P_i^m(\eta') P_{j}^m(\eta') \frac{d\eta'}{\eta - \eta'} \right.$$

$$\left. - \int_{0}^{1} \varrho^m(\eta', \zeta) \, P_i^m(\eta') \frac{\eta' d\eta'}{\eta - \eta'} + \frac{P_i^m(-\zeta)}{2(\eta + \zeta)} \right]. \tag{7.14}$$

We shall use the well known representation of the associated Legendre polynomials

$$P_i^m(\eta) = (1 - \eta^2)^{m/2} D_i^m(\eta), \tag{7.15}$$

where $D_i^m(\eta)$ is the polynomial of order $i - m$,

$$D_i^m(\eta) = \frac{d^m P_i^m(\eta)}{d\eta^m}. \tag{7.16}$$

With the aid of (7.15) we find instead of (7.14)

$$u_i^m(\eta, \zeta) = \frac{\lambda}{2} c_i^m \left[\eta \sum_{j=m}^{n} u_j^m(\eta, \zeta) D_j^m(\eta) \int_{-1}^{1} P_i^m(\eta')(1 - \eta'^2)^{m/2} \frac{d\eta'}{\eta - \eta'} \right.$$

$$\left. - \eta \sum_{j=i+1}^{n} u_j^m(\eta, \zeta) \beta_{ij}^m(\eta) - \int_{0}^{1} \varrho^m(\eta', \zeta) \, P_i^m(\eta') \frac{\eta' d\eta'}{\eta - \eta'} + \frac{P_i^m(-\zeta)}{2(\eta + \zeta)} \right], \tag{7.17}$$

where we have defined

$$\beta_{ij}^m(\eta) = \int_{-1}^{1} \frac{D_j^m(\eta) - D_j^m(\eta')}{\eta - \eta'} \, D_i^m(\eta') (1 - \eta'^2)^m \, d\eta' \tag{7.18}$$

and have used the fact that $\beta_{ij}^m(\eta) = 0$ if $j \le i$. Equation (7.17) may be rewritten in the form

$$u_i^m(\eta, \zeta) = U_i^m(\eta, \zeta) - \frac{\lambda}{2} c_i^m \eta \sum_{j=i+1}^{n} u_j^m(\eta, \zeta) \beta_{ij}^m(\eta), \tag{7.19}$$

where

$$U_i^m(\eta, \zeta) = \frac{\lambda}{2} c_i^m \left[\frac{\eta \varrho^m(\eta, \zeta)}{(1 - \eta^2)^{m/2}} \int_{-1}^{1} P_i^m(\eta')(1 - \eta'^2)^{m/2} \frac{d\eta'}{\eta - \eta'} \right.$$

$$\left. - \int_{0}^{1} \varrho^m(\eta', \zeta) P_i^m(\eta') \frac{\eta' d\eta'}{\eta - \eta'} + \frac{P_i^m(-\zeta)}{2(\eta + \zeta)} \right]. \tag{7.20}$$

Substitution of the solution $u_i^m(\eta, \zeta)$ to equation (7.19) into equation (7.13) will produce an expression of the form

$$\varrho^m(\eta, \zeta) = \sum_{i=m}^{n} U_i^m(\eta, \zeta) R_i^m(\eta), \tag{7.21}$$

where $R_i^m(\eta)$ are certain functions to be determined. To find $R_i^m(\eta)$ we substitute $U_i^m(\eta, \zeta)$ from (7.19) into (7.21). This yields

$$\varrho^m(\eta, \zeta) = \sum_{i=m}^{n} u_i^m(\eta, \zeta) R_i^m(\eta) + \frac{\lambda}{2} \eta \sum_{i=m}^{n} u_i^m(\eta, \zeta) \sum_{j=m}^{i-1} c_j^m R_j^m(\eta) \beta_{ji}^m(\eta). \tag{7.22}$$

From a comparison of (7.22) and (7.13) we find

$$R_i^m(\eta) = P_i^m(\eta) - \frac{\lambda}{2} \eta \sum_{j=m}^{i-1} c_j^m \beta_{ji}^m(\eta) R_j^m(\eta). \tag{7.23}$$

Equation (7.23) is a recurrence relation for the determination of the $R_i^m(\eta)$. Clearly, $R_m^m(\eta) = P_m^m(\eta)$.

Another recurrence relation for the functions $R_i^m(\eta)$ also follows from equation (7.23). In order to find it, we utilize the recurrence relation (5.6) for the associated Legendre polynomials. We set i in equation (7.23) first equal to $i+1$ and then equal to $i-1$, multiply the resulting equations by $i-m+1$ and $i+m$, respectively, multiply (7.23) by $-(2i+1)\eta$, and add the resulting three equations. With the aid of equation (5.6) we find

$$(i-m+1) R_{i+1}^m(\eta) + (i+m) R_{i-1}^m(\eta) = (2i+1)\eta R_i^m(\eta)$$

$$- \frac{\lambda}{2} \eta c_i^m (i-m+1) R_i^m(\eta) \beta_{i, i+1}^m(\eta). \tag{7.24}$$

Since, however,

$$\beta_{i, i+1}^m(\eta) = \int_{-1}^{1} \frac{D_{i+1}^m(\eta) - D_{i+1}^m(\eta')}{\eta - \eta'} \, D_i^m(\eta')(1 - \eta'^2)^m \, d\eta' = \frac{2}{i-m+1} \frac{(i+m)!}{(i-m)!}, \tag{7.25}$$

10*

we may write (7.24) as

$$(i-m+1) R_{i+1}^m(\eta)+(i+m) R_{i-1}^m(\eta) = (2i+1-\lambda x_i)\eta R_i^m(\eta).$$ (7.26)

In the last equation it must be observed that

$$R_m^m(\eta) = P_m^m(\eta), \quad R_{m+1}^m(\eta) = (2m+1-\lambda x_m)\eta P_m^m(\eta),$$ (7.27)

which follows from (7.23).

Substitution of (7.20) into (7.21) yields the desired equation for the function $\varrho^m(\eta, \zeta)$:

$$\varrho^m(\eta, \zeta) T^m(\eta) = \frac{\lambda}{2} \int_0^1 \varrho^m(\eta', \zeta) A^m(\eta, \eta') \frac{\eta' \, d\eta'}{\eta'-\eta} + \frac{\lambda}{4} \frac{A^m(\eta, -\zeta)}{\eta+\zeta},$$ (7.28)

where

$$T^m(\eta) = 1 - \frac{\lambda}{2} \frac{\eta}{(1-\eta^2)^{m/2}} \int_{-1}^1 A^m(\eta, \eta')(1-\eta'^2)^{m/2} \frac{d\eta'}{\eta-\eta'},$$ (7.29)

and

$$A^m(\eta, \eta') = \sum_{i=m}^n c_i^m R_i^m(\eta) P_i^m(\eta').$$ (7.30)

The expression for $T^m(\eta)$ may be significantly simplified. Substituting (7.30) into (7.29) we obtain

$$T^m(\eta) = 1 - \frac{\lambda}{2} \eta \sum_{i=m}^n c_i^m \int_{-1}^1 R_i^m(\eta') P_i^m(\eta') \frac{d\eta'}{\eta-\eta'}$$

$$- \frac{\lambda}{2} \eta \sum_{i=m}^n c_i^m \int_{-1}^1 \left[\frac{R_i^m(\eta)}{(1-\eta^2)^{m/2}} - \frac{R_i^m(\eta')}{(1-\eta'^2)^{m/2}} \right] P_i^m(\eta') (1-\eta'^2)^{m/2} \frac{d\eta'}{\eta-\eta'}.$$ (7.31)

But all the integrals in the second sum vanish because of the orthogonality properties of the $P_i^m(\eta)$, since $R_i^m(\eta)/(1-\eta^2)^{m/2}$ is a polynomial of order $i-m$. We thus find

$$T^m(\eta) = 1 - \eta \int_{-1}^1 \frac{\Psi^m(\eta')}{\eta-\eta'} \, d\eta',$$ (7.32)

where

$$\Psi^m(\eta) = \frac{\lambda}{2} A^m(\eta, \eta).$$ (7.33)

We note that the function $T^m(\eta)$ was already introduced in Chapter 5. The quantities $R_i^m(\eta)$ and $A^m(\eta, \zeta)$ are easily seen to be related to the quantities $R_{im}^m(\eta)$ and $A_m^m(\eta, \zeta)$ introduced in that chapter by the relations

$$R_i^m(\eta) = P_m^m(\eta) R_{im}^m(\eta), \quad A^m(\eta, \zeta) = P_m^m(\eta) A_m^m(\eta, \zeta).$$ (7.34)

For the determination of $\varrho^m(\eta, \zeta)$ we have thus obtained the linear integral equation (7.28), in which $A^m(\eta, \eta')$ is given by equation (7.30) and $T^m(\eta)$ by equation (7.32). We emphasize that the variable ζ enters equation (7.28) as a parameter.

As is well known, V. A. Ambartsumyan expressed $\varrho^m(\eta, \zeta)$ in terms of the auxiliary functions $\varphi_i^m(\eta)$ and obtained a system of nonlinear integral equations (2.42) for the latter functions. It is possible, however, to find linear integral equations for these functions. This was initially accomplished for the simplest phase functions in [3].

In order to obtain these equations for the auxiliary functions, we shall use equation (7.28) and the following expression for $\varphi_i^m(\eta)$ in terms of $\varrho^m(\eta, \zeta)$:

$$\varphi_i^m(\eta) = P_i^m(\eta) + 2\eta(-1)^i \int_0^1 P_i^m(\zeta)\varrho^m(\eta, \zeta)\, d\zeta, \qquad (7.35)$$

which was derived earlier (see Section 2.2). Multiplying (7.28) by $P_i^m(\zeta)$, integrating over ζ between 0 and 1, and using (7.35), we find

$$[\varphi_i^m(\eta) - P_i^m(\eta)]T^m(\eta) = \frac{\lambda}{2}\,\eta \int_0^1 \frac{\varphi_i^m(\eta')}{\eta' - \eta} A^m(\eta, \eta')\, d\eta' - \frac{\lambda}{2}\,\eta \int_{-1}^1 \frac{P_i^m(\eta')}{\eta' - \eta} A^m(\eta, \eta')\, d\eta'. \qquad (7.36)$$

This equation may be rewritten with the aid of (7.29) and (7.30) in the form

$$\varphi_i^m(\eta)T^m(\eta) = \frac{\lambda}{2}\,\eta \int_0^1 \frac{\varphi_i^m(\eta')}{\eta' - \eta} A^m(\eta, \eta')\, d\eta' + P_i^m(\eta)$$

$$-\frac{\lambda}{2}\,\eta \sum_{j=m}^n c_j^m R_j^m(\eta) \int_{-1}^1 P_j^m(\eta') \left[P_i^m(\eta) \left(\frac{1 - \eta'^2}{1 - \eta^2}\right)^{m/2} - P_i^m(\eta') \right] \frac{d\eta'}{\eta - \eta'}. \qquad (7.37)$$

From equations (7.15) and (7.18), however, we see that the integrals under the summation sign in (7.37) are just the quantities $\beta_{ji}^m(\eta)$. Using equation (7.23), we now obtain in place of (7.37)

$$\varphi_i^m(\eta)T^m(\eta) = \frac{\lambda}{2}\,\eta \int_0^1 \frac{\varphi_i^m(\eta')}{\eta' - \eta} A^m(\eta, \eta')\, d\eta' + R_i^m(\eta). \qquad (7.38)$$

This is the desired linear integral equation for $\varphi_i^m(\eta)$. We see that each of the functions $\varphi_i^m(\eta)$ satisfies a separate equation. This consitutes the great advantage of equation (7.38) over the system of nonlinear integral equations (2.42).

7.2. The Radiation Intensity Averaged over Azimuth

We shall now apply the results obtained above to the case $m = 0$, corresponding to the radiation field averaged over azimuth. For example, equation (7.28) for $m = 0$ specifies the reflection coefficient averaged over azimuth. For simplicity in the present case we shall omit the index "0" of all these quantities, so that we shall write ϱ instead of ϱ^0, T instead of T^0, etc.

The equation for the reflection coefficient then takes the form

$$\varrho(\eta, \zeta)T(\eta) = \frac{\lambda}{2} \int_0^1 \varrho(\eta', \zeta)A(\eta, \eta') \frac{\eta' \, d\eta'}{\eta' - \eta} + \frac{\lambda}{4} \frac{A(\eta, -\zeta)}{\eta + \zeta}, \tag{7.39}$$

where, on the basis of (7.29) and (7.32),

$$T(\eta) = 1 - \frac{\lambda}{2} \eta \int_{-1}^1 A(\eta, \eta') \frac{d\eta'}{\eta - \eta'} = 1 - \frac{\lambda}{2} \eta \int_{-1}^1 A(\eta', \eta') \frac{d\eta'}{\eta - \eta'}, \tag{7.40}$$

and, as follows from (7.30),

$$A(\eta, \eta') = \sum_{i=0}^n x_i R_i(\eta) P_i(\eta'). \tag{7.41}$$

The polynomial $R_i(\eta)$ which enters (7.41) is determined by the recurrence relation

$$(i+1) R_{i+1}(\eta) + iR_{i-1}(\eta) = (2i+1 - \lambda x_i) \eta R_i(\eta), \tag{7.42}$$

which follows from (7.26). Note that $R_0(\eta) = 1$ and $R_1(\eta) = (1-\lambda)\eta$. An alternative recurrence relation for $R_i(\eta)$ may be obtained from (7.23). It has the form

$$R_i(\eta) = P_i(\eta) - \frac{\lambda}{2} \eta \sum_{j=0}^{i-1} x_j \beta_{ji}(\eta) R_j(\eta), \tag{7.43}$$

where

$$\beta_{ji}(\eta) = \int_{-1}^1 \frac{P_i(\eta) - P_i(\eta')}{\eta - \eta'} P_j(\eta') \, d\eta'. \tag{7.44}$$

We may also obtain an integral equation satisfied by $A(\eta, \eta')$. Let us substitute (7.44) into (7.42) and make use of (7.41). We find

$$R_i(\eta) = P_i(\eta) - \frac{\lambda}{2} \eta \int_{-1}^1 A(\eta, \eta') \frac{P_i(\eta) - P_i(\eta')}{\eta - \eta'} \, d\eta'. \tag{7.45}$$

Multiplying (7.45) by $x_i P_i(\zeta)$ and summing over i we obtain

$$A(\eta, \zeta) = p(\eta, \zeta) - \frac{\lambda}{2} \eta \int_{-1}^1 A(\eta, \eta') \frac{p(\eta, \zeta) - p(\eta', \zeta)}{\eta - \eta'} \, d\eta', \tag{7.46}$$

where

$$p(\eta, \zeta) = \sum_{i=0}^n x_i P_i(\eta) P_i(\zeta). \tag{7.47}$$

We recall (see Section 2.1) that $p(\eta, \zeta)$ may also be written in the form

$$p(\eta, \zeta) = \frac{1}{2\pi} \int_0^{2\pi} x(\gamma) \, d\varphi, \tag{7.48}$$

where

$$\cos \gamma = \eta\zeta + \sqrt{(1-\eta^2)(1-\zeta^2)} \cos \varphi. \tag{7.49}$$

We thus have the integral equation (7.46) for the determination of the function $A(\eta, \zeta)$, with the quantity $p(\eta, \zeta)$ given by equation (7.48). This equation may be used for the numerical computation of $A(\eta, \zeta)$ without the necessity of expanding the phase function in Legendre polynomials.

It is known that study of the diffusion of radiation in a semi-infinite medium leads to the "Milne problem" (see Section 2.3), which consists of determining the radiation field when the energy sources are located at infinitely large optical depth. In particular, one wishes to find the relative angular distribution of radiation emerging from the medium. This last problem is equivalent to the determination of the transmission coefficient (in relative units) for an atmosphere of optical thickness τ_0 as $\tau_0 \to \infty$. This relative transmission coefficient does not depend upon azimuth. As previously, we shall denote it by $u(\eta)$, where η is the cosine of the angle between the emergent radiation and the normal to the lower atmospheric boundary.

An equation for determining $u(\eta)$ may be found from equation (7.39). This is a result of the fact that the Milne problem reduces to the solution of the equations (7.1) and (7.2) for $m = 0$ without the inhomogeneous term. The equation for $u(\eta)$ is thus obtained from (7.39) by omitting the inhomogeneous term, so that

$$u(\eta)T(\eta) = \frac{\lambda}{2} \int_0^1 u(\eta') A(\eta, \eta') \frac{\eta' \, d\eta'}{\eta' - \eta}. \tag{7.50}$$

The functions $T(\eta)$ and $A(\eta, \eta')$ are given by equations (7.40) and (7.41), as previously.

The analysis of equations (7.39) and (7.50) (and also of equations (7.28) and (7.38) obtained in the previous section) may be carried out on the basis of the well-developed theory of singular integral equations (see the book by N. I. Muskhelishvili [4]). This theory shows that for $A(\eta, \eta) > 0$ equation (7.50) has a unique solution (within a multiplicative constant), while the solution of equation (7.39) includes a single arbitrary constant, which may be determined through use of the relation (2.76). Equations (7.39) and (7.50) may be solved explicitly. This was initially done for equation (7.50) by the method of Carleman [2]; that equation was subsequently studied in detail by I. B. Russman [5].

7.3. Expressions in Terms of the Functions $H^m(\eta)$

The important role played by the functions $H^m(\eta)$ in the theory of radiative transfer was explained in Chapter 5. Each of these functions satisfies both the nonlinear integral equation (5.48) and the linear integral equation (5.66). Once $H^m(\eta)$ is known, it is easy to find all of the auxiliary functions $\varphi_i^m(\eta)$ $(i = m, m+1, \ldots, n)$. It was shown in Section 5.2 that the $\varphi_i^m(\eta)$ are expressed in terms of $H^m(\eta)$ by

$$\varphi_i^m(\eta) = q_i^m(\eta) P_m^m(\eta) H^m(\eta), \tag{7.51}$$

where $q_i^m(\eta)$ is a polynomial in η of order $n-m$. These polynomials are determined by the system of algebraic equations (5.40).

We shall now present another method for finding the polynomials $q_i^m(\eta)$ which uses the linear integral equations (7.38) (see [6]). Substituting (7.51) into (7.38) and recalling (7.34), we obtain

$$q_i^m(\eta) H^m(\eta) T^m(\eta) = \frac{\lambda}{2}\, \eta \int_0^1 \frac{q_i^m(\eta') P_m^m(\eta') H^m(\eta')}{\eta' - \eta}\, A_m^m(\eta,\, \eta')\, d\eta' + R_{im}^m(\eta). \tag{7.52}$$

We shall need the linear integral equation for $H^m(\eta)$, which may be written in the form

$$H^m(\eta) T^m(\eta) = 1 + \frac{\lambda}{2}\, \eta \int_0^1 \frac{P_m^m(\eta') H^m(\eta')}{\eta' - \eta}\, A_m^m(\eta',\, \eta)\, d\eta'. \tag{7.53}$$

Substituting (7.53) into (7.52), we find

$$q_i^m(\eta) = \frac{\lambda}{2}\, \eta \int_0^1 P_m^m(\eta')\, H^m(\eta')\, \frac{q_i^m(\eta') A_m^m(\eta,\, \eta') - q_i^m(\eta) A_m^m(\eta',\, \eta)}{\eta' - \eta}\, d\eta' + R_{im}^m(\eta), \tag{7.54}$$

or

$$q_i^m(\eta) = \frac{\lambda}{2}\, \eta \int_0^1 P_m^m(\eta') H^m(\eta') A_m^m(\eta',\, \eta)\, \frac{q_i^m(\eta) - q_i^m(\eta')}{\eta - \eta'}\, d\eta'$$

$$- \frac{\lambda}{2}\, \eta \int_0^1 P_m^m(\eta') H^m(\eta') q_i^m(\eta')\, \frac{A_m^m(\eta,\, \eta') - A_m^m(\eta',\, \eta)}{\eta - \eta'}\, d\eta' + R_{im}^m(\eta). \tag{7.55}$$

Each of the polynomials $q_i^m(\eta)$ may thus be found from either equation (7.54) or (7.55).

If we had not known beforehand that the function $\varphi_i^m(\eta)$ can be represented in the form (7.51), where $q_i^m(\eta)$ is a polynomial of order $n - m$, we could have deduced that fact from equation (7.55), since $R_{im}^m(\eta)$ is a polynomial of order $i - m$ and $A_m^m(\eta, \eta')$ is a polynomial in η of order $n - m$.

In order to solve equation (7.55), we may represent $q_i^m(\eta)$ as

$$q_i^m(\eta) = R_{im}^m(0) + \sum_{k=1}^n a_{ik}^m \eta^k. \tag{7.56}$$

Substitution of (7.56) into (7.55) leads to a system of algebraic equations for the determination of the coefficients a_{ik}^m (separate systems for each pair of values of i and m). The solution of this system expresses the coefficients a_{ik}^m in terms of moments of the functions $H^m(\eta)$.

The quantity $\varrho^m(\eta, \zeta)$ may also be expressed in terms of $H^m(\eta)$. For this purpose we may substitute equation (7.51) into equation (2.41) and obtain

$$\varrho^m(\eta, \zeta) = \frac{\lambda}{4}\, \frac{H^m(\eta) H^m(\zeta)}{\eta + \zeta}\, P_m^m(\eta) P_m^m(\zeta) G^m(\eta, \zeta), \tag{7.57}$$

where $G^m(\eta, \zeta)$ is a polynomial of order $n - m$ in each of the variables η and ζ. Specifically,

$$G^m(\eta, \zeta) = \sum_{i=m}^n c_i^m (-1)^{i+m} q_i^m(\eta) q_i^m(\zeta). \tag{7.58}$$

It is not necessary to determine $G^m(\eta, \zeta)$ from equation (7.58) and the previously determined form of $q_i^m(\eta)$. By substituting equation (7.57) into (7.28) we may obtain an integral equation which will determine $G^m(\eta, \zeta)$ directly. Carrying out this substitution and setting

$$A^m(\eta, \eta') = P_m^m(\eta) P_m^m(\eta') a^m(\eta, \eta'),$$ (7.59)

we find

$$G^m(\eta, \zeta) H^m(\eta) T^m(\eta) = \frac{\lambda}{2} \eta \int_0^1 H^m(\eta') \frac{G^m(\eta', \zeta) a^m(\eta, \eta')}{\eta' - \eta} [P_m^m(\eta')]^2 \, d\eta'$$ (7.60)

$$+ \frac{\lambda}{2} \zeta \int_0^1 H^m(\eta') \frac{G^m(\eta', \zeta) a^m(\eta, \eta')}{\eta' + \zeta} [P_m^m(\eta')]^2 \, d\eta' + \frac{a^m(\eta, -\zeta)}{H^m(\zeta)}.$$

We next make use of equation (7.53) and also the nonlinear integral equation (5.48), which may be rewritten in the form

$$\frac{1}{H^m(\zeta)} = 1 - \frac{\lambda}{2} \zeta \int_0^1 H^m(\eta') \frac{a^m(\eta', \eta')}{\eta' + \zeta} [P_m^m(\eta')]^2 \, d\eta'.$$ (7.61)

We then arrive at the following equation for the determination of $G^m(\eta, \zeta)$:

$$G^m(\eta, \zeta) = \frac{\lambda}{2} \eta \int_0^1 H^m(\eta') \frac{G^m(\eta', \zeta) a^m(\eta, \eta') - G^m(\eta, \zeta) a^m(\eta', \eta')}{\eta' - \eta} [P_m^m(\eta')]^2 \, d\eta'$$

$$+ \frac{\lambda}{2} \zeta \int_0^1 H^m(\eta') \frac{G^m(\eta', \zeta) a^m(\eta, \eta') - a^m(\eta, -\zeta) a^m(\eta', \eta')}{\eta' + \zeta} [P_m^m(\eta')]^2 \, d\eta' + a^m(\eta, -\zeta).$$ (7.62)

Equation (7.62) may be solved numerically for any given value of ζ. It may also be reduced to a system of linear algebraic equations which determine the coefficients of the polynomial $G^m(\eta, \zeta)$.

We note that the polynomial $G^m(\eta, \zeta)$ obeys the two relations

$$G^m(\eta, \zeta) = G^m(\zeta, \eta), \quad G^m(-\zeta, \zeta) = a^m(\zeta, \zeta).$$ (7.63)

The first of these is obvious, while the second follows from the condition that the second integral in (7.62) must be a polynomial. The existence of these relations significantly simplifies the determination of the coefficients of the polynomial $G^m(\eta, \zeta)$.

We now turn to the expression of the relative transmission coefficient $u(\eta)$ in terms of the function $H(\eta)$ (once again, we shall write $H(\eta)$ instead of $H^0(\eta)$, recalling that $u(\eta)$ is independent of azimuth). For this purpose we substitute (7.51) into equation (2.68), which yields

$$u(\eta) = \frac{H(\eta)}{1 - k\eta} Q(\eta),$$ (7.64)

where $Q(\eta)$ is a polynomial of order n equal to

$$Q(\eta) = \frac{\lambda}{2} \sum_{i=0}^{n} x_i a_i q_i(\eta).$$ (7.65)

An equation for the direct determination of $Q(\eta)$ may be obtained by substitution of (7.64) into (7.50). Proceeding in this manner and using equation (7.53) for $m = 0$, we find

$$Q(\eta) = \frac{\lambda}{2} \eta \int_0^1 H(\eta') \frac{Q(\eta')A(\eta, \eta') - Q(\eta)A(\eta', \eta)}{\eta' - \eta} d\eta' + \frac{\lambda}{2} \int_0^1 \frac{H(\eta')}{1 - k\eta'} Q(\eta') A(\eta, \eta') d\eta'.$$ (7.66)

In the case of pure scattering ($\lambda = 1$, $k = 0$) equation (7.66) may be put in the form

$$Q(\eta) = \frac{1}{2} \int_0^1 H(\eta') \frac{\eta' Q(\eta') A(\eta, \eta') - \eta Q(\eta) A(\eta', \eta)}{\eta' - \eta} d\eta'.$$ (7.67)

Systems of linear algebraic equations for the determination of the coefficients of the polynomial $Q(\eta)$ follow from equations (7.66) and (7.67).

7.4. The Case of a Three-term Phase Function

We shall now examine the problem of diffuse light reflection from a semi-infinite atmosphere characterized by a three-term phase function. We have previously considered this problem in Section 5.5. We shall now apply to it the method of solution presented in the previous section [6].

We should point out that the present problem was examined by Horak and Chandrasekhar [7]. They did not, however, have a general method for obtaining the quantities $\varphi_i^m(\eta)$ and $\varrho^m(\eta, \zeta)$ in terms of $H^m(\eta)$. As a result their solution turned out to be extremely complicated.

For the three-term phase function

$$x(\gamma) = 1 + x_1 \cos \gamma + x_2 P_2 (\cos \gamma),$$ (7.68)

the reflection coefficient has the form

$$\varrho(\eta, \zeta, \varphi) = \varrho^0(\eta, \zeta) + 2\varrho^1(\eta, \zeta) \cos \varphi + 2\varrho^2(\eta, \zeta) \cos 2\varphi,$$ (7.69)

where φ is the azimuthal angle as measured from the plane of incidence (that is, taking the azimuthal angle of the incident solar radiation as 0). The quantities ϱ^0, ϱ^1, and ϱ^2 are expressed in terms of the functions $H^0(\eta)$, $H^1(\eta)$, and $H^2(\eta)$, respectively. These functions are determined by the integral equation (5.48), in which the characteristic functions $\Psi^0(\eta)$, $\Psi^1(\eta)$, and $\Psi^2(\eta)$ are given by equations (5.121), (5.127), and (5.130). We shall assume that the functions $H^m(\eta)$ are known.

Let us first find the quantity $\varrho^0(\eta, \zeta)$. It may be expressed in terms of the three auxiliary functions $\varphi_0^0(\eta)$, $\varphi_1^0(\eta)$, and $\varphi_2^0(\eta)$. They in turn are expressed in terms of $H^0(\eta)$ through equation (7.51), which takes the form

$$\varphi_i^0(\eta) = q_i^0(\eta)\, H^0(\eta),\tag{7.70}$$

where $q_i^0(\eta)$ is a quadratic polynomial. For the determination of $q_i^0(\eta)$ we have the equation

$$q_i^0(\eta) = \frac{\lambda}{2}\, \eta \int_0^1 H^0(\eta')\, \frac{q_i^0(\eta')\, A_0^0(\eta,\,\eta') - q_i^0(\eta)\, A_0^0(\eta',\,\eta')}{\eta' - \eta}\, d\eta' + R_{i0}^0(\eta),\tag{7.71}$$

which follows from (7.54). We recall from Section 5.5 that

$$\left.\begin{array}{l} R_{00}^0(\eta) = 1, \quad R_{10}^0(\eta) = (1-\lambda)\eta, \\[4pt] R_{20}^0(\eta) = \frac{1}{2}[(1-\lambda)(3-\lambda x_1)\eta^2 - 1], \end{array}\right\}\tag{7.72}$$

and that the quantity $A_0^0(\eta,\,\eta')$ is given on the basis of (7.30) and (7.34) as

$$A_0^0(\eta,\,\eta') = 1 + x_1(1-\lambda)\eta\eta' + \frac{x_2}{2}[(1-\lambda)(3-\lambda x_1)\eta^2 - 1]\, P_2(\eta').\tag{7.73}$$

We shall seek the polynomials $q_i^0(\eta)$ in the form

$$q_i^0(\eta) = R_{i0}^0(0) + a_{i1}\eta + a_{i2}\eta^2,\tag{7.74}$$

where a_{i1} and a_{i2} are coefficients to be determined. Substitution of (7.73) and (7.74) into (7.71) gives

$$a_{i1} + a_{i2}\eta = \frac{\lambda}{2}\int_0^1 H^0(\eta')\, A_0^0(\eta',\,\eta')\,[a_{i1} + a_{i2}(\eta + \eta')]\, d\eta'$$

$$-\frac{\lambda}{2}\int_0^1 H^0(\eta')\, q_i^0(\eta')\left[x_1(1-\lambda)\eta' + \frac{x_2}{2}(1-\lambda)(3-\lambda x_1)(\eta+\eta')\, P_2(\eta')\right] d\eta' + m_i + n_i\eta,\tag{7.75}$$

where m_i and n_i are the expansion coefficients of

$$R_{i0}^0(\eta) = R_{i0}^0(0) + m_i\eta + n_i\eta^2.\tag{7.76}$$

From (7.75) we obtain the following system of equations which determines the quantities a_{ik}:

$$\left.\begin{array}{l} M_1 a_{i1} + M_2 a_{i2} = M_0 R_{i0}^0(0) + m_i, \\[4pt] N_1 a_{i1} + N_2 a_{i2} = N_0 R_{i0}^0(0) + n_i. \end{array}\right\}\tag{7.77}$$

The coefficients of (7.77) are independent of i and are determined by the expressions

$$M_0 = -\frac{\lambda}{2}(1-\lambda)\int_0^1 H^0(\eta)\left[x_1 + \frac{x_2}{2}(3-\lambda x_1)\, P_2(\eta)\right]\eta\, d\eta,\tag{7.78}$$

$$M_1 = 1 - \frac{\lambda}{2}\int_0^1 H^0(\eta)\left[1 - \frac{x_2}{2}\, P_2(\eta)\right] d\eta,\tag{7.79}$$

$$M_2 = -\frac{\lambda}{2} \int_0^1 H^0(\eta) \left[1 - \frac{x_2}{2} P_2(\eta)\right] \eta \, d\eta, \tag{7.80}$$

$$N_0 = -\frac{\lambda}{4} x_2(1-\lambda)(3-\lambda x_1) \int_0^1 H^0(\eta) P_2(\eta) \, d\eta, \tag{7.81}$$

$$N_1 = \frac{\lambda}{4} x_2(1-\lambda)(3-\lambda x_1) \int_0^1 H^0(\eta) P_2(\eta) \eta \, d\eta, \tag{7.82}$$

$$N_2 = 1 - \frac{\lambda}{2} \int_0^1 H^0(\eta) \left[1 + x_1(1-\lambda)\eta^2 - \frac{x_2}{2} P_2(\eta)\right] d\eta. \tag{7.83}$$

The solution of the system of equations (7.77) has the form

$$\begin{aligned} \Delta a_{i1} &= N_2[M_0 R_{i0}^0(0) + m_i] - M_2[N_0 R_{i0}^0(0) + n_i], \\ \Delta a_{i2} &= M_1[N_0 R_{i0}^0(0) + n_i] - N_1[M_0 R_{i0}^0(0) + m_i], \end{aligned} \right\} \tag{7.84}$$

where

$$\Delta = M_1 N_2 - M_2 N_1. \tag{7.85}$$

Setting in (7.84) $i = 1, 2, 3$ consecutively, we obtain all the coefficients a_{ik}. After substituting them into equation (7.74), we obtain the following expressions for the polynomials we are seeking:

$$q_0^0(\eta) = 1 + \frac{N_2 M_0 - M_2 N_0}{\Delta} \eta + \frac{M_1 N_0 - N_1 M_0}{\Delta} \eta^2, \tag{7.86}$$

$$q_1^0(\eta) = \frac{1-\lambda}{\Delta} (N_2\eta - N_1\eta^2), \tag{7.87}$$

$$q_2^0(\eta) = -\frac{1}{2} q_0^0(\eta) - \frac{(1-\lambda)(3-\lambda x_1)}{2\Delta} (M_2\eta - M_1\eta^2). \tag{7.88}$$

When $\lambda = 1$ we may use the integral properties of the H-functions (see Section 5.3) to obtain in place of (7.86) and (7.88)

$$q_0^0(\eta) = 1 - \frac{h_0^0 - 2}{h_1^0} \eta, \tag{7.89}$$

$$q_2^0(\eta) = -\frac{1}{2} + \frac{h_0^0}{2h_1^0} \eta, \tag{7.90}$$

where h_0^0 and h_1^0 are the zeroth and first moments of $H^0(\eta)$, respectively. The relation $q_1^0(\eta) = 0$ which follows from (7.87) for $\lambda = 1$ is known to be valid for arbitrary phase function.

After the determination of the polynomials $q_i^0(\eta)$, the quantity $\varrho^0(\eta, \zeta)$ may be found from the equation

$$\varrho^0(\eta, \zeta) = \frac{\lambda}{4} \frac{H^0(\eta) H^0(\zeta)}{\eta+\zeta} [q_0^0(\eta) q_0^0(\zeta) - x_1 q_1^0(\eta) q_1^0(\zeta) + x_2 q_2^0(\eta) q_2^0(\zeta)], \tag{7.91}$$

which follows from (7.57) and (7.58).

The quantities $\varrho^1(\eta, \zeta)$ and $\varrho^2(\eta, \zeta)$ may be found in an analogous manner. We shall not present the derivation, but shall give only the final results. The auxiliary functions $\varphi_i^1(\eta)$ are expressed in terms of the function $H^1(\eta)$ by the equation

$$\varphi_i^1(\eta) = q_i^1(\eta) H^1(\eta) \sqrt{1-\eta^2},\tag{7.92}$$

where $q_1^1(\eta)$ and $q_2^1(\eta)$ are polynomials of first order. They are given explicitly as

$$q_1^1(\eta) = 1 - \eta \, \frac{\dfrac{\lambda}{4} x_2 (3 - \lambda x_1) \displaystyle\int_0^1 (1-\eta'^2) H^1(\eta')\eta' \, d\eta'}{1 - \dfrac{\lambda}{4} x_1 \displaystyle\int_0^1 (1-\eta'^2) H^1(\eta') \, d\eta'},\tag{7.93}$$

and

$$q_2^1(\eta) = \eta \, \frac{3 - \lambda x_1}{1 - \dfrac{\lambda}{4} x_1 \displaystyle\int_0^1 (1-\eta'^2) H^1(\eta') \, d\eta'}.\tag{7.94}$$

For the quantity $\varrho^1(\eta, \zeta)$ we have

$$\varrho^1(\eta, \zeta) = \frac{\lambda}{8} \frac{H^1(\eta) H^1(\zeta)}{\eta + \zeta} \sqrt{\{(1-\eta^2)(1-\zeta^2)\}}$$
$$\times \left[x_1 q_1^1(\eta) q_1^1(\zeta) - \frac{x_2}{3} q_2^1(\eta) q_2^1(\zeta) \right].\tag{7.95}$$

The auxiliary function $\varphi_2^2(\eta)$ and the quantity $\varrho^2(\eta, \zeta)$ are determined by the equations

$$\varphi_2^2(\eta) = 3(1 - \eta^2) H^2(\eta),\tag{7.96}$$

$$\varrho^2(\eta, \zeta) = \frac{3}{32} \lambda x_2 \frac{H^2(\eta) H^2(\zeta)}{\eta + \zeta_1} (1 - \eta^2)(1 - \zeta^2).\tag{7.97}$$

Equations (7.69), (7.91), (7.95), and (7.97), together with the expressions for the polynomials $q_i^m(\eta)$ which have been derived, give the complete solution to the problem of diffuse reflection of light by an atmosphere with a three-term phase function. Various particular problems may also be solved with these equations.

It is possible, for example, to find an expression for the albedo of the atmosphere. We recall that for this purpose it is sufficient to know only the function $\varphi_1^0(\zeta)$. Using equations (3.27), (7.70), and (7.87) we obtain for the plane albedo

$$A(\zeta) = 1 - \frac{1-\lambda}{\Delta} H^0(\zeta) (N_2 - N_1 \zeta).\tag{7.98}$$

For the spherical albedo we find with the aid of equations (1.84) and (7.98)

$$A_s = 1 - 2 \frac{1-\lambda}{\Delta} (N_2 h_1^0 - N_1 h_2^0),\tag{7.99}$$

where h_1^0 and h_2^0 are the first and second moments of $H^0(\zeta)$.

Knowledge of the polynomials $q_0^0(\eta)$, $q_1^0(\eta)$, and $q_2^0(\eta)$ allows us also to find the relative transmission coefficient of the semi-infinite atmosphere. For this purpose we must use equations (7.64) and (7.65). In particular, when $\lambda = 1$ this transmission coefficient is

$$u(\eta) = \frac{H^0(\eta)}{2h_1^0}, \qquad (7.100)$$

which may be obtained most simply through use of equations (2.144), (7.70), (7.89), and (7.90).

7.5. Numerical Results

The equations we have obtained in the previous section allow us to calculate very easily various quantities which characterize the radiation field for a three-term phase function. These calculations were first carried out for the case $m = 0$, and subsequently extended to the cases $m = 1$ and $m = 2$ (see [8] and [9]). Two phase functions were employed for the computations. They are labeled according to the values of the parameters chosen: (A) $x_1 = 1$, $x_2 = 1$; (B) $x_1 = 3/2$, $x_2 = 1$. Values of λ equal to 1, 0.99, 0.95, and 0.90 were used. We shall now present a portion of the results obtained.

Table 7.1 contains values of the functions $H^0(\eta)$, $H^1(\eta)$, and $H^2(\eta)$ which were obtained by numerical solution of equation (5.48) for the appropriate characteristic functions $\Psi^m(\eta)$ and for the case of pure scattering ($\lambda = 1$). The table also gives the first five moments of the functions $H^m(\eta)$. The values of $H^0(\eta)$ and $H^2(\eta)$ coincide for the phase functions A and B, since the characteristic functions $\Psi^0(\eta)$ and $\Psi^2(\eta)$ do not depend on x_1 when $\lambda = 1$.

TABLE 7.1. THE FUNCTION $H^m(\eta)$ AND ITS MOMENTS FOR $\lambda = 1$
AND TWO SAMPLE THREE-TERM PHASE FUNCTIONS

η	$H^0(\eta)$	$H^1(\eta)$		$H^2(\eta)$
	A, B	A	B	A, B
0	1.0000	1.0000	1.0000	1.0000
0.1	1.2833	1.0712	1.1052	1.0359
0.2	1.5127	1.1106	1.1645	1.0515
0.3	1.7290	1.1393	1.2081	1.0615
0.4	1.9388	1.1615	1.2424	1.0687
0.5	2.1447	1.1796	1.2705	1.0742
0.6	2.3481	1.1946	1.2939	1.0785
0.7	2.5496	1.2073	1.3139	1.0821
0.8	2.7499	1.2182	1.3312	1.0850
0.9	2.9491	1.2276	1.3463	1.0875
1.0	3.1475	1.2360	1.3597	1.0896
h_0	2.1294	1.1637	1.2469	1.0675
h_1	1.2385	0.5982	0.6487	0.5392
h_2	0.8823	0.4031	0.4391	0.3608
h_3	0.6870	0.3040	0.3319	0.2711
h_4	0.5630	0.2440	0.2668	0.2171

TABLE 7.2. VALUES OF THE FUNCTION $\bar{\varphi}_i^m(\eta)$ FOR $\lambda = 1$ FOR TWO SAMPLE THREE-TERM PHASE FUNCTIONS

η	$\varphi_0^0(\eta)$	$\varphi_1^0(\eta)$	$\varphi_2^0(\eta)$	$\bar{\varphi}_1^1(\eta)$		$\bar{\varphi}_2^1(\eta)$	
	A, B	A, B	A, B	A	B	A	B
0	1.0000	0	−0.5000	1.0000	1.0000	0.0000	0.0000
0.1	1.2699	0	−0.5313	1.0517	1.0864	0.2645	0.2372
0.2	1.4811	0	−0.4962	1.0703	1.1248	0.5486	0.5012
0.3	1.6748	0	−0.4186	1.0772	1.1463	0.8441	0.7799
0.4	1.8578	0	−0.3027	1.0771	1.1577	1.1474	1.0694
0.5	2.0327	0	−0.1504	1.0724	1.1622	1.4566	1.3669
0.6	2.2009	0	0.0372	1.0644	1.1616	1.7701	1.6706
0.7	2.3631	0	0.2595	1.0537	1.1572	2.0870	1.9792
0.8	2.5200	0	0.5163	1.0411	1.1498	2.4067	2.2917
0.9	2.6717	0	0.8073	1.0269	1.1399	2.7286	2.6075
1.0	2.8186	0	1.1322	1.0114	1.1280	3.0524	2.9259

Table 7.2 presents values of the functions $\bar{\varphi}_i^m(\eta)$, which are related to the functions $\varphi_i^m(\eta)$ by

$$\varphi_i^m(\eta) = (1 - \eta^2)^{m/2} \bar{\varphi}_i^m(\eta).$$

Obviously $\varphi_i^0(\eta) = \bar{\varphi}_i^0(\eta)$. Values of the function $\bar{\varphi}_2^2(\eta)$ are not given, since it equals $3H^2(\eta)$.

After the functions $\varphi_i^m(\eta)$ have been determined, it is easy to find the quantities ϱ^0, ϱ^1, and ϱ^2, as well as the complete reflection coefficient $\varrho(\eta, \zeta, \varphi)$. We shall not present here tables of these quantities, but shall give only values of the atmospheric albedo $A(\zeta)$. Values of $A(\zeta)$ are given in Table 7.3 as a function of the angle of incidence arccos ζ of the solar radiation and the albedo for single scattering λ. The values were computed from equation (7.98).

TABLE 7.3. PLANE ALBEDO $A(\zeta)$ OF A SEMI-INFINITE ATMOSPHERE WITH SINGLE SCATTERING ALBEDO λ FOR TWO SAMPLE THREE-TERM PHASE FUNCTIONS

ζ	λ							
	1.00	0.99		0.95		0.90	
	A, B	A	B	A	B	A	B
0	1	0.8890	0.8763	0.7591	0.7410	0.6656	0.6479
0.1	1	0.8601	0.8438	0.7038	0.6802	0.5971	0.5735
0.2	1	0.8378	0.8186	0.6638	0.6356	0.5498	0.5212
0.3	1	0.8177	0.7957	0.6293	0.5967	0.5103	0.4768
0.4	1	0.7989	0.7742	0.5984	0.5615	0.4760	0.4378
0.5	1	0.7810	0.7536	0.5702	0.5292	0.4458	0.4030
0.6	1	0.7640	0.7341	0.5445	0.4993	0.4187	0.3713
0.7	1	0.7477	0.7151	0.5206	0.4714	0.3943	0.3414
0.8	1	0.7321	0.6969	0.4985	0.4454	0.3722	0.3165
0.9	1	0.7170	0.6792	0.4780	0.4211	0.3523	0.2926
1.0	1	0.7024	0.6621	0.4590	0.3982	0.3342	0.2706

Knowledge of the functions $\varphi_i^0(\eta)$ allows us to determine the relative transmission coefficient $u(\eta)$. Values of $u(\eta)$ are given in Table 7.3 as a function of η and λ for the phase functions A and B. The table was computed according to equation (2.68) with the coefficients a_i found from equation (2.87).

TABLE 7.4. THE FUNCTION $u(\eta)$ FOR SINGLE SCATTERING ALBEDO λ AND TWO SAMPLE THREE-TERM PHASE FUNCTIONS

η	λ						
	1.00	0.99		0.95		0.90	
	A, B	A	B	A	B	A	B
0	0.4037	0.3388	0.3324	0.2531	0.2455	0.1906	0.1849
0.1	0.5181	0.4346	0.4266	0.3244	0.3151	0.2440	0.2375
0.2	0.6107	0.5128	0.5034	0.3842	0.3736	0.2904	0.2834
0.3	0.6980	0.5871	0.5764	0.4428	0.4309	0.3375	0.3297
0.4	0.7827	0.6598	0.6478	0.5022	0.4887	0.3871	0.3782
0.5	0.8659	0.7320	0.7185	0.5635	0.5479	0.4405	0.4299
0.6	0.9480	0.8041	0.7891	0.6275	0.6093	0.4990	0.4857
0.7	1.0294	0.8767	0.8599	0.6948	0.6734	0.5638	0.5466
0.8	1.1102	0.9496	0.9310	0.7665	0.7408	0.6365	0.6136
0.9	1.1906	1.0234	1.0028	0.8430	0.8121	0.7191	0.6882
1.0	1.2707	1.0983	1.0753	0.9255	0.8877	0.8142	0.7718

We have already stated (see Sections 2.5 and 2.6) that in the case of small true absorption $(1-\lambda \ll 1)$ the quantities $\varphi_i^0(\eta)$, $\varrho^0(\eta, \zeta)$, $A(\zeta)$, $u(\eta)$ for arbitrary phase function may be written with the aid of asymptotic expressions in terms of the analogous quantities for the case of pure scattering ($\lambda = 1$). These quantities have been calculated by the asymptotic expressions for the three-term phase functions A and B, and the results may be compared with the corresponding exact values by Kolesov and Sobolev ([7] of Chapter 6). Such a comparison characterizes the precision of the asymptotic expressions.

The reflection and transmission coefficients for a semi-infinite atmosphere for arbitrary phase function may be computed by using the equations introduced in Section 7.3. Such computations require previous knowledge of the functions $H^m(\eta)$. Once they are known, the quantities $\varrho^m(\eta, \zeta)$ and $u(\eta)$ are given by equations (7.57) and (7.64), respectively, with the polynomials $G^m(\eta, \zeta)$ and $Q(\eta)$ which enter those equations being given by (7.62) and (7.66). Such computations were carried out by A. K. Kolesov [10] for the Henyey–Greenstein phase function given in equation (1.16). The polynomials $G^m(\eta, \zeta)$ and $Q(\eta)$ were approximated with ten terms, and systems of linear algebraic equations for the coefficients of these polynomials were obtained from equations (7.62) and (7.66). Part of his results for $\lambda = 1$ are contained in Tables 7.5 and 7.6. Table 7.5 illustrates values of $u_0(\eta)$ for various values of the parameter g, which characterizes the elongation of the phase function. It is obvious from the table that the function $u_0(\eta)$ has a noticeable dependence on g only for small values of η. Table 7.6 gives values of the function $\varrho^0(\eta, \zeta)$; that is, the reflection coefficient averaged over azimuth. The upper portion of the table, above the main diagonal,

TABLE 7.5. VALUES OF THE FUNCTION $u_0(\eta)$ FOR THE HENYEY–GREENSTEIN PHASE FUNCTION (1.16)

η	g					
	0	0.3	0.5	0.6	0.7	0.75
0	0.4333	0.4202	0.3951	0.3754	0.3489	0.3326
0.1	0.5401	0.5314	0.5160	0.5051	0.4917	0.4842
0.2	0.6280	0.6218	0.6118	0.6054	0.5982	0.5940
0.3	0.7112	0.7069	0.7009	0.6975	0.6939	0.6921
0.4	0.7921	0.7894	0.7862	0.7847	0.7833	0.7826
0.5	0.8716	0.8702	0.8691	0.8688	0.8687	0.8686
0.6	0.9501	0.9499	0.9503	0.9508	0.9515	0.9518
0.7	1.0280	1.0287	1.0303	1.0313	1.0325	1.0331
0.8	1.1054	1.1070	1.1093	1.1107	1.1121	1.1129
0.9	1.1824	1.1847	1.1876	1.1892	1 1908	1.1917
1.0	1.2591	1.2620	1.2653	1.2670	1.2688	1.2697

TABLE 7.6. REFLECTION COEFFICIENT $\varrho^0(\eta, \zeta)$ FOR THE HENYEY–GREENSTEIN PHASE FUNCTIONS WITH $g = 0.25$ (ABOVE MAIN DIAGONAL) AND $g = 0.50$ (BELOW DIAGONAL)

η	0.0	0.2	0.4	0.6	0.8	1.0	ζ
	∞	1.922	1.174	0.909	0.707	0.683	0.0
		1.373	1.130	0.996	0.910	0.851	0.2
0.0	∞		1.058	1.005	0.966	0.937	0.4
0.2	2.320	1.560		1.003	1.000	0.995	0.6
0.4	1.245	1.185	1.080		1.022	1.038	0.8
0.6	0.861	0.986	1.005	1.006		1.072	1.0
0.8	0.677	0.869	0.952	1.001	1.032		
1.0	0.577	0.795	0.914	0.995	1.053	1.095	
ζ / η	0.0	0.2	0.4	0.6	0.8	1.0	

refers to the case $g = 0.25$, while the lower part refers to the case $g = 0.5$. This form of the table is possible because of the symmetry of the functions $\varrho^m(\eta, \zeta)$.

7.6. Atmospheres of Finite Optical Thickness

The results obtained above for a semi-infinite atmosphere may be easily generalized to atmospheres of finite optical thickness. We shall now present linear integral equations for the reflection and transmission coefficients, and also for the auxiliary functions $\varphi_i^m(\eta, \tau_0)$ and $\psi_i^m(\eta, \tau_0)$. Derivation of these equations for the simplest cases may be found in the references

11

[1, 3] pointed out previously, while the derivation for arbitrary phase function is contained in more recent work [11, 12].

We shall continue to assume that the reflection coefficient $\varrho(\eta, \zeta, \varphi, \tau_0)$ and transmission coefficient $\sigma(\eta, \zeta, \varphi, \tau_0)$ are expanded in cosine series in azimuth according to equations (3.2) and (3.3). Our problem consists of determining the quantities $\varrho^m(\eta, \zeta, \tau_0)$ and $\sigma^m(\eta, \zeta, \tau_0)$ which enter these expansions. In order to simplify the notation in what follows, we shall not include τ_0 among the arguments of the relevant quantities. We shall thus simply write $\varrho^m(\eta, \zeta)$ and $\sigma^m(\eta, \zeta)$ instead of $\varrho^m(\eta, \zeta, \tau_0)$ and $\sigma^m(\eta, \zeta, \tau_0)$, $\varphi_i^m(\eta)$ and $\psi_i^m(\eta)$ instead of $_i^m(\eta, \tau_0)$ and $\psi_i^m(\eta, \tau_0)$, etc.

φ For a semi-infinite atmosphere the reflection coefficient is determined by equation (7.28). By generalizing this equation we find that the quantities $\varrho^m(\eta, \zeta)$ and $\sigma^m(\eta, \zeta)$ are determined by the following system of linear integral equations:

$$\varrho^m(\eta, \zeta)\, T^m(\eta) = \frac{\lambda}{2} \int_0^1 \varrho^m(\eta', \zeta)\, A^m(\eta, \eta')\, \frac{\eta'\, d\eta'}{\eta' - \eta}$$

$$-\frac{\lambda}{2} e^{-\tau_0/\eta} \int_0^1 \sigma^m(\eta', \zeta)\, A^m(\eta, -\eta')\, \frac{\eta'\, d\eta'}{\eta + \eta'} + A^m(\eta, -\zeta)\, \varrho_1(\eta, \zeta), \qquad (7.101)$$

$$\sigma^m(\eta, \zeta)\, T^m(\eta) = \frac{\lambda}{2} \int_0^1 \sigma^m(\eta', \zeta)\, A^m(\eta, \eta')\, \frac{\eta'\, d\eta'}{\eta' - \eta}$$

$$-\frac{\lambda}{2} e^{-\tau_0/\eta} \int_0^1 \varrho^m(\eta', \zeta)\, A^m(\eta, -\eta')\, \frac{\eta'\, d\eta'}{\eta + \eta'} + A^m(\eta, \zeta)\, \sigma_1(\eta, \zeta), \qquad (7.102)$$

where

$$\varrho_1(\eta, \zeta) = \frac{\lambda}{4}\, \frac{1 - e^{-\tau_0\left(\frac{1}{\eta} + \frac{1}{\zeta}\right)}}{\eta + \zeta}, \qquad \sigma_1(\eta, \zeta) = \frac{\lambda}{4}\, \frac{e^{-\tau_0/\eta} - e^{-\tau_0/\zeta}}{\eta - \zeta}, \qquad (7.103)$$

and the functions $A^m(\eta, \zeta)$ and $T^m(\eta)$ are given by equations (7.30) and (7.32).

We have shown in Chapter 3 that the quantities $\varrho^m(\eta, \zeta)$ and $\sigma^m(\eta, \zeta)$ may be expressed in terms of the auxiliary functions $\varphi_i^m(\eta)$ and $\psi_i^m(\eta)$. In turn, these latter functions are expressed through the quantities $\varrho^m(\eta, \zeta)$ and $\sigma^m(\eta, \zeta)$ with the aid of

$$\varphi_i^m(\eta) = P_i^m(\eta) + 2\eta \int_0^1 P_i^m(-\zeta)\, \varrho^m(\eta, \zeta)\, d\zeta, \qquad (7.104)$$

$$\psi_i^m(\eta) = P_i^m(\eta)\, e^{-\tau_0/\eta} + 2\eta \int_0^1 P_i^m(\zeta)\, \sigma^m(\eta, \zeta)\, d\zeta. \qquad (7.105)$$

By using equations (7.104) and (7.105), we may obtain from equations (7.102) and (7.103) the following system of linear integral equations for the determination of $\varphi_i^m(\eta)$ and $\psi_i^m(\eta)$:

$$\varphi_i^m(\eta)\, T^m(\eta) = \frac{\lambda}{2}\,\eta \int_0^1 \varphi_i^m(\eta')\, A^m(\eta,\eta')\frac{d\eta'}{\eta'-\eta}$$

$$-\frac{\lambda}{2}\,\eta e^{-\tau_0/\eta}(-1)^{1+m}\int_0^1 \psi_i^m(\eta')\, A^m(\eta,-\eta')\frac{d\eta'}{\eta+\eta'} + R_i^m(\eta), \tag{7.106}$$

$$\psi_i^m(\eta)\, T^m(\eta) = \frac{\lambda}{2}\,\eta \int_0^1 \psi_i^m(\eta')\, A^m(\eta,\eta')\frac{d\eta'}{\eta'-\eta}$$

$$-\frac{\lambda}{2}\,\eta e^{-\tau_0/\eta}(-1)^{i+m}\int_0^1 \varphi_i^m(\eta')\, A^m(\eta,-\eta')\frac{d\eta'}{\eta+\eta'} + R_i^m(\eta)\, e^{-\tau_0/\eta}, \tag{7.107}$$

where the function $R_i^m(\eta)$ is given by the recurrence relation (7.26).

The system of equations (7.101) and (7.102), and also the system (7.106) and (7.107), may not have unique solutions. In order to obtain solutions which are physically realistic, we must specify some additional constraints. Let us do this for the case $m = 0$.

As we know, the function $T^0(\eta)$ has the root $\eta = 1/k$ (see Section 5.4). Thus, if we set $\eta = 1/k$ in equation (7.101) for $m = 0$, we find

$$\frac{\lambda}{2}k\int_0^1 \varrho^0(\eta,\zeta)\, A^0\left(\frac{1}{k},\eta\right)\frac{\eta\, d\eta}{1-k\eta}$$

$$+\frac{\lambda}{2}ke^{-k\tau_0}\int_0^1 \sigma^0(\eta,\zeta)\, A^0\left(\frac{1}{k},-\eta\right)\frac{\eta\, d\eta}{1+k\eta} = A^0\left(\frac{1}{k},-\zeta\right)\varrho_1\left(\frac{1}{k},\zeta\right). \tag{7.108}$$

This relation may be rewritten by making use of several expressions for the quantities which characterize the radiation field in deep layers of a semi-infinite medium. When we compare expression (2.10) for the function $b(\eta)$ and expression (7.30) for the function $A^0(\eta,\zeta)$, and also the recurrence relations (2.20) and (7.26) for the quantities b_i and $R_i^0(\eta)$, we see that

$$b(\eta) = A^0\left(\frac{1}{k},\eta\right). \tag{7.109}$$

We thus find on the basis of (2.6) that

$$i(\eta) = \frac{A^0\left(\frac{1}{k},\eta\right)}{1-k\eta}. \tag{7.110}$$

Using (7.110) and (7.103), we obtain instead of (7.108)

$$2\int_0^1 \varrho^0(\eta,\zeta)\, i(\eta)\eta\, d\eta + 2e^{-k\tau_0}\int_0^1 \sigma^0(\eta,\zeta)\, i(-\eta)\eta\, d\eta = [1-e^{-\tau_0(k+1/\zeta)}]\, i(-\zeta). \tag{7.111}$$

11*

By setting $\eta = 1/k$ in equation (7.102) for $m = 0$, we find in analogous manner

$$2 \int_0^1 \sigma^0(\eta, \zeta) i(\eta) \eta \, d\eta + 2e^{-k\tau_0} \int_0^1 \varrho^0(\eta, \zeta) i(-\eta) \eta \, d\eta = (e^{-k\tau_0} - e^{-\tau_0/\zeta}) i(\zeta). \qquad (7.112)$$

Equations (7.111) and (7.112) may serve as the desired constraints on equations (7.101) and (7.102) for $m = 0$.

In a completely similar manner we may obtain constraint relations for equations (7.106) and (7.107) for $m = 0$. Setting $\eta = 1/k$ in those equations and using equation (7.110), we find

$$\frac{\lambda}{2} \int_0^1 \varphi_i^0(\eta) i(\eta) \, d\eta + (-1)^i \frac{\lambda}{2} e^{-k\tau_0} \int_0^1 \psi_i^0(\eta) i(-\eta) \, d\eta = R_i^0\left(\frac{1}{k}\right), \qquad (7.113)$$

$$\frac{\lambda}{2} \int_0^1 \psi_i^0(\eta) i(\eta) \, d\eta + (-1)^i \frac{\lambda}{2} e^{-k\tau_0} \int_0^1 \varphi_i^0(\eta) i(-\eta) \, d\eta = e^{-k\tau_0} R_i^0\left(\frac{1}{k}\right). \qquad (7.114)$$

Let us examine the form of the above constraints in the case of pure scattering ($\lambda = 1$, $k = 0$). In this process we utilize the fact that for small k the quantity $i(\eta)$ is given by

$$i(\eta) = 1 + \frac{3k\eta}{3 - x_1}, \qquad (7.115)$$

which was obtained in Section 2.1. Setting $k = 0$ in equations (7.112) and (7.111), we have

$$2 \int_0^1 \varrho^0(\eta, \zeta) \eta \, d\eta + 2 \int_0^1 \sigma^0(\eta, \zeta) \eta \, d\eta = 1 - e^{-\tau_0/\zeta}. \qquad (7.116)$$

Differentiating (7.111) and (7.112) with respect to k and setting $k = 0$, we obtain

$$\left(1 - \frac{x_1}{3}\right) \tau_0 \left[1 - 2 \int_0^1 \varrho^0(\eta, \zeta) \eta \, d\eta\right] - 2 \int_0^1 \varrho^0(\eta, \zeta) \eta^2 \, d\eta + 2 \int_0^1 \sigma^0(\eta, \zeta) \eta^2 \, d\eta = \zeta(1 - e^{-\tau_0/\zeta}).$$

$$(7.117)$$

We recall that relations (7.116) and (7.117) follow from the flux integral and the so-called K-integral, which are valid for $\lambda = 1$ (see Section 1.6).

Likewise, in the case of pure scattering we find from equations (7.113) and (7.114) with the aid of (7.115) that

$$\int_0^1 [\varphi_i^0(\eta) + (-1)^i \psi_i^0(\eta)] \, d\eta = 2\delta_{i0}, \qquad (7.118)$$

$$\int_0^1 \varphi_i^0(\eta) \left[\left(1 - \frac{x_1}{3}\right) \tau_0 + \eta\right] d\eta - \int_0^1 \psi_i^0(\eta) \eta \, d\eta = 2\left(1 - \frac{x_1}{3}\right) \tau_0 \delta_{i0} \quad \text{(for even } i\text{)}, \qquad (7.119)$$

$$\int_0^1 \varphi_i^0(\eta) \left[\left(1 - \frac{x_1}{3}\right) \tau_0 + \eta\right] d\eta + \int_0^1 \psi_i^0(\eta) \eta \, d\eta = \frac{2}{3} \delta_{i1} \quad \text{(for odd } i\text{)}. \qquad (7.120)$$

Equations (7.118)–(7.120) may also be easily obtained by use of equations (7.116), (7.117), (7.104), and (7.105) (with $m = 0$).

We note the important investigations in the theory of anisotropic light scattering which have been carried out by Mullikin [13, 14]. He obtained a singular integral equation for the source function, from which he obtained equations (7.106) and (7.107) for the functions $\varphi_i^m(\eta)$ and $\psi_i^m(\eta)$. Mullikin also examined in great detail the question of the uniqueness of the solutions of these equations.

7.7. Expressions in Terms of the Functions $X^m(\eta)$ and $Y^m(\eta)$

During the consideration of the diffusion of radiation in a semi-infinite medium in Section 7.3, we expressed all the auxiliary functions $\varphi_i^m(\eta)$ in terms of the single function $H^m(\eta)$. This result may be generalized to atmospheres of finite optical thickness, so that all the functions $\varphi_i^m(\eta)$ and $\psi_i^m(\eta)$ ($i = m, m+1, \ldots, n$) may be expressed in terms of the two functions $X^m(\eta)$ and $Y^m(\eta)$. We shall now carry out this generalization (for more detail, see [11] and [12]).

By using equations (7.106) and (7.107) for $\varphi_i^m(\eta)$ and $\psi_i^m(\eta)$ and equations (6.32) and (6.33) for $X^m(\eta)$ and $Y^m(\eta)$, it is not difficult to obtain the following interrelations between these functions:

$$\varphi_i^m(\eta) = [X^m(\eta)q_i^m(\eta)+(-1)^{i+m}Y^m(\eta)s_i^m(-\eta)]\,P_m^m(\eta), \tag{7.121}$$

$$\psi_i^m(\eta) = [X^m(\eta)s_i^m(\eta)+(-1)^{i+m}Y^m(\eta)q_i^m(-\eta)]\,P_m^m(\eta), \tag{7.122}$$

where $q_i^m(\eta)$ and $s_i^m(\eta)$ are polynomials in η of order $n-m$. These polynomials are determined by the equations

$$q_i^m(\eta) = \frac{\lambda}{2}\,\eta \int_0^1 X^m(\eta')\,[q_i^m(\eta')A_m^m(\eta, \eta')-q_i^m(\eta)\,A_m^m(\eta', \eta)]\,\frac{P_m^m(\eta')}{\eta'-\eta}\,d\eta'$$

$$+(-1)^{i+m}\frac{\lambda}{2}\,\eta \int_0^1 Y^m(\eta')\,[s_i^m(-\eta')\,A_m^m(\eta, \eta')$$

$$-s_i^m(-\eta)\,A_m^m(\eta', \eta)]\,\frac{P_m^m(\eta')}{\eta'-\eta}\,d\eta'+R_{im}^m(\eta), \tag{7.123}$$

$$s_i^m(\eta) = \frac{\lambda}{2}\,\eta \int_0^1 X^m(\eta')\,[s_i^m(\eta')\,A_m^m(\eta, \eta')-s_i^m(\eta)A_m^m(\eta', \eta)]\,\frac{P_m^m(\eta')}{\eta'-\eta}\,d\eta'$$

$$+(-1)^{i+m}\frac{\lambda}{2}\,\eta \int_0^1 Y^m(\eta')\,[q_i^m(-\eta')\,A_m^m(\eta, \eta')-q_i^m(-\eta)\,A_m^m(\eta', \eta)]\,\frac{P_m^m(\eta')}{\eta'-\eta}\,d\eta'. $$

$$\tag{7.124}$$

from which we may obtain a system of linear algebraic equations for the coefficients of the polynomials $q_i^m(\eta)$ and $s_i^m(\eta)$.

We note that equations (7.121) and (7.122) which express $\varphi_i^m(\eta)$ and $\psi_i^m(\eta)$ in terms of $X^m(\eta)$ and $Y^m(\eta)$ were obtained earlier (see Section 6.3). We also presented an alternative method for the determination of the polynomials $q_i^m(\eta)$ and $s_i^m(\eta)$.

Using equations (7.121) and (7.122) we may express the reflection and transmission coefficients in terms of the functions $X^m(\eta)$ and $Y^m(\eta)$ and the polynomials $q_i^m(\eta)$ and $s_i^m(\eta)$. Substituting (7.121) and (7.122) into equations (3.16) and (3.17), we obtain

$$\varrho^m(\eta,\,\zeta) = \frac{\lambda}{4}\,\frac{P_m^m(\eta)\,P_m^m(\zeta)}{\eta+\zeta}\,[X^m(\eta)\,X^m(\zeta)\,M^m(\eta,\,\zeta)-Y^m(\eta)\,Y^m(\zeta)\,M^m(-\eta,\,-\zeta)$$
$$+X^m(\eta)\,Y^m(\zeta)\,N^m(\eta,\,\zeta)+X^m(\zeta)\,Y^m(\eta)\,N^m(\eta,\,\zeta)], \qquad (7.125)$$

and

$$\sigma^m(\eta,\,\zeta) = \frac{\lambda}{4}\,\frac{P_m^m(\eta)\,P_m^m(\zeta)}{\eta-\zeta}\,[X^m(\eta)\,X^m(\zeta)\,N^m(\zeta,\,-\eta)+Y^m(\eta)\,Y^m(\zeta)\,N^m(-\eta,\,\zeta)$$
$$+X^m(\zeta)\,Y^m(\eta)\,M^m(-\eta,\,\zeta)-X^m(\eta)\,Y^m(\zeta)\,M^m(\eta,\,-\zeta)], \qquad (7.126)$$

where we have defined

$$M^m(\eta,\,\zeta) = \sum_{i=m}^{n}(-1)^{i+m}c_i^m\,[q_i^m(\eta)\,q_i^m(\zeta)-s_i^m(\eta)\,s_i^m(\zeta)], \qquad (7.127)$$

$$N^m(\eta,\,\zeta) = \sum_{i=m}^{n}c_i^m\,[q_i^m(\eta)\,s_i^m(-\zeta)-s_i^m(\eta)\,q_i^m(-\zeta)]. \qquad (7.128)$$

The quantities $M^m(\eta,\,\zeta)$ and $N^m(\eta,\,\zeta)$ are polynomials of order $n-m$ in each of the variables η and ζ. They obviously satisfy the conditions

$$M^m(\eta,\,\zeta) = M^m(\zeta,\,\eta), \quad N^m(\eta,\,\zeta) = -N^m(-\zeta,\,-\eta). \qquad (7.129)$$

The quantities $\varrho^m(\eta,\,\zeta)$ and $\sigma^m(\eta,\,\zeta)$ may also be found from equations (7.125) and (7.126) without first obtaining the polynomials $q_i^m(\eta)$ and $s_i^m(\eta)$. In order to do this, we must find equations which determine the polynomials $M^m(\eta,\,\zeta)$ and $N^m(\eta,\,\zeta)$ directly. Such equations are easily obtained upon substitution of equations (7.125) and (7.126) into equations (7.101) and (7.102). This procedure, following use of (6.30)–(6.33), leads to

$$M^m(\eta,\,\zeta) = \frac{\lambda}{2}\,\eta\int_0^1 X^m(\eta')\,\frac{a^m(\eta,\,\eta')\,M^m(\eta',\,\zeta)-a^m(\eta',\,\eta)\,M^m(\eta,\,\zeta)}{\eta'-\eta}\,[P_m^m(\eta')]^2\,d\eta'$$

$$+\frac{\lambda}{2}\,\eta\int_0^1 Y^m(\eta')\frac{a^m(\eta,\,\eta')\,N^m(\zeta,\,\eta')-a^m(\eta',\,\eta)\,N^m(\zeta,\,\eta)}{\eta'-\eta}\,[P_m^m(\eta')]^2\,d\eta'$$

$$+\frac{\lambda}{2}\,\zeta\int_0^1 X^m(\eta')\,\frac{a^m(\eta,\,\eta')\,M^m(\eta',\,\zeta)-a^m(\eta',\,\eta)\,a^m(\eta,\,-\zeta)}{\eta'+\zeta}\,[P_m^m(\eta')]^2\,d\eta'$$

$$+\frac{\lambda}{2}\,\zeta\int_0^1 Y^m(\eta')\,a^m(\eta,\,\eta')\frac{N^m(\zeta,\,\eta')}{\eta'+\zeta}[P_m^m(\eta')]^2\,d\eta'+a^m(\eta,\,-\zeta), \qquad (7.130)$$

$$N^m(\eta, \zeta) = \frac{\lambda}{2}\eta \int_0^1 X^m(\eta') \frac{a^m(\eta, \eta')N^m(\eta', \zeta) - a^m(\eta', \eta')N^m(\eta, \zeta)}{\eta' - \eta} [P_m^m(\eta')]^2 d\eta'$$

$$-\frac{\lambda}{2}\eta \int_0^1 Y^m(\eta') \frac{a^m(\eta, \eta')M^m(-\eta', -\zeta) - a^m(\eta', \eta')M^m(-\eta, -\zeta)}{\eta' - \eta} [P_m^m(\eta')]^2 d\eta'$$

$$+\frac{\lambda}{2}\zeta \int_0^1 Y^m(\eta') \frac{a^m(\eta, \eta')M^m(-\eta', -\zeta) - a^m(\eta', \eta')M^m(-\eta, -\zeta)}{\eta' + \zeta} [P_m^m(\eta')]^2 d\eta'$$

$$+\frac{\lambda}{2}\zeta \int_0^1 X^m(\eta')a^m(\eta, \eta') \frac{N^m(\eta', \zeta)}{\eta' + \zeta} [P_m^m(\eta')]^2 d\eta'. \tag{7.131}$$

where we recall the notation defined by equation (7.59). Equations (7.130) and (7.131) reduce to a system of linear algebraic equations for the coefficients of the polynomials $M^m(\eta, \zeta)$ and $N^m(\eta, \zeta)$. These coefficients are expressed in terms of moments of the functions $X^m(\eta)$ and $Y^m(\eta)$. It is easily seen that (7.130) and (7.131) imply the relations

$$M^m(-\zeta, \zeta) = a^m(\zeta, \zeta), \quad N^m(-\zeta, \zeta) = 0, \tag{7.132}$$

which, together with equation (7.129), considerably simplify the task of finding the coefficients of the polynomials $M^m(\eta, \zeta)$ and $N^m(\eta, \zeta)$.

The equations introduced in the present section have been used to determine the reflection and transmission coefficients for atmospheres with phase functions of two terms [11] and three terms [15]. In the latter case certain numerical computations were carried out. We shall examine below the case of a two-term phase function.

7.8. The Case of a Two-term Phase Function

We shall now find the reflection and transmission coefficients when the phase function $x(\gamma) = 1 + x_1 \cos \gamma$. This problem was first considered by Chandrasekhar (see [9] in the References to Chapter 1). We shall now solve it as an example of the use of the general method presented in the previous section.

For the given phase function the reflection and transmission coefficients take the form

$$\varrho(\eta, \zeta, \varphi) = \varrho^0(\eta, \zeta) + 2\varrho^1(\eta, \zeta) \cos \varphi, \tag{7.133}$$

$$\sigma(\eta, \zeta, \varphi) = \sigma^0(\eta, \zeta) + 2\sigma^1(\eta, \zeta) \cos \varphi. \tag{7.134}$$

In order to find the quantities $\varrho^0(\eta, \zeta)$ and $\sigma^0(\eta, \zeta)$, it is first necessary to determine the functions $X^0(\eta)$ and $Y^0(\eta)$. Equations (6.30) and (6.31) for $m = 0$ may be used for this purpose. The characteristic function in the present case is given by equation (5.102). We shall consider the functions $X^0(\eta)$ and $Y^0(\eta)$ as known.

The next step consists of determining the linear polynomials $q_i^0(\eta)$ and $s_i^0(\eta)$. According to (7.123) and (7.124), they have the form

$$q_0^0(\eta) = 1 + C_0\eta, \quad s_0^0(\eta) = D_0\eta, \tag{7.135}$$

$$q_1^0(\eta) = C_1\eta, \quad s_1^0(\eta) = D_1\eta. \tag{7.136}$$

Substituting these expressions into equations (7.123) and (7.124) for $m = 0$ and taking into consideration that

$$A_0^0(\eta, \eta') = 1 + x_1(1-\lambda)\eta\eta', \tag{7.137}$$

we obtain

$$C_0 = -\frac{\lambda}{\Delta}(1-\lambda)x_1[(2-\lambda\alpha_0)\alpha_1 - \lambda\beta_0\beta_1], \tag{7.138}$$

$$D_0 = -\frac{\lambda}{\Delta}(1-\lambda)x_1[(2-\lambda\alpha_0)\beta_1 - \lambda\beta_0\alpha_1], \tag{7.139}$$

$$C_1 = \frac{2}{\Delta}(1-\lambda)(2-\lambda\alpha_0), \tag{7.140}$$

$$D_1 = \frac{2}{\Delta}(1-\lambda)\lambda\beta_0, \tag{7.141}$$

where

$$\Delta = (2-\lambda\alpha_0)^2 - (\lambda\beta_0)^2. \tag{7.142}$$

We denote by α_k and β_k the moments of the functions $X^0(\eta)$ and $Y^0(\eta)$, so that

$$\alpha_k = \int_0^1 X^0(\eta)\eta^k\, d\eta, \quad \beta_k = \int_0^1 Y^0(\eta)\eta^k\, d\eta. \tag{7.143}$$

By introducing (7.135) and (7.136) into equations (7.121) and (7.122), we find the following expressions for the auxiliary functions:

$$\varphi_0^0(\eta) = (1+C_0\eta)X^0(\eta) - D_0\eta Y^0(\eta), \tag{7.144}$$
$$\psi_0^0(\eta) = D_0\eta X^0(\eta) + (1-C_0\eta)Y^0(\eta), \tag{7.145}$$
$$\varphi_1^0(\eta) = C_1\eta X^0(\eta) + D_1\eta Y^0(\eta), \tag{7.146}$$
$$\psi_1^0(\eta) = D_1\eta X^0(\eta) + C_1\eta Y^0(\eta). \tag{7.147}$$

If we substitute these equations into equations (3.16) and (3.17), we obtain the desired expressions for the quantities $\rho^0(\eta, \zeta)$ and $\sigma^0(\eta, \zeta)$.

The quantities $\rho^0(\eta, \zeta)$ and $\sigma^0(\eta, \zeta)$ may also be found from equations (7.125) and (7.126), once the polynomials $M^0(\eta, \zeta)$ and $N^0(\eta, \zeta)$ have been determined. Substituting (7.135) and (7.136) into equations (7.127) and (7.128), we have

$$M^0(\eta, \zeta) = 1 + C_0(\eta+\zeta) + [C_0^2 - D_0^2 - x_1(C_1^2 - D_1^2)]\eta\zeta, \tag{7.148}$$
$$N^0(\eta, \zeta) = -D_0(\eta+\zeta). \tag{7.149}$$

The polynomials $M^0(\eta, \zeta)$ and $N^0(\eta, \zeta)$ may also be determined without previously finding the quantities $q_i^0(\eta)$ and $s_i^0(\eta)$. Using the relations (7.129) and (7.132), we obtain immediately

$$M^0(\eta, \zeta) = 1 + C_0(\eta+\zeta) - x_1(1-\lambda)\eta\zeta, \tag{7.150}$$
$$N^0(\eta, \zeta) = -D_0(\eta+\zeta) + D_2(\eta^2-\zeta^2), \tag{7.151}$$

where C_0, D_0, and D_2 are certain constants. Substitution of (7.150) and (7.151) into equations (7.130) and (7.131) (for $\eta = 0$ or $\zeta = 0$) gives $D_2 = 0$, and the constants C_0 and D_0 are determined by equations (7.138) and (7.139).

Introducing (7.149) and (7.150) into equations (7.125) and (7.126) for $m = 0$, we obtain the expressions which we have been seeking for the quantities $\varrho^0(\eta, \zeta)$ and $\sigma^0(\eta, \zeta)$ in terms of the functions $X^0(\eta)$ and $Y^0(\eta)$:

$$\varrho^0(\eta, \zeta) = \frac{\lambda}{4}\left\{[1-x_1(1-\lambda)\eta\zeta]\,\frac{X^0(\eta)\,X^0(\zeta)-Y^0(\eta)\,Y^0(\zeta)}{\eta+\zeta}\right.$$

$$\left.+C_0[X^0(\eta)\,X^0(\zeta)+Y^0(\eta)\,Y^0(\zeta)]-D_0[X^0(\eta)\,Y^0(\zeta)+X^0(\zeta)\,Y^0(\eta)]\right\}, \qquad (7.152)$$

$$\sigma^0(\eta, \zeta) = \frac{\lambda}{4}\left\{[1-x_1(1-\lambda)\eta\zeta]\,\frac{X^0(\zeta)\,Y^0(\eta)-X^0(\eta)\,Y^0(\zeta)}{\eta-\zeta}\right.$$

$$\left.+D_0[X^0(\eta)\,X^0(\zeta)+Y^0(\eta)\,Y^0(\zeta)]-C_0[X^0(\eta)\,Y^0(\zeta)-X^0(\zeta)\,Y^0(\eta)]\right\}. \qquad (7.153)$$

In order to find the quantities $\varrho^1(\eta, \zeta)$ and $\sigma^1(\eta, \zeta)$, it is necessary to know the functions $X^1(\eta)$ and $Y^1(\eta)$. The latter functions are determined from equations (6.30) and (6.31) for $m = 1$, with the characteristic function $\Psi^1(\eta)$ given by equation (5.105). It follows from relations (7.121) and (7.122) that the auxiliary functions are

$$\varphi_1^1(\eta) = X^1(\eta)\,\sqrt{1-\eta^2}, \quad \psi_1^1(\eta) = Y^1(\eta)\,\sqrt{1-\eta^2}. \qquad (7.154)$$

Therefore, on the basis of equations (3.16) and (3.17), we have

$$\varrho^1(\eta, \zeta) = \frac{\lambda}{8}\,x_1\,\frac{X^1(\eta)\,X^1(\zeta)-Y^1(\eta)\,Y^1(\zeta)}{\eta+\zeta}\,\sqrt{(1-\eta^2)(1-\zeta^2)}, \qquad (7.155)$$

$$\sigma^1(\eta, \zeta) = \frac{\lambda}{8}\,x_1\,\frac{X^1(\zeta)\,Y^1(\eta)-X^1(\eta)\,Y^1(\zeta)}{\eta-\zeta}\,\sqrt{(1-\eta^2)(1-\zeta^2)}. \qquad (7.156)$$

Knowledge of the functions $X^0(\eta)$ and $Y^0(\eta)$ allows us to determine quite simply the atmospheric albedo and the illumination of the underlying surface. For the albedo, we obtain following use of equations (3.27) and (7.146)

$$A(\zeta) = 1-2\,\frac{1-\lambda}{\Delta}\,[(2-\lambda\alpha_0)\,X^0(\zeta)+\lambda\beta_0 Y^0(\zeta)]. \qquad (7.157)$$

The ratio of the illumination of the surface to the solar flux incident on the atmosphere we shall again denote by $V(\zeta)$. On the basis of equations (3.30) and (7.147), this quantity is just

$$V(\zeta) = 2\,\frac{1-\lambda}{\Delta}\,[\lambda\beta_0 X^0(\zeta)+(2-\lambda\alpha_0)\,Y^0(\zeta)]. \qquad (7.158)$$

When $\lambda = 1$ the indeterminancy in these equations may be resolved with the aid of the relation

$$\frac{\Delta}{1-\lambda} = 4\left(1-\frac{\lambda}{3}\,x_1+\lambda x_1\alpha_2\right)-\lambda^2 x_1\,[2(\alpha_0\alpha_2-\beta_0\beta_2)+(1-\lambda)\,x_1(\alpha_2^2-\beta_2^2)], \qquad (7.159)$$

which is easily obtained from equation (6.30).

References

1. V. V. SOBOLEV, On the diffuse reflection and transmission of light by a plane layer of a turbid medium, *Dokl. Akad. Nauk SSSR* **69**, 353 (1949).
2. V. V. SOBOLEV, On the distribution of brightness across the disc of a star, *Astron. Zh.* **26**, 22 (1949).
3. V. V. SOBOLEV, On the problem of diffuse reflection and transmission of light, *Dokl. Akad. Nauk SSSR* **69**, 547 (1949).
4. N. I. MUSKHELISHVILI, *Singular Integral Equations*, Fizmatgiz, Moscow, 1962.
5. I. B. RUSSMAN, On the solution of the equation for the law of limb darkening, *Zh. Vychisl. Matem. i Matem. Fiz.* **5**, 1130 (1965).
6. V. V. SOBOLEV, Anistropic scattering of light in a semi-infinite atmosphere. II, *Astron. Zh.* **45**, 528 (1968) [*Sov. Astron. A.J.* **12**, 420 (1968)].
7. H. G. HORAK and S. CHANDRASEKHAR, Diffuse reflection by a semi-infinite atmosphere, *Astrophys. J.* **134**, 45 (1961).
8. A. K. KOLESOV and V. V. SOBOLEV, Transfer of radiation in a semi-infinite medium for a three term phase function, *Uch. Zapiski Leningrad Gos. University*, No. 347 (*Trudy Astron. Obs., Leningrad Gos. Univ.* **26**, 1969).
9. V. M. LOSKUTOV, Diffuse reflection of light from a semi-infinite atmosphere for a three term phase function, *Trudy Astron. Obs., Leningrad Gos. Univ.* **29** (1972).
10. A. K. KOLESOV, Reflection and transmission of light by a semi-infinite atmosphere for anisotropic scattering, *Trudy Astron. Obs., Leningrad Gos. Univ.*, **29** (1972).
11. V. V. SOBOLEV, Diffuse reflection and transmission of light by an atmosphere for anisotropic scattering, *Astrofizika* **5**, 5 (1969) [*Astrophysics* **5**, 1 (1969)].
12. V. V. SOBOLEV, Diffuse reflection and transmission of light by an atmosphere for an arbitrary phase function, *Astron. Zh.* **46**, 1137 (1969) [*Sov. Astron. A.J.* **13**, 893 (1970)].
13. T. W. MULLIKIN, Radiative transfer in finite homogeneous atmospheres with anisotropic scattering. I. Linear singular equation, *Astrophys. J.* **139**, 379 (1964).
14. T. W. MULLIKIN, Radiative transfer in finite homogeneous atmospheres with anisotropic scattering. II. The uniqueness problem for Chandrasekhar's ψ_i and φ_i equations, *Astrophys. J.* **139**, 1267 (1964).
15. A. K. KOLESOV and O. I. SMOKTII, Diffuse reflection and transmission of light by a planetary atmosphere with a three term phase function, *Astron. Zh.* **47**, 397 (1970) [*Sov. Astron. A.J.* **14**, 319 (1970)].

Chapter 8

APPROXIMATE FORMULAS

IN THE previous chapters we have considered the exact solution of the problem of light scattering in planetary atmospheres. We shall now present a number of approximate methods for the solution of this problem. Some of them have long been used in astrophysics and geophysics, while others have begun to be applied only recently.

With the aid of each of the methods discussed, we shall solve a particular sample problem. We obtain as a result an approximate expression for the quantity being sought. Other quantities may be obtained by the same method, but we give only a brief indication of the procedure or direct the reader to the References.

We begin the derivation of approximate expressions by making use of some integral relations, followed by use of certain inequalities. We then describe the method based on "similarity principles", with whose aid anisotropic scattering may be approximately reduced to isotropic scattering. We examine in most detail the method of directional averaging of the radiation intensity, which has been used often in the past. At the end of the chapter we give an approximate solution for the case of a very forward-directed phase function.

8.1. The Use of Integral Relations

In previous chapters we have obtained exact equations for the various quantities characterizing the radiation field. These equations in most cases are quite complex. As a result of their solution, however, it turns out that the desired quantities may often be represented with great accuracy by simple expressions. Consequently, it would seem natural to seek initially these quantities in the form of simple expressions with unknown coefficients. In order to find these coefficients, it is convenient to use various integral relations which are satisfied by the quantities under examination.

As an example we shall find (following [1]) by this procedure an approximate expression for the function $u_0(\eta)$, which we have called the transmission function of a semi-infinite atmosphere for the case that the single scattering albedo $\lambda = 1$. It gives the relative angular distribution of the radiation transmitted by an atmosphere of optical thickness τ_0 as $\tau_0 \to \infty$. This function is interesting not only because it gives the brightness distribution over the cloudy sky, but more importantly because of its presence in the asymptotic expressions for many other quantities (see Sections 2.5, 2.6, 3.3 and 3.4).

One of the integral relations for the function $u_0(\eta)$ is obtained from the linear integral equation (7.50). Setting $\eta = 0$ and $\lambda = 1$ in that equation and using (7.46) for $\eta = 0$, we find

$$u_0(0) = \frac{1}{2} \int_0^1 u_0(\eta) p(0, \eta) \, d\eta. \tag{8.1}$$

The normalization condition

$$2 \int_0^1 u_0(\eta) \eta \, d\eta = 1 \tag{8.2}$$

will serve as another integral relation satisfied by $u_0(\eta)$.

Let us seek $u_0(\eta)$ in the form

$$u_0(\eta) = u_0(0)(1+\beta\eta), \tag{8.3}$$

where $u_0(0)$ and β are constants to be determined. With the aid of (8.1) and (8.2) we obtain

$$u_0(\eta) = \frac{1+\beta\eta}{1+\frac{2}{3}\beta}, \tag{8.4}$$

where

$$\frac{1}{\beta} = \int_0^1 p(0, \eta) \eta \, d\eta. \tag{8.5}$$

The quantity β may be expressed directly in terms of the phase function $x(\gamma)$. For this purpose we must substitute equation (7.48) for $\zeta = 0$ into (8.5). Transforming from the variables of integration η and φ to the variables γ and ψ according to $\eta = \sin \gamma \cos \psi$, $\cos \gamma = \sqrt{1-\eta^2} \cos \varphi$, we then find

$$\frac{1}{\beta} = \frac{1}{\pi} \int_0^\pi x(\gamma) \sin^2 \gamma \, d\gamma. \tag{8.6}$$

We have thus obtained the approximate expression (8.4) for the function $u_0(\eta)$ with the constant β given by equation (8.6).

As an example, we shall apply these results to the case of the Henyey–Greenstein phase function given by equation (1.16). Substituting that equation into (8.6), we obtain

$$\frac{1}{\beta} = \frac{1-g^2}{g^2} \frac{2}{\pi} [K(g)-E(g)], \tag{8.7}$$

where $K(g)$ and $E(g)$ are complete elliptic integrals. For small g we may expand $K(g)$ and $E(g)$ in powers of g^2. We then find from (8.7)

$$\beta = 2+\tfrac{5}{4}g^2. \tag{8.8}$$

In that case equation (8.4) takes the form

$$u_0(\eta) = \tfrac{3}{7}[1-\tfrac{5}{14}g^2+(2+\tfrac{15}{28}g^2)\eta], \tag{8.9}$$

where we have dropped terms of order g^4. For $g = 0$ (isotropic scattering) it follows from (8.9) that

$$u_0(\eta) = \tfrac{3}{7}(1+2\eta).\tag{8.10}$$

These approximate values of $u_0(\eta)$ may be compared with exact values from Section 6.5. As an example, Fig. 8.1 presents both the curve plotted using the exact values and the straight line given by equations (8.4) and (8.7) for $g = 0.75$. The analogous figures for other values of g have a similar appearance. We may conclude that the use of equations (8.4) and (8.6) gives values of $u_0(\eta)$ with a precision that is sufficient for many applications.

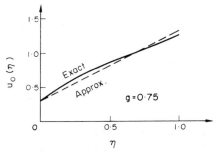

FIG. 8.1. Exact and approximate (eq. (8.4)) results for the relative transmission coefficient of a semi-infinite atmosphere $u_0(\eta)$ for pure scattering with a Henyey–Greenstein phase function with $g = 0.75$.

A more precise formula for the determination of $u_0(\eta)$ may be obtained if we use an additional integral relation. In order to find it, we shall make use of equation (2.37) for $m = 0$. Multiplying this equation by $\eta\zeta$, integrating over η and ζ between 0 and 1, and taking account of (2.106) and (2.110), we find

$$4 \int_0^1 u_0(\eta)\, d\eta \int_0^1 u_0(\zeta)p(-\eta,\,\zeta)\, d\zeta = 3-x_1.\tag{8.11}$$

The function $u_0(\eta)$ may now be sought, for example, in the form

$$u_0(\eta) = a+b\sqrt{\eta}+c\eta\tag{8.12}$$

with the constants to be determined from the relations (8.1), (8.2), and (8.11).

Approximate expressions for other quantities may also be obtained with the aid of integral relations. In particular, approximate expressions for the quantity $\varrho_0(\eta,\,\zeta)$, the reflection coefficient averaged over azimuth for $\tau_0 = \infty$ and $\lambda = 1$, are of great interest. Let us point out several relations which the quantity $\varrho_0(\eta,\,\zeta)$ satisfies. One of these is

$$2 \int_0^1 \varrho_0(\eta,\,\zeta)\eta\, d\eta = 1,\tag{8.13}$$

which expresses the fact that the albedo of a semi-infinite atmosphere in the case of pure scattering is unity. In addition, from equation (2.37) for $m = 0$ we have

$$\eta \varrho_0(\eta, 0) = \tfrac{1}{2}\eta \int_0^1 \varrho_0(\eta, \zeta)p(0, \zeta) \, d\zeta + \tfrac{1}{4}p(\eta, 0), \tag{8.14}$$

$$\lim_{\eta,\, \zeta \to 0} \varrho_0(\eta, \zeta)(\eta + \zeta) = \tfrac{1}{4}p(0, 0). \tag{8.15}$$

From equation (2.110) with the aid of (8.13) we also find

$$u_0(\eta) = \tfrac{3}{2} \int_0^1 \varrho_0(\eta, \zeta)(\eta + \zeta)\zeta \, d\zeta. \tag{8.16}$$

We may then take, for example, the following approximate expression for $\varrho_0(\eta, \zeta)$:

$$\varrho_0(\eta, \zeta) = \frac{a + b(\eta + \zeta) + c\eta\zeta}{\eta + \zeta}, \tag{8.17}$$

where a, b, c are certain constants. The values obtained for these constants may be somewhat different depending on which of the relations obtained above are used for their determination. Consequently, several approximate expressions for the quantity $\varrho_0(\eta, \zeta)$ may be obtained which differ from each other. We shall not, however, spend further time on this question.

8.2. Some Inequalities

In some cases, we may easily obtain from the exact relations certain inequalities which determine limiting values of the quantity in which we are interested. These inequalities are of interest for their own sake, since it is often possible to draw significant conclusions from them. In addition, it is possible to obtain with their aid approximate values for the quantities under discussion.

We shall first find inequalities for the function $u_0(\eta)$, for which an approximate expression was given in the preceding section. For this purpose we shall use the relation

$$u_0(\eta) = \tfrac{3}{4}\left[\eta + 2\int_0^1 \varrho_0(\eta, \zeta)\zeta^2 \, d\zeta\right], \tag{8.18}$$

which was derived in Section 2.5. From (8.18) we easily obtain with the aid of (8.13) that $u_0(\eta)$ is always confined within the limits

$$\tfrac{3}{4}\eta < u_0(\eta) < \tfrac{3}{4}(\eta + 1). \tag{8.19}$$

From (8.19) we find

$$0 < u_0(0) < \tfrac{3}{4}, \tag{8.20}$$

and

$$\tfrac{3}{4} < u_0(1) < \tfrac{3}{2}. \tag{8.21}$$

We thus have

$$u_0(0) < u_0(1). \tag{8.22}$$

The inequality (8.22) shows that the intensity of radiation diffusely transmitted by a semi-infinite atmosphere is always greater for the transmission angle 0 than for the transmission angle $\pi/2$. This conclusion holds for any phase function. In particular, it remains true (although it may seem surprising) for a phase function which is strongly elongated to the sides rather than in the forward direction.

If we assume that $u_0(\eta)$ has its maximum for $\eta = 1$, we may obtain a more stringent inequality than (8.22). Indeed, from (8.1) and the fact that

$$\int_0^1 p(0, \eta) \, d\eta = 1, \tag{8.23}$$

we find

$$2u_0(0) < u_0(1). \tag{8.24}$$

The rather crude inequalities (8.19)–(8.21) may be made significantly more precise if the phase function is given. For example, in order to find a lower bound for $u_0(\eta)$ we may substitute in equation (8.18) instead of $\varrho_0(\eta, \zeta)$ the value of the latter quantity including only first order scattering. We then obtain

$$u_0(0) > \tfrac{3}{8} \int_0^1 p(0, \zeta)\zeta \, d\zeta. \tag{8.25}$$

Making the same transformation as in the derivation of (8.6) from (8.5), we find

$$u_0(0) > \frac{3}{8\pi} \int_0^{\zeta} x(\gamma) \sin^2 \gamma \, d\gamma. \tag{8.26}$$

We may also obtain inequalities satisfied by the quantity $V(\zeta, \tau_0)$ for the case of pure scattering. As previously, it represents the ratio of the illumination of the lower atmospheric boundary to the external illumination of the upper boundary. $V(\zeta, \tau_0)$ may be expressed in terms of the transmission coefficient averaged over azimuth with the aid of the equation

$$V(\zeta, \tau_0) = e^{-\tau_0/\zeta_0} + 2 \int_0^1 \sigma(\eta, \zeta, \tau_0)\eta \, d\eta. \tag{8.27}$$

In the present case ($\lambda = 1$), $V(\zeta, \tau_0)$ may also be expressed in the form

$$V(\zeta, \tau_0) = 1 - A(\zeta, \tau_0), \tag{8.28}$$

where $A(\zeta, \tau_0)$ is the atmospheric albedo

$$A(\zeta, \tau_0) = 2 \int_0^1 \varrho(\eta, \zeta, \tau_0) \, \eta \, d\eta. \tag{8.29}$$

We shall make use of equations (7.116) and (7.117), which follow from the flux integral and the "K-integral". With the aid of the first of these equations, we may rewrite the second in the form

$$\left(1 - \frac{x_1}{3}\right) \tau_0 V(\zeta, \tau_0) + 2 \int_0^1 \sigma(\eta, \zeta, \tau_0)\eta^2 \, d\eta + \zeta e^{-\tau_0/\zeta} = 2 \int_0^1 \varrho(\eta, \zeta, \tau_0)\eta^2 \, d\eta + \zeta. \tag{8.30}$$

If we neglect the last two terms on the left side of this equation and replace the first term on the right side by $A(\zeta, \tau_0)$, we obtain, following use of (8.28),

$$\left[\left(1-\frac{x_1}{3}\right)\tau_0+1\right]V(\zeta, \tau_0) < 1+\zeta. \tag{8.31}$$

Alternatively, we may neglect the first term on the right side of equation (8.30) and replace the two last terms on the left side by $V(\zeta, \tau_0)$. This yields

$$\left[\left(1-\frac{x_1}{3}\right)\tau_0+1\right]V(\zeta, \tau_0) > \zeta. \tag{8.32}$$

It follows from (8.31) and (8.32) that $V(\zeta, \tau_0)$ will lie between the following limits:

$$\frac{\zeta}{\left(1-\frac{x_1}{3}\right)\tau_0+1} < V(\zeta, \tau_0) < \frac{1+\zeta}{\left(1-\frac{x_1}{3}\right)\tau_0+1}. \tag{8.33}$$

The inequality (8.33) is valid for arbitrary phase function and any optical thickness of the atmosphere. When $\tau_0 \gg 1$ it yields the following asymptotic expression for the decrease in $V(\zeta, \tau_0)$ with increasing τ_0:

$$V(\zeta, \tau_0) \sim \frac{1}{\left(1-\frac{x_1}{3}\right)\tau_0}. \tag{8.34}$$

This expression was obtained earlier by another method (see Sections 3.3 and 4.3).

If we take for $V(\zeta, \tau_0)$ the arithmetic mean of the limiting values in the inequality (8.33), we obtain an expression close to the approximate expression (8.93) found below. Inequalities have, however, a great advantage over approximate formulas, since the latter are usually deduced without an evaluation of their precision. Inequalities, in contrast, permit such an evaluation. For example, we may assert that any value of $V(\zeta, \tau_0)$ satisfying the inequality (8.33) with $\zeta = 1$ differs from the exact value by less than a factor of 2.

8.3. Similarity Relations

In determining the radiation field in the atmosphere we have taken the phase function $x(\gamma)$, the optical thickness τ_0, and the single scattering albedo λ as given. A change in these quantities of course produces a change in the radiation field. We may, however, pose the following question: Is it possible to change these quantities in a manner such that the radiation field remains approximately the same? The relation between a given case and an atmosphere with isotropic scattering has particular interest, since the solution of the transfer problem in the latter case is quite simple.

It was observed some time ago that the latter transformation is possible. A number of the equations derived above testify to this possibility. As an example, consider equation (8.34), which shows that the illumination of the lower boundary of an atmosphere for anisotropic scattering with an optical thickness τ_0 has approximately the same value as that for an atmosphere with isotropic scattering and an optical thickness $(1-x_1/3)\,\tau_0$.

It must be understood that the approximate similarity of the radiation field in an atmosphere with anisotropic scattering to the corresponding field in an atmosphere with isotropic scattering will take place only after a large number of scatterings (that is, in the case of large optical thickness τ_0 and for values of λ close to 1). It is also clear that we may speak of similarity only for the radiation fields averaged over azimuth, since for isotropic scattering the intensity does not depend upon azimuth.

We know that the radiation field in an atmosphere consisting of plane-parallel layers satisfies the following integro-differential equation:

$$\cos\vartheta\,\frac{dI}{d\tau} = -I + \lambda \int Ix(\gamma)\,\frac{d\omega}{4\pi}, \tag{8.35}$$

which may be deduced from equations (1.26) and (1.27) (we neglect for simplicity the inhomogeneous term). In order to relate anisotropic scattering to isotropic scattering, we shall write the phase function approximately as

$$x(\gamma) = r + (1-r)\,\delta, \tag{8.36}$$

where δ is the Dirac δ-function. We thus assume that a fraction r of the radiation is scattered isotropically, and a fraction $1-r$ is re-emitted in the direction of the incident radiation. Substituting (8.36) into (8.35), we obtain

$$\cos\vartheta\,\frac{dI}{d\tau} = -[1-(1-r)\lambda]I + r\lambda \int I\,\frac{d\omega}{4\pi}. \tag{8.37}$$

This equation may be rewritten in the form

$$\cos\vartheta\,\frac{dI}{d\tau^*} = -I + \lambda^* \int I\,\frac{d\omega}{4\pi}, \tag{8.38}$$

where we have defined

$$\lambda^* = \frac{r\lambda}{1-(1-r)\lambda}, \tag{8.39}$$

and

$$\tau^* = [1-(1-r)\lambda]\tau. \tag{8.40}$$

In this way we have approximately transformed the transfer equation (8.35) for anisotropic scattering into the transfer equation (8.38) for isotropic scattering. A new single scattering albedo λ^* is given by equation (8.39), and a new optical depth τ^* by equation (8.40), with an analogous expression for the new optical thickness of the atmosphere τ_0^*. We shall refer to equations (8.39) and (8.40) as "similarity relations". Note that elimination of the quantity r from (8.39) and (8.40) leads to the following equality:

$$(1-\lambda^*)\tau^* = (1-\lambda)\tau. \tag{8.41}$$

The quantity r entering the expressions derived above is still undetermined, and the precision of the approximation depends upon its choice. Clearly, the more forward-directed

12

is the phase function, the larger will be the quantity $1-r$. We know that the forward elonga-tion of a phase function is usually characterized by the parameter x_1, the first coefficient of its expansion in Legendre polynomials. It thus seems natural to choose the quantity r such that the value of the parameter x_1 for the actual phase function and for the phase function (8.36) will coincide. This condition gives

$$r = 1 - \frac{x_1}{3}. \tag{8.42}$$

With this choice of r, the similarity relations (8.39) and (8.40) take the form

$$\lambda^* = \frac{(3-x_1)\lambda}{3-x_1\lambda}, \tag{8.43}$$

and

$$\tau^* = \left(1 - \frac{x_1}{3}\lambda\right)\tau. \tag{8.44}$$

The replacement of equation (8.35) by equation (8.38) with the determination of the quantity r by equation (8.42) has been used in the theory of neutron diffusion and is called the "transport approximation" (see the book of Davison cited in Chapter 1). Recently, van de Hulst [2] proposed using the similarity relations in the following form:

$$k^*\tau^* = k\tau \tag{8.45}$$

and

$$\frac{1-\lambda^*}{k^*} = \frac{1-\lambda}{k}, \tag{8.46}$$

where k is the smallest root of the characteristic equation (2.22) and k^* is the root of equa-tion (2.14). The basis for the relation (8.45) is the fact that in the deep layers of a homo-geneous semi-infinite medium the radiation intensity falls off with increasing τ as $e^{-k\tau}$, while if the scattering is isotropic it falls off as $e^{-k^*\tau^*}$. The relation (8.46) follows from (8.45) and (8.41).

In the most interesting case of small true absorption ($1-\lambda \ll 1$) relations (8.45) and (8.46) reduce to (8.43) and (8.44). This may be confirmed from equation (2.24) for the quantities k and k^*.

We see that the approximate method under examination consists of transforming with the aid of the similarity relations from the quantities $x(\gamma)$, λ, and τ_0 to the quantities λ^* and τ_0^* for an isotropically scattering atmosphere, and then determining the radiation field in the latter atmosphere. The validity of the method has been confirmed by finding the reflection coefficient of a semi-infinite atmosphere for phase functions with different degrees of elongation (including the isotropic case) for the same value of the parameter $(1-\lambda)/k$. The agreement among the resulting reflection coefficients was satisfactory (see [2]).

We must note, however, that such agreement cannot be expected in all cases. For example, the similarity relation (8.39) shows that $\lambda = 1$ is always transformed to $\lambda^* = 1$; that is, pure scattering is always "similar" except for a scaling of optical thickness τ_0. In other words, in the case of pure scattering, the reflection coefficient of a semi-infinite atmosphere for an arbitrary phase function may be replaced according to the present method by the reflection coefficient for isotropic scattering. In reality, we know that these reflection coefficients may significantly differ from each other. The agreement between exact results and those obtained with the similarity relations will be greatest for integrated quantities such as the atmospheric albedo or the transmitted flux.

8.4. Directional Averaging of the Radiation Intensity

The approximate methods of Schwarzschild–Schuster and of Eddington, based on the directional averaging of the radiation intensity, have been widely applied in the theory of radiative transfer in stellar atmospheres. The light scattering in stellar atmospheres is isotropic. Such averaging may also be used, however, for the approximate solution of the equation of transfer in the case of anisotropic scattering (see [3] and also TRT, Chapter 10).

We shall begin with the expression (1.36) for the source function. It may be written approximately in the form

$$B(\tau, \eta, \zeta, \varphi) = \lambda \bar{I}(\tau, \zeta) + \lambda x_1 H(\tau, \zeta) \, \eta$$
$$+ \lambda x_1 \bar{G}(\tau, \zeta) \sqrt{1-\eta^2} \cos \varphi + \frac{\lambda}{4} Sx(\gamma) \, e^{-\tau/\zeta}, \qquad (8.47)$$

where

$$\bar{I}(\tau, \zeta) = \frac{1}{4\pi} \int_0^{2\pi} d\varphi \int_{-1}^{1} I(\tau, \eta, \zeta, \varphi) \, d\eta, \qquad (8.48)$$

$$\bar{H}(\tau, \zeta) = \frac{1}{4\pi} \int_0^{2\pi} d\varphi \int_{-1}^{1} I(\tau, \eta, \zeta, \varphi) \eta \, d\eta, \qquad (8.49)$$

$$\bar{G}(\tau, \zeta) = \frac{1}{4\pi} \int_0^{2\pi} d\varphi \int_{-1}^{1} I(\tau, \eta, \zeta, \varphi) \sqrt{1-\eta^2} \, d\eta, \qquad (8.50)$$

and x_1 is the first-order coefficient of the expansion of the phase function in Legendre polynomials, so that

$$x_1 = \tfrac{3}{2} \int_0^{\pi} x(\gamma) \cos \gamma \sin \gamma \, d\gamma. \qquad (8.51)$$

The quantity $\bar{I}(\tau, \zeta)$ is the mean intensity of diffuse radiation at an optical depth τ, $4\pi\bar{H}(\tau, \zeta)$ is the flux of diffuse radiation in the direction of increasing optical depth, and $4\pi\bar{G}(\tau, \zeta)$ is the flux of diffuse radiation in the direction $\varphi = 0$ in a horizontal plane.

In transforming from (1.36) to (8.47) we have kept the zeroth and first terms of the expansion (1.40) for the phase function and neglected the remainder. The error resulting from this procedure will be of the same order as that produced by the approximations introduced below.

We may say that scattering of first order is computed exactly by equation (8.47), while scattering of higher orders is approximated by the phase function

$$x(\gamma) = 1 + x_1 \cos \gamma. \qquad (8.52)$$

It is necessary, however, to bear in mind that the transformation from (1.36) to (8.47) is a mathematical procedure, and that consequently the parameter x_1 may take any value given by equation (8.51) (in particular, x_1 may be greater than 1, so that the quantity $x(\gamma)$ given by equation (8.52) may take negative values).

We may substitute the expression (8.47) into the equation of transfer (1.35). Before that, however, we replace the quantity $x(\gamma)$ in (8.47) by equations (8.52) and (1.38) (and then in the final result perform the inverse transformation). The radiation intensity may then be written in the form

$$I(\tau, \eta, \zeta, \varphi) = I^0(\tau, \eta, \zeta) + I^1(\tau, \eta, \zeta) \cos \varphi, \tag{8.53}$$

and we obtain separate integro-differential equations for the determination of each of the quantities I^0 and I^1.

We shall now find the quantity I^0, the radiation intensity averaged over azimuth. The quantity I^1 may be found in a similar manner (see TRT, § 4 of Chapter 10).

Writing for simplicity $I(\tau, \eta, \zeta)$ instead of $I^0(\tau, \eta, \zeta)$, we find for it the following equation:

$$\eta \frac{dI(\tau, \eta, \zeta)}{d\tau} = -I(\tau, \eta, \zeta) + \lambda \bar{I}(\tau, \zeta) + \lambda x_1 \bar{H}(\tau, \zeta)\, \eta + \frac{\lambda}{4} S(1 + x_1 \eta \zeta) e^{-\tau/\zeta}, \tag{8.54}$$

where

$$\bar{I}(\tau, \zeta) = \tfrac{1}{2} \int_{-1}^{1} I(\tau, \eta, \zeta)\, d\eta, \quad \bar{H}(\tau, \zeta) = \tfrac{1}{2} \int_{-1}^{1} I(\tau, \eta, \zeta)\eta\, d\eta. \tag{8.55}$$

We may obtain from equation (8.54) a system of approximate equations for the functions $\bar{I}(\tau, \zeta)$ and $\bar{H}(\tau, \zeta)$. Integrating (8.54) over η between -1 and 1, we find

$$\frac{d\bar{H}(\tau, \zeta)}{d\tau} = -(1 - \lambda)\, \bar{I}(\tau, \zeta) + \frac{\lambda}{4}\, S e^{-\tau/\zeta}. \tag{8.56}$$

Multiplying (8.54) by η, integrating over η between -1 and 1, and using the approximate relation

$$\tfrac{1}{2} \int_{-1}^{1} I(\tau, \eta, \zeta)\, \eta^2\, d\eta = \tfrac{1}{3} \bar{I}(\tau, \zeta), \tag{8.57}$$

we have

$$\frac{d\bar{I}(\tau, \zeta)}{d\tau} = -(3 - \lambda x_1)\bar{H}(\tau, \zeta) + x_1 \frac{\lambda}{4} S e^{-\tau/\zeta}\, \zeta. \tag{8.58}$$

We must specify the boundary conditions for equations (8.56) and (8.58). These conditions must express the fact that there is no diffuse radiation incident on the atmosphere from above or from below. For $\tau = 0$ we find approximately

$$\bar{H}(0, \zeta) = \tfrac{1}{2} \int_{-1}^{0} I(0, \eta, \zeta)\, \eta\, d\eta = -\tfrac{1}{4} \int_{-1}^{0} I(0, \eta, \zeta)\, d\eta, \tag{8.59}$$

so that

$$2\bar{H}(0, \zeta) = -\bar{I}(0, \zeta). \tag{8.60}$$

For $\tau = \tau_0$ we find analogously

$$2\bar{H}(\tau_0, \zeta) = \bar{I}(\tau_0, \zeta). \tag{8.61}$$

From equations (8.56) and (8.58) we deduce the following equation for $\bar{I}(\tau, \zeta)$:

$$\frac{d^2\bar{I}(\tau, \zeta)}{d\tau^2} = k^2\bar{I}(\tau, \zeta)-[3+(1-\lambda)x_1]\frac{\lambda}{4}Se^{-\tau/\zeta}, \tag{8.62}$$

where

$$k^2 = (1-\lambda)(3-\lambda x_1). \tag{8.63}$$

The solution of equation (8.62) has the form

$$\bar{I}(\tau, \zeta) = C_1e^{-k\tau}+C_2e^{k\tau}+De^{-\tau/\zeta}, \tag{8.64}$$

where

$$D = -\frac{3+(1-\lambda)x_1}{1-k^2\zeta^2}\zeta^2\frac{\lambda}{4}S, \tag{8.65}$$

and C_1 and C_2 are arbitrary constants. Substituting (8.64) into (8.58) and using equations (8.65) and (8.63), we find

$$\bar{H}(\tau, \zeta) = (C_1e^{-k\tau}-C_2e^{k\tau})\frac{k}{3-\lambda x_1}$$

$$-\frac{\lambda}{4}Se^{-\tau/\zeta}\frac{\zeta}{1-k^2\zeta^2}[1+(1-\lambda)x_1\zeta^2]. \tag{8.66}$$

In order to find the constants C_1 and C_2, we must use the boundary conditions (8.60) and (8.61). Substituting into them equations (8.64) and (8.66) for $\tau = 0$ and $\tau = \tau_0$, we obtain

$$C_1 = \frac{\lambda}{4}S\frac{\zeta}{1-k^2\zeta^2}\frac{1}{\Delta}\{[2+3\zeta+(1-\lambda)x_1\zeta(1+2\zeta)]$$

$$\times e^{k\tau_0}(1+b)+[2-3\zeta-(1-\lambda)x_1\zeta(1-2\zeta)]e^{-\tau_0/\zeta}(1-b)\}, \tag{8.67}$$

$$C_2 = -\frac{\lambda}{4}S\frac{\zeta}{1-k^2\zeta^2}\frac{1}{\Delta}\{[2+3\zeta+(1-\lambda)x_1\zeta(1+2\zeta)]$$

$$\times e^{-k\tau_0}(1-b)+[2-3\zeta-(1-\lambda)x_1\zeta(1-2\zeta)]e^{-\tau_0/\zeta}(1+b)\}, \tag{8.68}$$

where we have set

$$\Delta = e^{k\tau_0}(1+b)^2-e^{-k\tau_0}(1-b)^2 \tag{8.69}$$

and

$$b = \frac{2k}{3-\lambda x_1}. \tag{8.70}$$

Substitution of the values of C_1 and C_2 so obtained into equations (8.64) and (8.66) yields expressions for the functions $\bar{I}(\tau, \zeta)$ and $\bar{H}(\tau, \zeta)$. We may then determine the radiation intensity at any optical depth in the atmosphere with the aid of the equation of transfer (8.54).

It is easy to see that the albedo of the atmosphere and the illumination of the adjacent surface may be expressed directly in terms of the constants C_1, C_2, and D. For the atmospheric albedo we have

$$A(\zeta, \tau_0) = -\frac{4}{S\zeta}\bar{H}(0, \zeta) = \frac{2}{S\zeta}\bar{I}(0, \zeta), \tag{8.71}$$

or, recalling (8.64),

$$A(\zeta, \tau_0) = \frac{2}{S\zeta} (C_1 + C_2 + D).\tag{8.72}$$

The ratio of the illumination of the surface to the solar flux incident at the upper boundary of the atmosphere we have designated $V(\zeta, \tau_0)$. This quantity is clearly given by

$$V(\zeta, \tau_0) = \frac{4}{S\zeta} \bar{H}(\tau_0, \zeta) + e^{-\tau_0/\zeta} = \frac{2}{S\zeta} \bar{I}(\tau_0, \zeta) + e^{-\tau_0/\zeta},\tag{8.73}$$

where the last term represents the direct solar radiation transmitted through the atmosphere. Application of equation (8.64) yields

$$V(\zeta, \tau_0) = \frac{2}{S\zeta} (C_1 e^{-k\tau_0} + C_2 e^{k\tau_0} + D e^{-\tau_0/\zeta}) + e^{-\tau_0/\zeta}.\tag{8.74}$$

The above expressions are greatly simplified for a semi-infinite atmosphere. In that case we must set $\tau_0 = \infty$ and $C_2 = 0$. In particular, we find for the albedo of a semi-infinite atmosphere

$$A(\zeta) = \frac{\lambda}{2} \frac{2 - 3b\zeta + (1-\lambda) x_1 \zeta (2\zeta - b)}{(1+b)(1 - k^2\zeta^2)}.\tag{8.75}$$

In the case of small true absorption $(1-\lambda \ll 1)$ it follows from (8.75) that

$$A(\zeta) = 1 - \frac{4k}{3 - x_1} u_0(\zeta),\tag{8.76}$$

where

$$u_0(\zeta) = \tfrac{1}{2} + \tfrac{3}{4} \zeta.\tag{8.77}$$

Equation (8.76) does not differ in form from the asymptotic expression (2.111), except for the approximate form (8.77) for the function $u_0(\zeta)$.

We note that other formulas similar to the asymptotic expressions found earlier for a semi-infinite atmosphere for small true absorption (Section 2.6) and for atmospheres of large optical thickness (Section 3.3) may be obtained by the same method. The functions entering the expressions so derived will, however, be determined by approximate expressions.

8.5. The Case of Pure Scattering

We shall now apply the approximate method for the solution of radiative transfer problems presented in the previous section to the case of pure scattering. This case must be examined separately, since for pure scattering $k = 0$ and the previous solution is indeterminate.

Let us return to equations (8.56) and (8.58). Setting $\lambda = 1$ and integrating, we obtain

$$\bar{H}(\tau, \zeta) = F - \frac{S}{4} e^{-\tau/\zeta} \zeta,\tag{8.78}$$

$$\bar{I}(\tau, \zeta) = C - (3 - x_1) F\tau - \tfrac{3}{4} S e^{-\tau/\zeta} \zeta^2,\tag{8.79}$$

where C and F are arbitrary constants. Using the boundary conditions (8.60) and (8.61), we find

$$C = \frac{S\zeta}{2}(1 + \tfrac{3}{2}\zeta) - 2F, \tag{8.80}$$

$$[4 + (3 - x_1)\tau_0] F = \frac{S\zeta}{2} R(\zeta, \tau_0), \tag{8.81}$$

where we have defined

$$R(\zeta, \tau_0) = 1 + \tfrac{3}{2}\zeta + (1 - \tfrac{3}{2}\zeta) e^{-\tau_0/\zeta}. \tag{8.82}$$

Substitution of the expressions (8.78) and (8.79) into the equation of transfer (8.54) yields

$$\eta \frac{dI(\tau, \eta, \zeta)}{d\tau} = -I(\tau, \eta, \zeta) + C + x_1 F\eta - (3 - x_1) F\tau$$
$$-\frac{S}{4}(3\zeta^2 + x_1\eta\zeta) e^{-\tau/\zeta} + \frac{S}{4} p(\eta, \zeta) e^{-\tau/\zeta}. \tag{8.83}$$

Equation (8.83) determines the radiation intensity averaged over azimuth. In accordance with the method we have adopted, in the last term we have replaced the quantity $1 + x_1\eta\zeta$ by the function $p(\eta, \zeta)$; that is, by the actual phase function averaged over azimuth according to equation (7.48). The first order scattering is now given exactly by equation (8.83).[†]

The radiation intensity at any optical depth may now be found by integrating equation (8.83). We shall confine ourselves, however, to the intensity emerging from the atmosphere. Introducing the reflection and transmission coefficients defined by equations (1.74) and (1.75), we find

$$S\varrho(\eta, \zeta, \tau_0)\zeta = (C - 3F\eta)(1 - e^{-\tau_0/\eta}) + (3 - x_1) F\tau_0 e^{-\tau_0/\eta}$$
$$+ S[p(-\eta, \zeta) - 3\zeta^2 + x_1\eta\zeta] \varrho_1(\eta, \zeta)\zeta, \tag{8.84}$$

$$S\sigma(\eta, \zeta, \tau_0)\zeta = (C + 3F\eta)(1 - e^{\tau_0/\eta}) - (3 - x_1) F\tau_0$$
$$+ S[p(\eta, \zeta) - 3\zeta^2 - x_1\eta\zeta] \sigma_1(\eta, \zeta)\zeta, \tag{8.85}$$

where in the first of these equations η is the cosine of the angle of reflection, and in the second it is the cosine of the angle of transmission, so that in both cases $\eta > 0$. The quantities $\varrho_1(\eta, \zeta)$ and $\sigma_1(\eta, \zeta)$ entering (8.84) and (8.85) are determined by the expressions

$$\varrho_1(\eta, \zeta) = \frac{1}{4} \frac{1 - e^{-\tau_0(1/\eta + 1/\zeta)}}{\eta + \zeta}, \qquad \sigma_1(\eta, \zeta) = \frac{1}{4} \frac{e^{-\tau_0/\eta} - e^{-\tau_0/\zeta}}{\eta - \zeta}, \tag{8.86}$$

so that they represent the coefficients of reflection and transmission for single scattering with $x(\gamma) = 1$.

Introducing into (8.84) and (8.85) the expressions for the constants C and F from (8.80) and (8.81), we obtain the following final formulas for the reflection and transmission coeffi-

---

[†] If we replaced $p(\eta, \zeta)$ by $x(\gamma)$ in equation (8.83) then it will give the primary scattering exactly including the dependence upon azimuth, and the higher order scattering approximately (averaged over azimuth). Such an approximation is sometimes used in practice.

cients averaged over azimuth:

$$\varrho(\eta, \zeta, \tau_0) = 1 - \frac{R(\eta, \tau_0)\, R(\zeta, \tau_0)}{4 + (3 - x_1)\tau_0}$$

$$+ [(3 + x_1)\, \eta\zeta - 2(\eta + \zeta) + p(-\eta, \zeta)]\, \varrho_1(\eta, \zeta), \qquad (8.87)$$

$$\sigma(\eta, \zeta, \tau_0) = \frac{R(\eta, \tau_0)\, R(\zeta, \tau_0)}{4 + (3 - x_1)\, \tau_0} - \frac{1}{2} (e^{-\tau_0/\eta} + e^{-\tau_0/\zeta})$$

$$+ [p(\eta, \zeta) - (3 + x_1)\, \eta\zeta]\, \sigma_1(\eta, \zeta). \qquad (8.88)$$

It is known that the reflection and transmission coefficients must be symmetric functions of η and ζ. The approximate expressions (8.87) and (8.88) satisfy this requirement.

In the case of a semi-infinite atmosphere ($\tau_0 = \infty$) we find from (8.87) for the reflection coefficient

$$\varrho(\eta, \zeta) = \frac{1}{2} + \frac{(3 + x_1)\, \eta\zeta + p(-\eta, \zeta)}{4(\eta + \zeta)}. \qquad (8.89)$$

In the case of an atmosphere of large optical thickness ($\tau_0 \gg 1$) equations (8.87) and (8.88) take the form

$$\varrho(\eta, \zeta, \tau_0) = \varrho(\eta, \zeta) - \frac{4u_0(\eta)\, u_0(\zeta)}{4 + (3 - x_1)\tau_0}, \qquad (8.90)$$

$$\sigma(\eta, \zeta, \tau_0) = \frac{4u_0(\eta)\, u_0(\zeta)}{4 + (3 - x_1)\, \tau_0}, \qquad (8.91)$$

where the function $\varrho(\eta, \zeta)$ is given by equation (8.89) and the function $u_0(\zeta)$ by equation (8.77). We see that equations (8.90) and (8.91) are similar to the asymptotic expressions (3.65) and (3.66), except for the approximate expressions for $\varrho(\eta, \zeta)$ and $u_0(\zeta)$ entering the former equations.

The approximate expressions for the atmospheric albedo and illumination of the planetary surface in the present case are easily obtained. From (8.71), (8.78), and (8.81) we find for the atmospheric albedo

$$A(\zeta, \tau_0) = 1 - \frac{2R(\zeta, \tau_0)}{4 + (3 - x_1)\, \tau_0}. \qquad (8.92)$$

The illumination of the surface expressed in units of the illumination of the upper atmospheric boundary is, on the basis of (8.73), (8.78), and (8.81),

$$V(\zeta, \tau_0) = \frac{2R(\zeta, \tau_0)}{4 + (3 - x_1)\, \tau_0}. \qquad (8.93)$$

It is obvious from (8.92) and (8.93) that $A(\zeta, \tau_0) + V(\zeta, \tau_0) = 1$, as it should be in the case of pure scattering.

8.6. The Effect of the Reflection of Light by a Surface

The approximate formulas obtained in the two preceding sections relate to the case when the albedo of the planetary surface is zero. We shall now consider the reflection of light by the surface, assuming for simplicity that it is given by Lambert's Law (according to which the surface reflects light isotropically and its albedo a does not depend upon the angle of incidence).

The problem of light scattering in an atmosphere bounded by a reflecting surface has already been considered in detail in Chapter 4. In particular, for the case of Lambert reflection by the surface, we obtained exact expressions for the reflection and transmission coefficients for $a \neq 0$ in terms of the corresponding reflection and transmission coefficients for $a = 0$. We could now utilize these results. We prefer, however, to solve the present problem with the aid of the approximate method presented above.

It has been shown in Section 8.4 that the determination of the radiation field averaged over azimuth approximately reduces to the solution of equations (8.56) and (8.58) with the boundary conditions (8.60) and (8.61). These conditions assumed that no diffuse radiation was incident on the atmosphere either from above or from below. Now, with the presence of a reflecting surface, we must modify the boundary condition for $\tau = \tau_0$. The differential equations (8.56) and (8.58) and also the boundary condition for $\tau = 0$ clearly remain unchanged.

Let us find the new boundary condition for $\tau = \tau_0$. We shall denote the mean intensity of diffuse radiation falling on the surface by I_1, and the intensity reflected by the surface by I_2. If E is the illumination of the surface, we have

$$I_2 = \frac{a}{\pi} E. \tag{8.94}$$

The quantity E includes both the illumination of the surface by direct solar radiation, which is just $\pi S \zeta e^{-\tau_0/\zeta}$, and its illumination by diffuse radiation, which is πI_1. We thus have

$$I_2 = a(I_1 + S\zeta e^{-\tau_0/\zeta}). \tag{8.95}$$

We must relate the boundary values of $\bar{I}$ and $\bar{H}$ which enter equations (8.56) and (8.58). We have approximately

$$\bar{I} = \tfrac{1}{2}(I_1 + I_2), \quad \bar{H} = \tfrac{1}{4}(I_1 - I_2), \tag{8.96}$$

from which it follows that

$$I_1 = \bar{I} + 2\bar{H}, \quad I_2 = \bar{I} - 2\bar{H}. \tag{8.97}$$

Substitution of (8.97) into (8.95) leads to the desired boundary condition for $\tau = \tau_0$:

$$\bar{I} - 2\bar{H} = a(\bar{I} + 2\bar{H} + S\zeta e^{-\tau_0/\zeta}). \tag{8.98}$$

The functions $\bar{I}(\tau, \zeta)$ and $\bar{H}(\tau, \zeta)$ are given by equations (8.64) and (8.66) for $\lambda < 1$ and by equations (8.78) and (8.79) for $\lambda = 1$. The constants which enter these equations in the presence of a reflecting surface must be determined from the boundary conditions (8.60) and (8.98).

As an example, let us find the reflection and transmission coefficients for $a \neq 0$ in the case of pure scattering. The appropriate boundary conditions show that the constant C is expressed in terms of F by equation (8.80), and the constant F is

$$F = \frac{S\zeta}{2} \frac{(1-a) R(\zeta, \tau_0)}{4+(3-x_1)(1-a) \tau_0}. \tag{8.99}$$

These values of F and C must be substituted into equations (8.84) and (8.85). In addition, we must add to the right side of equation (8.84) a term

$$a(\bar{I}+2\bar{H}+S\zeta e^{-\tau_0/\zeta}) e^{-\tau_0/\eta},$$

which represents the radiation reflected by the surface which emerges from the atmosphere without being scattered. We find as a result that the reflection and transmission coefficients are given by

$$\varrho(\eta, \zeta, \tau_0) = 1 - \frac{(1-a) R(\eta, \tau_0) R(\zeta, \tau_0)}{4+(3-x_1)(1-a) \tau_0}$$

$$+ [(3+x_1) \eta\zeta - 2(\eta+\zeta) + p(-\eta, \zeta)] \varrho_1(\eta, \zeta), \tag{8.100}$$

$$\sigma(\eta, \zeta, \tau_0) = \frac{[(1-a) R(\eta, \tau_0)+2a] R(\zeta, \tau_0)}{4+(3-x_1)(1-a) \tau_0}$$

$$- \tfrac{1}{2}(e^{-\tau_0/\eta}+e^{-\tau_0/\zeta}) + [p(\eta, \zeta)-(3+x_1) \eta\zeta] \sigma_1(\eta, \zeta). \tag{8.101}$$

When $a = 0$ equations (8.100) and (8.101) reduce to (8.87) and (8.88).

Quite simple expressions may then be obtained for the albedo of the atmosphere and the illumination of the planetary surface. Using expressions (8.71), (8.78), and (8.99), we find for the albedo

$$A(\zeta, \tau_0) = 1 - \frac{2(1-a) R(\zeta, \tau_0)}{4+(3-x_1)(1-a) \tau_0}. \tag{8.102}$$

We may not, however, use equation (8.73) to determine the function $V(\zeta, \tau_0)$ when $a \neq 0$. In the present case the flux of radiation is the difference between the illumination of the surface and the energy reflected from it, so that we have

$$4\bar{H}(\tau_0, \zeta)+S\zeta e^{-\tau_0/\zeta} = (1-a) V(\zeta, \tau_0) S\zeta. \tag{8.103}$$

Equations (8.103), (8.78), and (8.99) then give

$$V(\zeta, \tau_0) = \frac{2R(\zeta, \tau_0)}{4+(3-x_1)(1-a) \tau_0}. \tag{8.104}$$

From (8.102) and (8.104) we conclude that

$$A(\zeta, \tau_0)+(1-a) V(\zeta, \tau_0) = 1, \tag{8.105}$$

which would be expected in the case of pure scattering.

The approximate formulas obtained in the last three sections have often been applied to the study of the atmospheres of the Earth and other planets. Some of these applications will be discussed in subsequent chapters. The accuracy of these expressions has been discussed by a number of authors ([4], [5] and others).

8.7. The Radiation Field for Highly Anisotropic Scattering

In the preceding chapters we have seen that the exact solution to the problem of light scattering in an atmosphere may be obtained without particular difficulty when the phase function is not very elongated in the forward direction. The difficulties increase, however, with increasing forward elongation. It is therefore of considerable interest to find a method of approximate solution of the light scattering problem for such phase functions. We shall now present one such method, which consists of reducing the integro-differential equation of radiative transfer to a second-order partial differential equation [6].

As previously, we shall assume that the atmosphere consists of plane-parallel layers. Then the transfer of radiation in the atmosphere is described by the integro-differential equation (8.35), which may be rewritten in the form

$$\cos \vartheta \, \frac{\partial I(\tau, \vartheta, \varphi)}{\partial \tau} = -I(\tau, \vartheta, \varphi) + \frac{\lambda}{4\pi} \int_0^{2\pi} d\varphi' \int_0^{\pi} x(\gamma) \, I(\tau, \vartheta', \varphi') \sin \vartheta' \, d\vartheta', \qquad (8.106)$$

where by $I(\tau, \vartheta, \varphi)$ we understand the radiation intensity at optical depth τ traveling in a direction characterized by the polar angle ϑ and the azimuthal angle φ.

We shall assume that the phase function is strongly elongated in the forward direction; that is, that the function $x(\gamma)$ has a sharp maximum in the direction characterized by the angles ϑ and φ (which corresponds to $\gamma = 0$). We may then approximately replace the quantity $I(\tau, \vartheta', \varphi')$ by its Taylor expansion in powers of $\vartheta' - \vartheta$ and $\varphi' - \varphi$. Limiting ourselves to the quadratic terms of this expansion, we find instead of equation (8.106)

$$\cos \vartheta \, \frac{\partial I}{\partial \tau} + (1 - \lambda) \, I = \lambda v \left[\frac{1}{\sin \vartheta} \frac{\partial}{\partial \vartheta} \left(\sin \vartheta \, \frac{\partial I}{\partial \vartheta} \right) + \frac{1}{\sin^2 \vartheta} \frac{\partial^2 I}{\partial \varphi^2} \right], \qquad (8.107)$$

where

$$v = \tfrac{1}{8} \int_0^{\pi} x(\gamma) \sin^3 \gamma \, d\gamma. \qquad (8.108)$$

Equation (8.107) may also be obtained in a different manner. We replace the actual phase function by the approximate expression

$$x(\gamma) = \begin{cases} c, & \gamma \leqslant \gamma_0, \\ 0, & \gamma > \gamma_0, \end{cases} \qquad (8.109)$$

where the quantities γ_0 and c are related by the normalization condition

$$c(1 - \cos \gamma_0) = 2. \qquad (8.110)$$

By substituting (8.109) into equation (8.106), expanding $I(\tau, \vartheta', \varphi')$ in a Taylor series and taking $\gamma_0 \ll 1$, instead of equation (8.106) we are again led to (8.107), in which

$$v = \frac{1}{2c}. \tag{8.111}$$

The greater is the degree of forward elongation of the phase function, the smaller is the angle γ_0 and the larger the quantity c. This elongation of the phase function is usually characterized by the parameter x_1, given by equation (8.51). Equating the value of x_1 for the real phase function to that found for the phase function (8.109), we obtain

$$c = \frac{3}{3 - x_1}. \tag{8.112}$$

Substituting (8.112) into (8.111) we have

$$v = \frac{3 - x_1}{6}. \tag{8.113}$$

The determination of the parameter v by equation (8.113) may be preferable to its determination by equation (8.108).

For the approximate determination of the intensity $I(\tau, \vartheta, \varphi)$ we have thus obtained equation (8.107) in which the quantity v is given by equation (8.108) or (8.113). The results of using this equation will be more exact, the more scatterings the photons undergo in the medium. This is understandable, since even for the phase function (8.109), after many scatterings a photon may have significantly changed its direction. Consequently, some photons will ultimately be scattered backwards. We note that equations of the type (8.107) are used in the theory of particle diffusion (see [7]).

To solve any particular problem we must add boundary conditions to equation (8.107). If the sources of radiation are external to the medium, these conditions must determine the radiation intensity incident on the boundaries from outside. If the sources are located within the medium, the boundary conditions will express the absence of external illumination. In the latter case we must introduce into equation (8.107) a term which characterizes the intrinsic emission of the medium.

As an example, we shall use equation (8.107) to solve the problem of the radiation field in the deep layers of a semi-infinite medium illuminated from outside. This problem may be solved exactly (see Section 2.1), and the results compared to the approximate solution.

In the given case, the intensity does not depend upon azimuth and may be written in the form

$$I(\tau, \vartheta) = y(\vartheta)\, e^{-k\tau}, \tag{8.114}$$

where k is a constant which depends only on the phase function and the single scattering albedo λ. Substituting (8.114) into (8.107), we obtain the following equation for the determination of the function $y(\vartheta)$:

$$\frac{1}{\sin \vartheta} \frac{d}{d\vartheta}\left(\sin \vartheta \frac{dy}{d\vartheta}\right) = (p - q \cos \vartheta)\, y, \tag{8.115}$$

where

$$p = \frac{1 - \lambda}{\lambda v}, \qquad q = \frac{k}{\lambda v}. \tag{8.116}$$

We shall seek a solution of equation (8.115) as an expansion in Legendre polynomials; that is, we set

$$\mathit{\Lambda}(\vartheta) = \sum_0^\infty y_n P_n(\cos\vartheta). \tag{8.17}$$

Substitution of (8.117) into (8.115) leads to the following recurrence relation for the ocefficients y_n:

$$[p+n(+1)]y_n = p\left(\frac{n}{2n-1}y_{n-1}+\frac{n+1}{2n+3}y_{n+1}\right). \tag{8.118}$$

In order for the system of homogeneous equations (8.118) to possess a solution, its determinant must vanish, from which condition we may find the relation between the quantities p and q. Then, with the aid of equation (8.116), we may obtain the relation between k, λ, and v. The determinant of the system (8.118) equals

$$\mathit{\Delta} = \begin{vmatrix} p & -\frac{1}{3}q & 0 & 0 & \cdots \\ -q & p+2 & -\frac{2}{5}q & 0 & \cdots \\ 0 & -\frac{2}{3}q & p+6 & -\frac{3}{7}q & \cdots \\ \multicolumn{5}{c}{\cdots\cdots\cdots\cdots\cdots\cdots} \end{vmatrix}. \tag{8.119}$$

From the condition $\mathit{\Delta}=0$ we may express p in terms of q as a continuing fraction. We denote by $\mathit{\Delta}_n$ the determinant obtained from (8.119) by crossing out the first n rows and columns. Then

$$\mathit{\Delta} = p\mathit{\Delta}_1 - \tfrac{1}{3}q^2\mathit{\Delta}_2, \tag{8.120}$$

$$\mathit{\Delta}_1 = (p+2)\,\mathit{\Delta}_2 - \tfrac{4}{15}q^2\mathit{\Delta}_3, \tag{8.121}$$

etc. Since $\mathit{\Delta}=0$, it follows from (8.120) that

$$p = \frac{1}{3}q^2\frac{\mathit{\Delta}_2}{\mathit{\Delta}_1}. \tag{8.122}$$

Substituting into (8.122) equation (8.121) and the succeeding expressions for $\mathit{\Delta}_2$, $\mathit{\Delta}_3$, $\cdots$, we obtain

$$p = \frac{1}{3}\cfrac{q^2}{p+2-\frac{4}{15}\cfrac{q^2}{p+6-\cfrac{9}{35}\cfrac{q^2}{p+12}-\cdots}}. \tag{8.123}$$

For small q, equation (8.123) gives $p = \frac{1}{6}q^2$. We then find on the basis of (8.116) and (8.113) that

$$\lambda = 1 - \frac{k^2}{3-x_1}. \tag{8.124}$$

Using equations (8.124), (8.117), and (8.118) we obtain

$$y(\vartheta) = y_0\left(1+\frac{3k}{3-x_1}\cos\vartheta\right). \tag{8.125}$$

We recall that the rigorous theory for small k also leads to equations (8.124) and (8.125). This in some measure justifies the determination of the parameter v by equation (8.113).

Table 8.1 presents values of p corresponding to values of q from 0 to 10. The results were computed from equation (8.123). The numerical relation between the quantities λ and k for three values of the parameter v is also given.

TABLE 8.1. THE RELATION BETWEEN THE PARAMETERS λ AND k IN THE APPROXIMATE THEORY

q	p	$v = 0.05$		$v = 0.10$		$v = 0.15$	
		λ	k	λ	k	λ	k
0	0	1	0	1	0	1	0
1	0.16	0.992	0.050	0.984	0.098	0.977	0.146
2	0.56	0.973	0.097	0.947	0.189	0.923	0.277
3	1.09	0.948	0.142	0.902	0.271	0.859	0.387
4	1.70	0.922	0.184	0.854	0.342	0.796	0.478
5	2.37	0.894	0.224	0.809	0.404	0.738	0.554
6	3.06	0.867	0.260	0.766	0.460	0.685	0.617
7	3.78	0.841	0.294	0.726	0.508	0.638	0.670
8	4.52	0.816	0.326	0.689	0.551	0.596	0.715
9	6.05	0.768	0.384	0.623	0.623	0.524	0.787
10	6.05	0.768	0.384	0.623	0.623	0.524	0.787

Having found this relation, the radiation intensity may be obtained from equations (8.114), (8.117), and (8.118) to within a multiplicative constant y_0.

We shall now apply these results to the case of the phase function (1.16). The exact relation between λ and k for this case was found earlier and presented in Table 2.2. Since for this phase function $x_1 = 3g$, equation (8.113) yields

$$v = \tfrac{1}{2}(1-g). \qquad (8.126)$$

This means that the approximate relation between λ and k presented in Table 8.1 corresponds to values of g equal to 0.9, 0.8, and 0.7. The two curves in Fig. 8.2 express the exact relation

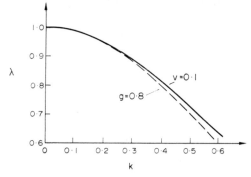

FIG. 8.2. The dependence of single scattering albedo λ on the parameter k as computed exactly for a Henyey-Greenstein phase function with $g = 0.8$ (dashed curve) and approximately, from the data in Table 8.1 for $v = 0.1$.

between λ and k for $g = 0.8$ and the approximate relation for $v = 0.1$. We see that for values of λ close to 1, the two curves are close to each other. Consequently, utilization of equation (8.107) leads in the present case to satisfactory results. We may expect that similar results will also be obtained in the solution of other problems with the aid of equation (8.107).

We should note that use of the so-called "small-angle approximation" (see, for example, [8]) has been proposed for the study of multiple scattering of particles in the case when the phase function is strongly elongated. L. M. Romanova [9] has developed a method for solving the equation of radiative transfer which consists of finding the difference between the solutions obtained using the small-angle approximation and the exact solution. Yu. N. Gnedin and A. Z. Dolginov [10] have also considered the theory of multiple scattering of particles and photons. They found an approximate solution to the transfer equation in explicit form when the phase function has a sharp maximum for small angles of scattering.

References

 1. V. V. SOBOLEV, The Milne problem for anisotropic scattering of radiation, *Astron. Zh.* **47**, 246 (1970) [*Sov. Astron. A.J.* **14**, 202 (1970)].
 2. H. C. VAN DE HULST and K. GROSSMAN, Multiple light scattering in planetary atmospheres, *The Atmospheres of Venus and Mars*, Gordon & Breach, New York, 1968.
 3. V. V. SOBOLEV, The approximate solution to the problem of light scattering in a medium with an arbitrary phase function, *Astron. Zh.* **20**, No. 5–6 (1943).
 4. O. A. AVASTE and V. S. ATROSHENKO, On the precision of V. V. Sobolev's method, *Izv. Akad. Nauk SSSR, Ser. Geofiz.*, No. 3, 507 (1960) [*Bull. (Izv.) Acad. Sci. USSR, Geophys. Ser.*, No. 3, 335 (1960)].
 5. I. W. BUSBRIDGE and S. E. ORCHARD, Reflection and transmission of light by a thick atmosphere according to a phase function $1 + x \cos \theta$, *Astrophys. J.* **149**, 655 (1967).
 6. V. V. SOBOLEV, The diffusion of radiation for highly anisotropic scattering, *Dokl. Akad. Nauk SSSR* **177**, 812 (1967) [*Sov. Physics—Doklady* **12**, 1083 (1968)].
 7. P. M. MORSE and H. FESHBACH, *Methods of Theoretical Physics*, McGraw-Hill, New York, 1953.
 8. M. C. WANG and E. GUTH, On the theory of multiple scattering, particulary of charged particles, *Phys. Rev.* **84**, 1092 (1951).
 9. L. M. ROMANOVA, Solution of the equation of radiative transfer when the phase function is strongly anisotropic, *Optika i Spektroskopiya* **13**, 429 (1962); **13**, 819 (1962) [*Optics & Spectroscopy* **13**, 238 (1962); **13**, 463 (1962)].
10. Yu. N. GNEDIN and A. Z. DOLGINOV, The theory of multiple scattering, *Zh. Eksp. Teor. Fiz.* **45**, 1136 (1963); **48**, 548 (1965) [*Sov. Physics JETP* **18**, 784 (1964); **21**, 364 (1965)].

Chapter 9

THE RADIATION EMERGING FROM A PLANET

IN ALL the preceding chapters we have examined the theory of light scattering in a plane layer illuminated by parallel radiation. Although this layer was frequently referred to as a "planetary atmosphere", the results obtained clearly have a more general significance, since they may be applied in other situations. We shall now apply them specifically to the objects of interest to us, the planets. More precisely, we shall use the previous results to find quantities which characterize the radiation from a planet as it depends upon the optical properties of the atmosphere and surface. By comparing theory and observation, we may then deduce these optical properties.

We begin this chapter by giving expressions for the quantities which are normally obtained through photometry of the disc of a planet. Section 9.1 presents equations giving the distribution of brightness across a planetary disc as a function of position of the Sun and Earth. The change of the total brightness of the planet as a function of changing phase angle is discussed in Section 9.2.

The theory of formation of planetary spectra is considered in the two following sections. In order to find the intensity and shape of absorption lines, we may use the previously obtained equations which determine the intensity diffusely reflected by the atmosphere. In these equations it is only necessary to assume that within the frequency range characteristic of a line, scattering takes place on aerosols, while true absorption is produced by both aerosols and molecules. Essential input to the theory is then the dependence on frequency of the absorption coefficient for molecules, as well as the relation between the number of molecules and the number of aerosol particles as a function of altitude in the atmosphere.

In the last section we briefly examine the problem of the polarization of light from a planet. This problem is very complicated, and a sufficiently complete solution exists at the present time only for the case of Rayleigh scattering. We have not included polarization in the equation of radiative transfer considered in previous chapters. We shall now give approximate expressions for the degree of polarization of light reflected by a planet.

The application of the equations obtained in this chapter to the determination of the optical properties of the atmospheres of the planets will be made in the following chapter.

9.1. The Distribution of Brightness Across a Planetary Disc

An observer on the Earth sees only that part of a planetary disc which is illuminated by the Sun. This part depends upon the phase angle α, which is the angle at the planet between the directions to the Sun and Earth. When $\alpha = 0$ we may see the whole disc, while with increasing α the visible part of the disc decreases, and for $\alpha = 180°$ the disc is completely invisible. Only for Mercury and Venus, however, is the full range $0 \leq \alpha \leq 180°$ observable from Earth. For the other planets, α oscillates between $0°$ and some limiting value ($47°$ for

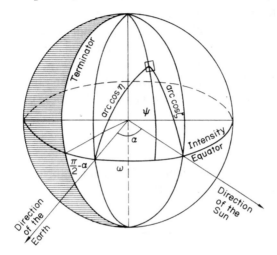

FIG. 9.1. The relation of the planetocentric coordinates ψ and ω to the angles of incidence arccos ζ and of reflection arccos η for a phase angle α.

Mars, $12°$ for Jupiter, etc.). It is clear that when the observable range of phase angles for a planet is large, more knowledge about the atmosphere may be obtained from photometric data. From this point of view the most favorable situation occurs for Venus, since for the planets more distant than Mars almost the whole disc is always visible. The methods and results of planetary photometry are described in detail in the books of V. V. Sharonov [1] and N. N. Sytinskaya [2].

In planetary photometry the location of a point on the disc is usually described by the planetocentric coordinates ψ and ω (Fig. 9.1). The angle ψ is the latitude computed from the intensity equator (the great circle passing through the sub-solar and sub-earth points), and ω is the corresponding longitude measured from the sub-Earth point. For a given phase angle α it is easy to find for every point the angle of incidence of solar radiation and the angle of reflection in the direction of a terrestrial observer, and also the difference of azimuth between the reflected and incident rays in the local horizontal plane. We have heretofore designated these angles arccos ζ, arccos η, and φ, respectively. With the aid of Fig. 9.1 we see that

$$\cos \alpha = \eta\zeta - \sqrt{(1-\eta^2)(1-\zeta^2)} \cos \varphi, \tag{9.1}$$

$$\zeta = \cos \psi \cos (\alpha - \omega), \tag{9.2}$$

$$\eta = \cos \psi \cos \omega. \tag{9.3}$$

The intensity of radiation diffusely reflected by the atmosphere is determined from theory as a function of η, ζ, and φ. We have written it previously in the form

$$I(\eta, \zeta, \varphi) = S\varrho(\eta, \zeta, \varphi)\zeta, \tag{9.4}$$

where πS is the flux through a small area oriented perpendicular to the solar radiation at the upper boundary of the atmosphere and $\varrho(\eta, \zeta, \varphi)$ is the reflection coefficient. Equations and tables for $\varrho(\eta, \zeta, \varphi)$ for various optical properties of the atmosphere have been given in Chapters 2, 3, and 7.

Using the computed values of $I(\eta, \zeta, \varphi)$ and applying equations (9.1)–(9.3), we may find the theoretical brightness distribution across the planetary disc for any phase angle α. As an example, let us find the brightness distribution along the intensity equator for a planet at exact opposition, so that $\alpha = 0$. In this case $\eta = \zeta$ and $\varphi = \pi$. Therefore, we must find the quantity $I(\eta, \eta, \pi)$. We shall assume that the planet has an atmosphere of infinitely large optical thickness. If the phase function is expanded in Legendre polynomials, then

$$I(\eta, \eta, \pi) = S\eta \left[\varrho^0(\eta, \eta) + 2 \sum_{m=1}^{n} (-1)^m \varrho^m(\eta, \eta) \right], \tag{9.5}$$

where, as follows from (2.41), the quantities $\varrho^m(\eta, \eta)$ are expressed in terms of the Ambartsumyan functions $\varphi_i^m(\eta)$ with the aid of the equation

$$\varrho^m(\eta, \eta) = \frac{\lambda}{8\eta} \sum_{i=m}^{n} c_i^m (-1)^{i+m} [\varphi_i^m(\eta)]^2, \tag{9.6}$$

while the coefficients c_i^m are given by equation (1.42). As an illustration, Table 9.1 presents the results of computations of $I(\eta, \eta, \pi)$ for the phase function $x(\gamma) = 1 + \frac{3}{2} \cos \gamma + P_2 (\cos \gamma)$ for various values of the single scattering albedo λ.

TABLE 9.1. VALUES OF THE QUANTITY $\dfrac{I(\eta, \eta, \pi)}{I(1, 1, \pi)}$ FOR A THREE-TERM PHASE FUNCTION

η	λ				η	λ			
	0.90	0.95	0.99	1.00		0.90	0.95	0.99	1.00
0	0.196	0.138	0.084	0.054	0.6	0.704	0.642	0.558	0.488
0.1	0.297	0.224	0.148	0.103	0.7	0.780	0.730	0.660	0.598
0.2	0.383	0.304	0.216	0.158	0.8	0.854	0.819	0.768	0.720
0.3	0.466	0.385	0.290	0.224	0.9	0.928	0.909	0.881	0.851
0.4	0.547	0.469	0.373	0.301	1.0	1.000	1.000	1.000	1.000
0.5	0.626	0.555	0.462	0.389	$\dfrac{I(1, 1, \pi)}{S}$	0.288	0.429	0.736	1.153

The table was constructed using values of the auxiliary functions $\varphi_i^m(\eta)$ calculated as discussed in Section 7.5. The quantity

$$\frac{I(\eta, \eta, \pi)}{I(1, 1, \pi)}$$

given in the main part of the table describes the brightness along the intensity equator relative to the brightness at the sub-solar point. The last row of the table contains values of the intensity $I(1, 1, \pi)$ divided by S, which equals the reflection coefficient $\varrho(1, 1, \pi)$ at the center of the planetary disc. It is clear from the table that the brightness at the center of the disc diminishes quickly with increasing true absorption (increasing values of $1-\lambda$), and that the ratio of the brightness at the center to that at the limb also diminishes in this process.

9.2. Dependence of the Planetary Brightness on Phase Angle

If the reflection coefficient $\varrho(\eta, \zeta, \varphi)$ is known, then the energy reaching an observer from the entire planetary disc may be easily determined for any phase angle α. This energy is normally described by the stellar magnitude of the planet m, which is a measure of the energy received from the planet relative to that received from a star of known brightness. Thus, our problem consists in finding the dependence of m on α.

If we represent the intensity emerging from the planet by equation (9.4), we find that the energy crossing an element of area $d\sigma$ into unit solid angle is just $S\varrho(\eta, \zeta, \varphi)\zeta \eta \, d\sigma$. Since $d\sigma = R^2 \cos \psi \, d\psi \, d\omega$, where R is the planetary radius, this energy may be written with the aid of equations (9.2) and (9.3) as

$$R^2 S \varrho(\eta, \zeta, \varphi) \cos (\alpha - \omega) \cos \omega \cos^3 \psi \, d\psi \, d\omega.$$

In order to find the total energy emerging from the planet in the direction of the Earth per unit solid angle, we must integrate the last expression over ψ between $-\pi/2$ and $\pi/2$ and over ω between $\alpha - \pi/2$ and $\pi/2$; that is, from the terminator to the limb. Calling the distance between the planet and the Earth Δ, we obtain for the illumination of the Earth

$$E_p(\alpha) = 2S \frac{R^2}{\Delta^2} H(\alpha), \tag{9.7}$$

where we have set

$$H(\alpha) = \int_{\alpha-\pi/2}^{\pi/2} \cos (\alpha - \omega) \cos \omega \, d\omega \int_0^{\pi/2} \varrho(\eta, \zeta, \varphi) \cos^3 \psi \, d\psi. \tag{9.8}$$

Equations (9.1)–(9.3) must be used to carry out the integration in (9.8).

The illumination of the Earth by the Sun is clearly

$$E_s = \pi S \left(\frac{r_1}{r_2}\right)^2, \tag{9.9}$$

where r_1 is the distance from the Sun to the planet and r_2 is the distance from the Sun to the Earth. Since

$$\frac{E_p}{E_s} = 2.512^{m_\odot - m}, \tag{9.10}$$

where $m_\odot$ is the stellar magnitude of the Sun, we obtain

$$2.512^{m_\odot - m} = \frac{2}{\pi} \left(\frac{r_2 R}{r_1 \Delta}\right)^2 H(\alpha). \tag{9.11}$$

13*

Equation (9.11) gives the desired theoretical dependence of the stellar magnitude of the planet m on phase angle α. The form of this dependence will be determined by the optical properties of the planetary atmosphere. Some knowledge of these properties may be obtained by comparing the theoretical and observational dependence of m on α. In the next chapter we shall do this for Venus.

We shall now present a number of further relations which are required in the photometric study of planets. First of all, from relation (9.7) we have

$$E_p(\alpha) = E_p(0)\Phi(\alpha), \tag{9.12}$$

where we have set

$$\Phi(\alpha) = \frac{H(\alpha)}{H(0)}. \tag{9.13}$$

The quantity $\Phi(\alpha)$ is usually called the phase function (this is a different use of the term from that employed throughout the earlier chapters of this book). Equation (9.12) assumes that the flux from the planet for each α has been corrected to the same distance Δ of the planet from the Earth. Multiplying (9.12) by $2\pi\Delta^2 \sin \alpha \, d\alpha$ and integrating over α between 0 and π, we obtain the total energy radiated by the planet in all directions,

$$2\pi\Delta^2 E_p(0) \int_0^\pi \Phi(\alpha) \sin \alpha \, d\alpha.$$

The corresponding energy incident on the planet from the Sun is $\pi S\pi R^2$. Dividing the first of these quantities by the second, we obtain the spherical albedo of the planet A_s Consequently,

$$A_s = q \frac{E_P(0) \Delta^2}{\pi S R^2}, \tag{9.14}$$

where

$$q = 2 \int_0^\pi \Phi(\alpha) \sin \alpha \, d\alpha. \tag{9.15}$$

The quantity q is called the phase integral.

In addition to the spherical albedo A_s, the geometric albedo A_g is used in the study of planets. The latter quantity is the ratio of the illumination at the Earth from the planet for phase angle $\alpha = 0$ to the illumination produced by a plane, absolutely white Lambert surface of the same radius as the planet placed at the same position. Since it follows from (9.7) that

$$E_P(0) = 2S \frac{R^2}{\Delta^2} H(0), \tag{9.16}$$

and the illumination from the corresponding Lambert disc is

$$E_L = \pi S \frac{R^2}{\Delta^2}, \tag{9.17}$$

we find for the geometric albedo

$$A_g = \frac{E_p(0)}{E_L} = \frac{2}{\pi} H(0).$$ (9.18)

This formula for A_g may also be written as

$$A_g = \frac{E_p(0) \, \Delta^2}{\pi S R^2}.$$ (9.19)

Comparing (9.19) with (9.14), we obtain the following relation between the quantities A and A_g:

$$A_s = q A_g.$$ (9.20)

If the planet reflects light according to Lambert's law, so that $\varrho = $ constant, it follows from (9.13) and (9.8) that

$$\Phi(\alpha) = \frac{1}{\pi} [\sin \alpha + (\pi - \alpha) \cos \alpha].$$ (9.21)

Substitution of (9.21) into (9.15) yields $q = \frac{3}{2}$.

Since we observe the phase angle change between 0 and 180° only for Mercury and Venus, it is only for these planets that we can determine directly the phase integral q and the spherical albedo A_s. Observations of the other planets give only the geometric albedo A_g, and the quantity A_s must then be determined from equation (9.20) on the basis of some assumption for the value of the phase integral. Hopefully, future spacecraft observations will provide data over a wider range of phase angles.

In order to estimate a value for the phase integral, we may use the results of theoretical computations for various optical properties of the atmosphere. For this purpose it is necessary to compute A_g and A_s for given atmospheric properties and then use equation (9.20) to find q.

If the reflection coefficient of the atmosphere is known, the spherical albedo is given by

$$A_s = 4 \int_0^1 \eta \, d\eta \int_0^1 \varrho^0(\eta, \zeta) \zeta \, d\zeta,$$ (9.22)

where $\varrho^0(\eta, \zeta)$ is the reflection coefficient averaged over azimuth (see Section 1.5). To find a corresponding geometric albedo we may use equation (9.18), in which we must substitute equation (9.8) for $\alpha = 0$. Since in that case $\eta = \zeta$ and $\varphi = \pi$, we have

$$A_g = \frac{2}{\pi} \int_{-\pi/2}^{\pi/2} \cos^2 \omega \, d\omega \int_0^{\pi/2} \varrho(\eta, \eta, \pi) \cos^3 \psi \, d\psi.$$ (9.23)

Making use of equation (9.3) and transforming from integration over ω and ψ to integration over η, we obtain

$$A_g = 2 \int_0^1 \varrho(\eta, \eta, \pi) \eta^2 \, d\eta.$$ (9.24)

As an example, Table 9.2 presents values of the spherical albedo A_s, the geometric albedo A_g, and the phase integral q as functions of λ (the single scattering albedo) for two types of three-term phase functions: (A) $x_1 = 1$, $x_2 = 1$, (B) $x_1 = \frac{3}{2}$, $x_2 = 1$. The table was compiled by V. M. Loskutov.

TABLE 9.2. VALUES OF THE QUANTITIES A_s, A_g AND q FOR A THREE-TERM PHASE FUNCTION

λ	A			B		
	A_s	A_g	q	A_s	A_g	q
0.90	0.410	0.305	1.35	0.360	0.216	1.67
0.95	0.534	0.394	1.36	0.486	0.302	1.61
0.99	0.755	0.556	1.36	0.724	0.473	1.53
1.00	1.000	0.743	1.35	1.000	0.683	1.46

It is clear from the table that the phase integral q depends rather strongly on the phase function.

9.3. Planetary Spectra for Different Points on the Disc

It is well known that planetary spectra exhibit absorption bands corresponding to different molecules. These bands have a very complicated structure and consist of separate, often overlapping, lines. The character of such spectra is well described in the book by R. Goody [3].

Spectra of planets are essentially different in their origin from stellar spectra. We know that stellar spectra arise, roughly speaking, as a result of the diffuse *transmission* of radiation through a layer of gas. The formation of planetary spectra results from diffuse *reflection* of radiation by a medium containing both gas and aerosols.

Photons incident on a planetary atmosphere are scattered by aerosol particles and molecules. At the same time, these photons may undergo true absorption, either immediately upon entering the atmosphere, or after a number of scatterings. The true absorption in the continuous spectrum is caused by the aerosol particles, and in the lines by both these particles and molecules.

As a result, the following quantities must be given in the theory of absorption line formation in planetary spectra: the true absorption coefficient for molecules $\varkappa_v$, the true absorption coefficient for aerosols $\varkappa$, the scattering coefficient for molecules plus aerosols σ, and the phase function $x(\gamma)$. Within a line the quantity $\varkappa_v$ depends strongly on frequency v, while the three other quantities may be taken as independent of frequency. The quantities $x(\gamma)$, σ, and $\varkappa$ may be found by comparing the theoretical and observational values of the photometric characteristics of the planet in the continuum (see Sections 9.1 and 9.2).

Generally speaking, the theory of planetary spectra is rather complicated. For simplicity we shall now examine isolated spectral lines. With the aid of certain assumptions, we may use the solution to the problem of diffuse reflection by an atmosphere given in previous chapters to determine the line profile and equivalent width (see [4]).

Let $I(\eta, \zeta, \varphi)$ be the intensity diffusely reflected by the atmosphere in the continuous spectrum close to the line. This intensity depends upon the phase function $x(\gamma)$, the single scattering albedo λ, the optical thickness of the atmosphere τ_0 and the albedo of the surface a. We recall that the optical thickness of the atmosphere is given by

$$\tau_0 = \int_R^\infty (\sigma + \varkappa)\, dr, \tag{9.25}$$

where R is the radius of the planet, and the albedo of single scattering is

$$\lambda = \frac{\sigma}{\sigma + \varkappa}. \tag{9.26}$$

We shall designate by $I_\nu(\eta, \zeta, \varphi)$ the intensity diffusely reflected by the atmosphere at a frequency ν within the line. This intensity depends on the same four quantities as the intensity $I(\eta, \zeta, \varphi)$, except that the optical thickness of the atmosphere is now

$$\tau_\nu^0 = \int_R^\infty (\sigma + \varkappa + \varkappa_\nu)\, dr, \tag{9.27}$$

and the single scattering albedo is

$$\lambda_\nu = \frac{\sigma}{\sigma + \varkappa + \varkappa_\nu}. \tag{9.28}$$

The continuum intensity $I(\eta, \zeta, \varphi)$ may be expressed in terms of the reflection coefficient $\varrho(\eta, \zeta, \varphi)$ through equation (9.4). In the same manner, we may express the intensity $I_\nu(\eta, \zeta, \varphi)$ as

$$I_\nu(\eta, \zeta, \varphi) = S\varrho_\nu(\eta, \zeta, \varphi)\zeta. \tag{9.29}$$

In the equations for the function $\varrho(\eta, \zeta, \varphi)$ obtained previously (see in particular Sections 2.2 2.6, 3.1–3, and 4.2) it was assumed that the quantities $x(\gamma)$ and λ were constant throughout the atmosphere. These same equations may be used for the function $\varrho_\nu(\eta, \zeta, \varphi)$ if we replace τ_0 by τ_ν^0 and λ by λ_ν, with the additional assumption that $\lambda_\nu = \text{constant}$.

Knowledge of the function $\varrho_\nu(\eta, \zeta, \varphi)$ for all frequencies within the line allows us to construct a line profile. It is usually characterized by the quantity

$$r_\nu(\eta, \zeta, \varphi) = \frac{I_\nu(\eta, \zeta, \varphi)}{I(\eta, \zeta, \varphi)} = \frac{\varrho_\nu(\eta, \zeta, \varphi)}{\varrho(\eta, \zeta, \varphi)}. \tag{9.30}$$

The equivalent width W is used to describe the integrated character of the absorption line. It represents the width of a neighboring part of the continuum whose energy equals the energy absorbed in the line. It is easy to see that

$$W(\eta, \zeta, \varphi) = \int_0^\infty [1 - r_\nu(\eta, \zeta, \varphi)]\, d\nu. \tag{9.31}$$

From equations (9.30) and (9.31) we may compute the profile and equivalent width of an absorption line as a function of position on the planetary disc. In order to transform from the variables η, ζ, and φ to the planetocentric coordinates ω, ψ, and the phase angle α, we must use equations (9.1)–(9.3).

Let us consider two examples of the application of the equations which we have obtained:

1. Let us first assume that the optical thickness of the atmosphere is much less than 1 in both the continuous spectrum and in the line ($\tau_0 \ll 1$ and $\tau_\nu^0 \ll 1$). Then we need include only first order scattering in the determination of ϱ and ϱ_ν. Let us also assume that the planetary surface reflects radiation according to Lambert's law with an albedo a. In this case the surface influences only the reflection coefficient averaged over azimuth, for the determination of which we may use equation (4.31). For the complete reflection coefficient with the inclusion of only first order scattering, we find

$$\varrho(\eta, \zeta, \varphi) = \frac{\lambda}{4} x(\gamma) \frac{1-e^{-\tau_0(1/\eta+1/\zeta)}}{\eta+\zeta} + \frac{aV(\eta, \tau_0) V(\zeta, \tau_0)}{1-aA_s(\tau_0)}. \tag{9.32}$$

The quantities $V(\zeta, \tau_0)$ and $A_s(\tau_0)$ are given by equations (4.28) and (4.29) in terms of the reflection and transmission coefficients averaged over azimuth. If we include in these coefficients only first-order scattering and limit ourselves to terms of first order in τ_0, we obtain

$$V(\zeta, \tau_0) = 1 - \frac{\tau_0}{\zeta}\left[1 - \frac{\lambda}{2}\int_0^1 p(\eta, \zeta)\, d\eta\right], \tag{9.33}$$

$$A_s(\tau_0) = \lambda\tau_0 \int_0^1 d\eta \int_0^1 p(-\eta, \zeta)\, d\zeta, \tag{9.34}$$

where $p(\eta, \zeta)$ is given by equation (2.3). Substituting (9.33) and (9.34) into (9.32) and again retaining only terms of first order in τ_0, we have

$$\varrho(\eta, \zeta, \varphi) = a\left[1-\tau_0\left(\frac{1}{\eta}+\frac{1}{\zeta}\right)\right] + \frac{\lambda}{4} x(\gamma)\frac{\tau_0}{\eta\zeta}$$
$$+a\frac{\lambda\tau_0}{2\zeta}\int_0^1 p(\eta', \zeta)\, d\eta' + a\frac{\lambda\tau_0}{2\eta}\int_0^1 p(\eta, \zeta')\, d\zeta' + a^2\lambda\tau_0\int_0^1 d\eta'\int_0^1 p(-\eta', \zeta')\, d\zeta'. \tag{9.35}$$

It is clear that equation (9.35) may be obtained directly from physical considerations. Each term on the right side of this equation corresponds to one of the following possibilities for solar radiation incident on the atmosphere: (1) radiation is transmitted by the atmosphere, reflected by the surface, and again transmitted through the atmosphere; (2) radiation is scattered by the atmosphere in the direction of the observer; (3) radiation is scattered by the atmosphere in the direction of the surface and reflected by it; (4) radiation is reflected by the surface and scattered by the atmosphere; (5) radiation is reflected by the surface, scattered by the atmosphere, and again reflected by the surface.

An expression analogous to (9.35) may also be written for the reflection coefficient in the spectral line, $\varrho_\nu(\eta, \zeta, \varphi)$. It is only necessary to replace λ by λ_ν and τ_0 by τ_ν^0 in equation (9.35). Recalling from equations (9.25)–(9.28) that

$$\lambda_\nu\tau_\nu^0 = \lambda\tau_0, \tag{9.36}$$

we find

$$\varrho(\eta, \zeta, \varphi) - \varrho_r(\eta, \zeta, \varphi) = a(\tau^0_\nu - \tau_0)\left(\frac{1}{\eta} + \frac{1}{\zeta}\right). \tag{9.37}$$

Dividing (9.37) by (9.35) and assuming that a is large in comparison with τ_0, we obtain

$$1 - r_\nu(\eta, \zeta, \varphi) = (\tau^0_\nu - \tau_0)\left(\frac{1}{\eta} + \frac{1}{\zeta}\right), \tag{9.38}$$

which determines the absorption line profile under the given assumptions.

It is clear that the quantity $\tau^0_\nu - \tau_0$ represents the optical thickness of the atmosphere resulting from true absorption by the molecules. Since the volume coefficient of molecular absorption may be written as $\varkappa_\nu = k_\nu n$, where k_ν is the absorption coefficient for a single molecule and n is the number of molecules per cubic centimeter, we have

$$\tau^0_\nu - \tau_0 = \int\limits_R^\infty \varkappa_\nu \, dr = k_\nu \int\limits_R^\infty n \, dr = k_\nu N, \tag{9.39}$$

where N is the number of molecules in a column of unit cross-sectional area above the surface of the planet. Substitution of (9.39) into (9.38) yields

$$1 - r_\nu(\eta, \zeta, \varphi) = k_\nu N \left(\frac{1}{\eta} + \frac{1}{\zeta}\right). \tag{9.40}$$

Equation (9.40) shows that in the present case the absorption line is basically formed by photons which are transmitted through the atmosphere, reflected by the planetary surface, and again transmitted through the atmosphere in the direction to the observer. In other words, the line has the same profile as an absorption line formed by transmission of radiation through a layer of gas with optical thickness $k_\nu N(1/\eta + 1/\zeta)$. It must be emphasized that the scattering processes enumerated above (that is, processes 2–5) do not play a significant role in the formation of the line.

Substituting (9.40) into (9.31), we obtain the following expression for the equivalent width of the absorption line:

$$W = N\left(\frac{1}{\eta} + \frac{1}{\zeta}\right)\int\limits_0^\infty k_\nu \, d\nu. \tag{9.41}$$

This equation may be used to compute the equivalent width W at different points on the planetary disc for a given value of the phase angle α. If, for example, $\alpha = 0$, then $\zeta = \eta$ and the equivalent width of the line increases from the center to the edge of the disc as $W \propto 1/\eta$. By using equation (9.41) and the observed values of W, we may determine the number of molecules N in the atmosphere of the planet.

2. Let us assume that the optical thickness of the atmosphere is very large (we shall take $\tau_0 = \infty$) and that the role of true absorption is small in both the continuum and in the line

$(1-\lambda \ll 1$ and $1-\lambda_{\nu} \ll 1)$. In this case, as in the preceding one, the absorption line will be weak.

The reflection coefficient in the continuum for small true absorption is given by the following asymptotic expression:

$$\varrho(\eta, \zeta, \varphi) = \varrho_0(\eta, \zeta, \varphi) - 4 \sqrt{\frac{1-\lambda}{3-x_1}}\, u_0(\eta)\, u_0(\zeta), \qquad (9.42)$$

where $\varrho_0(\eta, \zeta, \varphi)$ and $u_0(\eta)$ are the reflection coefficient and relative transmission coefficient of the semi-infinite atmosphere for $\lambda = 1$. In equation (9.42) we have retained the term of order $\sqrt{1-\lambda}$ and dropped terms of order $1-\lambda$ or higher. In this approximation the azimuthal dependence of $\varrho(\eta, \zeta, \varphi)$ enters only through the zeroth-order term of the expansion in powers of $\sqrt{1-\lambda}$, since the first-order term of this expansion does not depend upon azimuth. As a result, the term of order $\sqrt{1-\lambda}$ in equation (9.42) may be written on the basis of equation (2.129), which gives the reflection coefficient averaged over azimuth.

An equation for the quantity $\varrho_{\nu}(\eta, \zeta, \varphi)$ may be obtained from (9.42) by replacing λ by λ_{ν}. We then find, recalling equation (9.30), that

$$1 - r_{\nu}(\eta, \zeta, \varphi) = C(\eta, \zeta, \varphi)\, \{\sqrt{1-\lambda_{\nu}} - \sqrt{1-\lambda}\}, \qquad (9.43)$$

where

$$C(\eta, \zeta, \varphi) = \frac{4}{\sqrt{3-x_1}}\, \frac{u_0(\eta)\, u_0(\zeta)}{\varrho_0(\eta, \zeta, \varphi)}. \qquad (9.44)$$

We see that the quantity $1 - r_{\nu}(\eta, \zeta, \varphi)$ consists of two factors, one of which depends only on the location on the disc, and the second only on frequency. Consequently, as a function of frequency, the line has the same shape (in relative units) at every point on the disc.

Substituting (9.43) into (9.31), we find the following expression for the equivalent width of the line:

$$W(\eta, \zeta, \varphi) = C(\eta, \zeta, \varphi) \int_0^\infty \{\sqrt{1-\lambda_{\nu}} - \sqrt{1-\lambda}\}\, d\nu. \qquad (9.45)$$

In order to determine the change of W with position on the disc, we must find the function $C(\eta, \zeta, \varphi)$. This function depends only on the phase function $x(\gamma)$ which we have thus far considered to be arbitrary.

Let us take isotropic scattering as an example. The quantities $u_0(\eta)$ and $\varrho_0(\eta, \zeta)$ are then expressed in terms of the Ambartsumyan function $\varphi(\eta)$ by equations (2.150) and (2.43) for $\lambda = 1$. It then follows from equation (9.44) that

$$C(\eta, \zeta) = \sqrt{3}\,(\eta + \zeta). \qquad (9.46)$$

We may also easily find the function $C(\eta, \zeta, \varphi)$ for a three-term phase function by using equation (9.44), since the quantities $u_0(\eta)$ and $\varrho_0(\eta, \zeta, \varphi)$ for this phase function have already been determined (see Sections 7.4 and 7.5).

Table 9.3 presents values of the function $C(\eta, \eta, \pi)$ which characterizes the variation of the equivalent line width from the center of the disc to the limb for zero phase angle. The first row of the table corresponds to isotropic scattering, and the second and third rows to the three-term phase functions A (with Legendre expansion coefficients $x_1 = 1$, $x_2 = 1$) and $B(x_1 = \frac{3}{2}, x_2 = 1)$.

TABLE 9.3. VALUES OF THE FUNCTION $C(\eta, \eta, \pi)$
FOR THREE PHASE FUNCTIONS

η	0	0.2	0.4	0.6	0.8	1.0
Isotropic scattering	0	0.69	1.39	2.08	2.77	3.46
A	0	0.77	1.57	2.37	3.17	3.96
B	0	1.33	2.30	3.13	3.88	4.57

It is clear from the table that the function $C(\eta, \eta, \pi)$ and, consequently, also the equivalent width of the line W, vanishes at the limb. This behavior is also evident from equation (9.44) for arbitrary phase function, since we always have $\varrho_0(\eta, \eta, \pi) \to \infty$ for $\eta \to 0$.

Thus, the nature of the variation of the equivalent width with position on the disc is very different in the two cases we have considered. In particular, at opposition ($\alpha = 0$), in going from the center of the disc to the limb, W increases in the first case (an atmosphere of small optical thickness) and decreases in the second case (a semi-infinite atmosphere with small true absorption). We must emphasize that the first case may also occur with a semi-infinite atmosphere when a layer of small optical thickness consisting of molecules and aerosol particles lies above a dense cloud layer which acts as a reflecting surface. Consequently, one may determine the structure of the upper layers of the atmosphere (or more precisely, the distribution of molecules and aerosol particles with altitude) from the behavior of absorption lines at various points on the planetary disc.

9.4. Planetary Spectra for Different Phase Angles

In the previous section we determined the profile and equivalent width of an absorption line for an arbitrary position on the planetary disc. We shall now find the profile and equivalent width of a line in the spectrum of the entire planet as a function of phase angle α.

We shall characterize the line profile by the quantity $r_\nu(\alpha)$, which is the ratio of the flux from the planet at the frequency ν within the line to the flux from the planet in a neighboring portion of the continuous spectrum (per unit frequency interval). The flux in the continuous spectrum has been shown above to be given by equation (9.7). Writing the analogous expression for the flux at a frequency ν within the line, we obtain

$$r_\nu(\alpha) = \frac{H_\nu(\alpha)}{H(\alpha)}, \tag{9.47}$$

where $H(\alpha)$ is given by equation (9.8) and

$$H_\nu(\alpha) = \int_{\alpha-\pi/2}^{\pi/2} \cos(\alpha-\omega) \cos\omega \, d\omega \int_0^{\pi/2} \varrho_\nu(\eta, \zeta, \varphi) \cos^3\psi \, d\psi. \tag{9.48}$$

With the aid of (9.47), (9.8), and (9.48) we then find

$$1 - r_\nu(\alpha) = \frac{1}{H(\alpha)} \int_{\alpha - \pi/2}^{\pi/2} \cos(\alpha - \omega) \cos \omega \, d\omega \int_0^{\pi/2} [1 - r_\nu(\eta, \zeta, \varphi)] \varrho(\eta, \zeta, \varphi) \cos^3 \psi \, d\psi, \quad (9.49)$$

where $r_\nu(\eta, \zeta, \varphi)$ is given by equation (9.30).

For the equivalent width of the line, we have in analogy to (9.31)

$$W(\alpha) = \int_0^\infty [1 - r_\nu(\alpha)] \, d\nu. \quad (9.50)$$

Substituting (9.49) into (9.50) and using (9.31), we obtain

$$W(\alpha) = \frac{1}{H(\alpha)} \int_{\alpha - \pi/2}^{\pi/2} \cos(\alpha - \omega) \cos^2 \omega \, d\omega \int_0^{\pi/2} W(\eta, \zeta, \varphi) \varrho(\eta, \zeta, \varphi) \cos^3 \psi \, d\psi. \quad (9.51)$$

Thus, the computation of $r_\nu(\alpha)$ and $W(\alpha)$ from equations (9.49) and (9.51) requires knowledge of $r_\nu(\eta, \zeta, \varphi)$ and $W(\eta, \zeta, \varphi)$, respectively, and also the reflection coefficient in the continuum $\varrho(\eta, \zeta, \varphi)$. We shall now apply these equations to the two cases considered in the preceding section.

1. Let the optical thickness of the atmosphere be small in both the continuous spectrum and in the line. Then $r_\nu(\eta, \zeta, \varphi)$ is given by equation (9.40) and the reflection coefficient in the continuum, on the basis of (9.35), is $\varrho(\eta, \zeta, \varphi) = a$, provided that a is much larger than τ_0. Substituting these values of $r_\nu(\eta, \zeta, \varphi)$ and $\varrho(\eta, \zeta, \varphi)$ into equation (9.49) and carrying out the integration, we obtain

$$1 - r_\nu(\alpha) = \tfrac{3}{2}\pi \frac{1 + \cos \alpha}{\sin \alpha + (\pi - \alpha) \cos \alpha} k_\nu N. \quad (9.52)$$

Introducing equation (9.52) into (9.50), we have

$$W(\alpha) = \tfrac{3}{2}\pi \frac{1 + \cos \alpha}{\sin \alpha + (\pi - \alpha) \cos \alpha} N \int_0^\infty k_\nu \, d\nu. \quad (9.53)$$

Equations (9.52) and (9.53) determine the line profile and equivalent width at a given phase angle α.

The quantity k_ν entering these equations is the absorption coefficient per molecule. As an example, we may consider the Lorentz profile

$$k_\nu = \frac{k_0}{1 + \xi^2}, \quad (9.54)$$

where k_0 is the absorption coefficient at the center of the line and ξ is the dimensionless frequency parameter $\xi = (\nu - \nu_0)/\Delta\nu$, with ν_0 equal to the frequency at the line center and $\Delta\nu$ equal to the natural half-width. We then obtain for the equivalent width of the line

$$W(\alpha) = \tfrac{3}{2}\pi^2 \frac{1 + \cos \alpha}{\sin \alpha + (\pi - \alpha) \cos \alpha} N k_0 \, \Delta\nu. \quad (9.55)$$

2. Let us assume that the optical thickness of the atmosphere is infinitely large and that the role of true absorption is small in both the continuous spectrum and in the line. In this case the line profile for given α is determined by substituting equation (9.43) into (9.49). Use of equation (9.44) and replacement of ϱ by ϱ_0 then gives

$$1-r_\nu(\alpha) = \{\sqrt{1-\lambda_\nu}-\sqrt{1-\lambda}\}$$

$$\times \frac{4}{H(\alpha)\sqrt{3-x_1}} \int_{\alpha-\pi/2}^{\pi/2} \cos(\alpha-\omega) \cos \omega \, d\omega \int_0^{\pi/2} u_0(\eta) u_0(\zeta) \cos^3 \psi \, d\psi. \qquad (9.56)$$

Substitution of (9.56) into (9.50) yields

$$W(\alpha) = Q \frac{4}{H(\alpha)\sqrt{3-x_1}} \int_{\alpha-\pi/2}^{\pi/2} \cos(\alpha-\omega) \cos \omega \, d\omega \int_0^{\pi/2} u_0(\eta) u_0(\zeta) \cos^3 \psi \, d\psi, \qquad (9.57)$$

where we have set

$$Q = \int_0^\infty \{\sqrt{1-\lambda_\nu}-\sqrt{1-\lambda}\} \, d\nu. \qquad (9.58)$$

In order to find $r_\nu(\alpha)$ and $W(\alpha)$ from equations (9.56) and (9.57), we must know the functions $\varrho_0(\eta, \zeta, \varphi)$ and $u_0(\eta)$, the reflection coefficient and relative transmission coefficient of the semi-infinite atmosphere in the case of pure scattering (the first of these functions enters equation (9.8) for $H(\alpha)$). These functions depend only on the phase function. As a result, the double integrals entering equations (9.56) and (9.57) also depend only on the phase function.

Computation of the quantity Q requires that the dependence of the absorption coefficient k_ν on frequency be given. As in the previous case, we shall assume that this dependence is given by equation (9.54). Using equations (9.26) and (9.28), we then obtain

$$1-\lambda_\nu = 1-\lambda+\frac{\lambda b}{1+\xi^2+b}, \qquad (9.59)$$

where we have set

$$b = \frac{\varkappa_0}{\sigma+\varkappa}, \qquad (9.60)$$

and $\varkappa_0$ is the molecular absorption coefficient at the center of the line, so that $\varkappa_0 = n k_0$. Substituting (9.59) into (9.58), we obtain

$$Q = 2\Delta\nu \int_0^\infty \left\{\sqrt{1-\lambda+\frac{\lambda b}{1+\xi^2+b}}-\sqrt{1-\lambda}\right\} d\xi. \qquad (9.61)$$

Transforming to a new integration variable $v = \xi/\sqrt{1+b}$ and introducing the notation

$$p = \frac{\lambda b}{(1+b)(1-\lambda)}, \qquad (9.62)$$

we find

$$Q = 2\Delta\nu \sqrt{(1+b)(1-\lambda)} \int_0^\infty \left\{ \sqrt{1 + \frac{p}{1+v^2}} - 1 \right\} dv. \tag{9.63}$$

This equation may also be written in the form

$$Q = 2\Delta\nu \sqrt{(1+b)(1-\lambda)(1+p)} \left[K\left(\sqrt{\frac{p}{1+p}} \right) - E\left(\sqrt{\frac{p}{1+p}} \right) \right], \tag{9.64}$$

where K and E are complete elliptic integrals.

Since we have assumed that $1 - \lambda \ll 1$ and $1 - \lambda_\nu \ll 1$, we must take $b \ll 1$ in the equations we have obtained, so that we may write in place of (9.62)

$$p = \frac{b}{1-\lambda} = \frac{\varkappa_0}{\varkappa}. \tag{9.65}$$

In practice, both small and large values of p are encountered. When $p \ll 1$, then

$$Q = \frac{\pi}{2} \Delta\nu \frac{b}{\sqrt{1-\lambda}}, \tag{9.66}$$

while for $p \gg 1$ we have

$$Q = \Delta\nu \sqrt{b} \ln p. \tag{9.67}$$

We see that the quantity Q depends on the two variables b and $1 - \lambda$. The equivalent width W, which according to (9.57) is proportional to Q, will depend on these same variables. The quantity $1 - \lambda$ may be found from the brightness of the planet in the continuous spectrum. If it is known, the quantity b may be determined by comparing the theoretical and observational values of W. This allows the possibility of finding the relative number of molecules in the atmosphere, since $b = nk_0/(\sigma+\varkappa)$.

Both examples we have considered of the determination of the line profile and equivalent width in planetary spectra refer to weak lines. We have seen that a characteristic feature of these lines is the separation of the dependence upon frequency from that upon angle in the expressions for $1 - r_\nu(\eta, \zeta, \varphi)$. In the case of strong lines such a separation does not occur, and in order to find r_ν and W, it is necessary to know the reflection coefficient $\varrho(\eta, \zeta, \varphi)$ for different values of λ (in order to transform to the quantity $\varrho_\nu(\eta, \zeta, \varphi)$ for different λ_ν).

We must note that until recently astrophysicists have given very little attention to the theory of planetary spectra. A number of interesting papers on this subject have, however, recently appeared. First of all, we must point out the work of Chamberlain [5] who studied in detail the problem of the formation of planetary spectra for isotropic scattering. The work of Uesugi and Irvine [6] is also of considerable interest. These authors determined the diffusely reflected intensity by the method of successive scattering, which consists of finding the coefficients of an expansion of the intensity in powers of λ. Once these coefficients are known, it is easy to find the line profile by replacing λ by λ_ν. For values of λ close to 1, the required expansion converges slowly; the authors circumvented this problem by obtaining asymptotic expressions for the higher terms of the expansion.

9.5. Polarization of Light from a Planet

In our examination of multiple light scattering in previous chapters, we have neglected polarization. Numerical examples show that this procedure has little influence on the total radiation intensity. The determination of the degree of polarization for light emerging from a planetary atmosphere is a very complicated problem. If we had considered it, this book would have greatly increased in size and would have changed in character. We shall therefore examine the problem of polarization of planetary light only briefly.

Up to the present time, this problem has been studied in sufficient detail only for the case of Rayleigh scattering. Investigation of this field was begun by Chandrasekhar (see his book which is cited in Chapter 1) and Sobolev (see *TRT*, Chapter 5). Subsequently, important work was done by Sekera [7], Mullikin [8], Domke [9], and others. In addition, detailed tables were constructed of the functions which characterize the radiation emerging from a Rayleigh scattering atmosphere [10]. The investigation of the transfer of polarized light when scattering by small particles of arbitrary diameter occurs is only now beginning [11–15].

In practice, it is of great significance that the polarization of radiation which is diffusely reflected from a medium is caused in large measure by the single scattering. Thus, to evaluate the degree of polarization approximately, we may divide the difference of the two polarization components of the radiation field computed for single scattering by the total radiation intensity (obtained theoretically, without taking account of polarization, or found from observation). As an illustration of this procedure, consider Table 9.4, which is obtained by solving the problem of diffuse reflection of light by a semi-infinite atmosphere illuminated normally. It has been assumed that the atmosphere scatters according to Rayleigh's law with $\lambda = 1$. The first row of the table gives the exact degree of polarization and the second row gives the approximate values found by the method described here.

TABLE 9.4. EXACT AND APPROXIMATE DEGREE OF POLARIZATION FOR A RAYLEIGH
SCATTERING ATMOSPHERE

η	0	0.1	0.2	0.3	0.4	0.5	0.6	0.7	0.8	0.9	1.0
$p(\eta)$	0.45	0.37	0.31	0.25	0.20	0.16	0.12	0.09	0.06	0.03	0
$p_1(\eta)$	0.30	0.23	0.18	0.15	0.12	0.10	0.08	0.06	0.04	0.02	0

We see that in the present case the approximate values of the degree of polarization $p_1(\eta)$ are approximately two-thirds of the exact values $p(\eta)$. Similar results are obtained from an examination of the tables in reference [10] (with a few exceptions).

Let us introduce a few equations for the quantities characterizing the scattering of first order. Calculations using Mie theory usually give the intensity of radiation $i_1(\gamma)$ and $i_2(\gamma)$ for oscillations of the electric vector perpendicular and parallel to the plane of scattering as functions of the index of refraction m and the parameter $z = 2\pi a/\tilde{\lambda}$, where a is the radius of the particle and $\tilde{\lambda}$ is the wavelength of the radiation. Since an elementary volume of the atmosphere will contain particles of different sizes, the quantities $i_1(\gamma)$ and $i_2(\gamma)$ must be averaged according to the equation

$$I_k(\gamma) = \int_0^\infty i_k(\gamma)\, \Psi(a)\, da \qquad (k = 1, 2),$$ (9.68)

where $\Psi(a)\, da$ is the fraction of particles with radius between a and $a+da$.

The phase function $x(\gamma)$ is proportional to the sum $I_1(\gamma)+I_2(\gamma)$, so that

$$x(\gamma) = C[I_1(\gamma)+I_2(\gamma)],$$ (9.69)

where the constant C may be found from the normalization condition for the phase function. After C has been found, we may determine the quantity

$$y(\gamma) = C[I_1(\gamma)-I_2(\gamma)],$$ (9.70)

which, when divided by $x(\gamma)$, gives the fraction of polarization for scattering at an angle γ.

Table 9.5 presents as an example values of the quantities $x(\gamma)$ and $y(\gamma)$ for an index of refraction $m = 1.38$ (computations by V. M. Loskutov). The size distribution function for the particles was taken to be of the form (1.7) with $n = 2$.

TABLE 9.5. VALUES OF THE QUANTITIES $x(\gamma)$ AND $y(\gamma)$ FOR $m = 1.38$

z	1		2		5		10		20	
γ	x	y	x	y	x	y	x	y	x	y
0	7.81	0	18.6	0	52.2	0	188	0	715	0
10	7.24	0.05	15.0	−0.01	20.7	−0.20	15.3	−0.06	9.75	0.15
20	5.82	0.17	8.59	−0.01	6.14	−0.16	4.31	0.00	4.02	0.03
30	4.15	0.28	4.13	0.00	2.85	−0.14	2.33	−0.05	2.26	−0.06
40	2.72	0.33	1.94	0.01	1.51	−0.11	1.34	−0.07	1.27	−0.08
50	1.69	0.33	0.96	0.02	0.84	−0.08	0.78	−0.07	0.70	−0.07
60	1.03	0.29	0.52	0.02	0.49	−0.06	0.46	−0.06	0.39	−0.05
70	0.63	0.24	0.31	0.02	0.30	−0.04	0.28	−0.04	0.22	−0.03
80	0.40	0.20	0.20	0.02	0.20	−0.03	0.18	−0.03	0.13	−0.02
90	0.28	0.15	0.14	0.02	0.15	−0.02	0.13	−0.01	0.08	−0.01
100	0.20	0.12	0.10	0.02	0.12	−0.01	0.10	0.00	0.06	0.00
110	0.16	0.09	0.09	0.01	0.10	−0.01	0.08	0.00	0.06	0.02
120	0.14	0.06	0.08	0.01	0.10	−0.01	0.07	0.00	0.05	0.01
130	0.13	0.04	0.09	0.01	0.12	−0.01	0.09	−0.01	0.05	0.04
140	0.14	0.02	0.10	−0.05	0.16	0.00	0.16	0.00	0.12	0.02
150	0.15	0.01	0.11	−0.01	0.24	0.03	0.30	0.08	0.32	0.20
160	0.16	0	0.12	−0.03	0.35	0.07	0.36	0.14	0.26	0.03
170	0.18	0	0.13	−0.02	0.27	0.01	0.53	−0.01	0.40	−0.06
180	0.19	0	0.14	0	0.32	0	0.75	0	0.84	0

Knowledge of the function $y(\gamma)$ provides the possibility of determining (within the approximation described above) the degree of polarization for the radiation emerging from the planet. We shall assume that the atmosphere consists of molecules and aerosol particles, and examine the following two models: (i) the molecules and particles are mixed in a constant ratio throughout the atmosphere; (ii) a molecular layer of small optical thickness is situated above a cloud layer consisting of aerosol particles. In both models we will take the optical thickness of the atmosphere as infinitely great.

Model I

If the ratio of the number of molecules to the number of aerosol particles does not change with altitude in the atmosphere, the phase function may be written in the form

$$x(\gamma) = \delta x_M(\gamma) + (1-\delta) x_A(\gamma), \tag{9.71}$$

where $x_M(\gamma)$ and $x_A(\gamma)$ are the phase functions for molecular and particulate scattering, respectively, and $\delta = $ constant. The quantity δ is obviously the ratio of the molecular scattering coefficient to the total scattering coefficient. The phase function $x_M(\gamma)$ is, of course, given by the expression

$$x_M(\gamma) = \tfrac{3}{4}(1+\cos^2 \gamma), \tag{9.72}$$

while $x_A(\gamma)$ must be computed in the manner described above.

The quantity $y(\gamma)$, representing the polarized portion of the phase function, may be written in a manner analogous to (9.71), so that

$$y(\gamma) = \delta \tfrac{3}{4} \sin^2 \gamma + (1-\delta) y_A(\gamma). \tag{9.73}$$

The reflection coefficient for first-order scattering in a semi-infinite atmosphere with $\lambda = 1$ is

$$\varrho_1 = \frac{x(\gamma)}{4(\eta+\zeta)}. \tag{9.74}$$

If we replace $x(\gamma)$ by $y(\gamma)$, we obtain an expression determining the polarized portion of ϱ_1. But, according to our assumptions, this latter quantity characterizes also the polarized portion of the total reflection coefficient ϱ. We thus find for the degree of polarization of the radiation emerging from a given location in the atmosphere

$$p(\eta, \zeta, \varphi) = \frac{y(\gamma)}{4(\eta+\zeta)\, \varrho(\eta, \zeta, \varphi)}. \tag{9.75}$$

In order to determine the degree of polarization for radiation emerging from the entire planetary disc at phase angle α, it is necessary to substitute the quantity $y(\pi-\alpha)/4(\eta+\zeta)$ into (9.8) in place of $\varrho(\eta, \zeta, \varphi)$ and divide the resulting expression by $H(\alpha)$. This procedure gives

$$p(\alpha) = \frac{y(\pi-\alpha)}{4H(\alpha)} \int_{\alpha-\pi/2}^{\pi/2} \frac{\cos(\alpha-\omega)\cos\omega}{\cos(\alpha-\omega)+\cos\omega}\, d\omega \int_0^{\pi/2} \cos^2 \psi\, d\psi, \tag{9.76}$$

14

where we have used equations (9.2) and (9.3). Taking account of equation (9.73) and introducing the notation

$$F(\alpha) = \frac{1}{4} \int_{\alpha-\pi/2}^{\pi/2} \frac{\cos(\alpha-\omega)\cos\omega}{\cos(\alpha-\omega)+\cos\omega} \, d\omega \int_0^{\pi/2} \cos^2\psi \, d\psi$$

$$= \frac{\pi}{16}\left(1-\sin\frac{\alpha}{2}\tan\frac{\alpha}{2}\ln\cot\frac{\alpha}{4}\right), \tag{9.77}$$

we obtain in place of (9.76)

$$p(\alpha) = \frac{F(\alpha)}{H(\alpha)}[\delta\tfrac{3}{4}\sin^2\alpha+(1-\delta)y_A(\pi-\alpha)]. \tag{9.78}$$

The degree of polarization for Model I is thus determined by equations (9.75) and (9.78). The quantities $\varrho(\eta, \zeta, \varphi)$ and $H(\alpha)$ entering these equations may be found theoretically or from observation. In particular, equation (9.11) may be used to find $H(\alpha)$ from the observational data.

Model II

We assume that a purely molecular layer of optical thickness τ_0 lies on top of a cloudy layer. In this case the phase function will have the form (9.72) for optical depths between 0 and τ_0, and at greater depths will correspond to scattering by aerosol particles.

Limiting ourselves to first-order scattering, we find the following expression for the reflection coefficient:

$$\varrho_1 = \frac{1}{4(\eta+\zeta)}\left\{x_M(\gamma)[1-e^{-\tau_0(1/\eta+1/\zeta)}]+x_A(\gamma)e^{-\tau_0(1/\eta+1/\zeta)}\right\}, \tag{9.79}$$

where we have taken approximately $\lambda = 1$ in the second term. If we replace $x_M(\gamma)$ by $\frac{3}{4}\sin^2\gamma$ and $x_A(\gamma)$ by $y_A(\gamma)$ in this equation, we obtain the polarized portion of the reflection coefficient. Dividing the latter by the quantity $\varrho(\eta, \zeta, \varphi)$, we find the degree of polarization of the radiation emerging from the given location in the atmosphere as

$$p(\eta, \zeta, \varphi) = \frac{1}{4(\eta+\zeta)\varrho(\eta, \zeta, \varphi)}\left\{\tfrac{3}{4}\sin^2\gamma[1-e^{-\tau_0(1/\eta+1/\zeta)}]+y_A(\gamma)e^{-\tau_0(1/\eta+1/\zeta)}\right\}. \tag{9.80}$$

The degree of polarization of the radiation emerging from the entire planetary disc may be determined by substituting the quantity $p(\eta, \zeta, \varphi)\,\varrho(\eta, \zeta, \varphi)$ from equation (9.80) into equation (9.8) in place of $\varrho(\eta, \zeta, \varphi)$, and dividing the expression so obtained by $H(\alpha)$. Proceeding in this manner, we find

$$p(\alpha) = \frac{1}{H(\alpha)}\left\{\tfrac{3}{4}\sin^2\alpha\,[F(\alpha)-F_1(\tau_0, \alpha)]+y_A(\pi-\alpha)\,F_1(\tau_0, \alpha)\right\}, \tag{9.81}$$

where $F(\alpha)$ is given by equation (9.79) and

$$F_1(\tau_0, \alpha) = \frac{1}{4}\int_{\alpha-\pi/2}^{\pi/2}\frac{\cos(\alpha-\omega)\cos\omega}{\cos(\alpha-\omega)+\cos\omega}\,d\omega\int_0^{\pi/2}e^{-\tau_0(1/\eta+1/\zeta)}\cos^2\psi\,d\psi. \tag{9.82}$$

Equation (9.81) may be rewritten in the form

$$p(\alpha) = \frac{F(\alpha)}{H(\alpha)} \{\delta(\tau_0, \alpha)\tfrac{3}{4}\sin^2\alpha + [1 - \delta(\tau_0, \alpha)]\, y_A(\pi - \alpha)\}, \qquad (9.83)$$

where we have set

$$\delta(\tau_0, \alpha) = 1 - \frac{F_1(\tau_0, \alpha)}{F(\alpha)}. \qquad (9.84)$$

Equation (9.83) determines the degree of polarization for radiation from the planet as a function of phase angle α. Values of the quantity $\delta(\tau_0, \alpha)$ entering (9.83) are given in Table 9.6.

TABLE 9.6. VALUES OF THE FUNCTION $\delta(\tau_0, \alpha)$ ENTERING EQUATION (9.83)

τ_0 \ α	0°	15°	30°	45°	60°	75°	90°	105°	120°	135°	150°	165°	180°
0	0	0	0	0	0	0	0	0	0	0	0	0	0
0.01	0.03	0.03	0.03	0.04	0.04	0.04	0.05	0.05	0.06	0.08	0.12	0.22	1
0.02	0.07	0.07	0.07	0.07	0.08	0.08	0.09	0.10	0.12	0.16	0.22	0.38	1
0.03	0.10	0.10	0.10	0.10	0.11	0.12	0.13	0.15	0.18	0.22	0.31	0.50	1
0.04	0.13	0.13	0.13	0.13	0.14	0.15	0.17	0.19	0.22	0.28	0.38	0.60	1
0.05	0.15	0.16	0.16	0.16	0.17	0.18	0.20	0.23	0.27	0.34	0.45	0.68	1
0.06	0.18	0.18	0.19	0.19	0.20	0.22	0.24	0.27	0.31	0.38	0.51	0.74	1
0.07	0.21	0.21	0.21	0.22	0.23	0.24	0.27	0.30	0.35	0.43	0.56	0.79	1
0.08	0.23	0.24	0.24	0.24	0.25	0.27	0.30	0.34	0.39	0.47	0.60	0.82	1
0.09	0.26	0.26	0.26	0.27	0.28	0.30	0.33	0.37	0.43	0.51	0.64	0.86	1
0.10	0.28	0.28	0.29	0.29	0.31	0.33	0.36	0.40	0.46	0.55	0.68	0.88	1

TABLE 9.7. VALUES OF THE FUNCTIONS $G_0(\alpha, x_1)$, $G_1(\alpha)$ AND $F(\alpha)$

α \ x_1	$G_0(\alpha, x_1)$ 0	1.0	2.0	$G_1(\alpha)$	$F(\alpha)$
15°	0.855	0.925	0.900	4.67	0.187
30	0.773	0.840	0.838	4.18	0.169
45	0.650	0.708	0.731	3.46	0.146
60	0.512	0.561	0.604	2.64	0.122
75	0.375	0.413	0.466	1.84	0.0970
90	0.253	0.281	0.333	1.16	0.0739
105	0.153	0.171	0.213	0.626	0.0527
120	0.082	0.092	0.123	0.296	0.0345
135	0.034	0.040	0.056	0.107	0.0198
150	0.010	0.012	0.018	0.026	0.0089
165	0.0013	0.0016	0.0026	0.0024	0.0022

Values of the function $F(\alpha)$ are given in Table 9.7, which also includes two functions to be introduced in the following chapter. The function $H(\alpha)$, just as with Model I, may be found either theoretically or from observation.

14*

We see that equation (9.78) which determines $p(\alpha)$ for Model I is formally the same as equation (9.83) for Model II. However, in the first case $\delta = $ constant, while in the second case δ depends on τ_0 and α.

The formulas derived in this section for the degree of polarization of the light from a planet will be applied in the next chapter to the interpretation of polarimetric observations of Venus and Jupiter.

References

1. V. V. SHARONOV, *Nature of the Planets*, Fizmatgiz, Moscow, 1958.
2. N. N. SYTINSKAYA, *Absolute Photometry of Extended Celestial Objects*, Izd. Leningrad Univ., Leningrad, 1948.
3. R. M. GOODY, *Atmospheric Radiation*, Clarendon Press, Oxford, 1964.
4. V. V. SOBOLEV, On planetary spectra, *Astron. Zh.* **49**, 397 (1972) [*Sov. Astron. A.J.* **16**, 324 (1972)].
5. J. W. CHAMBERLAIN, Behavior of absorption lines in a hazy planetary atmosphere, *Astrophys. J.* **159**, 137 (1970).
6. A. UESUGI and W. M. IRVINE, Multiple scattering in a plane-parallel atmosphere, *Astrophys. J.* **159**, 127 (1970); **161**, 243 (1970).
7. Z. SEKERA, Radiative transfer in a planetary atmosphere with imperfect scattering, RAND Corporation, R-413-PR, 1963.
8. T. W. MULLIKIN, The complete Rayleigh-scattered field within a homogeneous plane-parallel atmosphere, *Astrophys. J.* **145**, 886 (1966).
9. H. DOMKE, Radiative transfer for Rayleigh scattering, *Astron. Zh.* **48**, 341 (1971); **48**, 777 (1971) [*Sov. Astron. A.J.* **15**, 266 (1971); **15**, 616 (1972)].
10. K. COULSON, J. DAVE and Z. SEKERA, *Tables Related to Radiation Emerging from Planetary Atmosphere with Rayleigh Scattering*, Univ. Calif. Press, 1960.
11. Z. SEKERA, Reciprocity relations for diffuse reflection and transmission of radiative transfer in the planetary atmosphere, *Astrophys. J.* **162**, 3 (1970).
12. J. W. HOVENIER, Multiple scattering of polarized light in planetary atmospheres, *Astronomy and Astrophysics* **13**, 7 (1971).
13. J. V. DAVE, Intensity and polarization of the radiation emerging from a plane-parallel atmosphere containing monodispersed aerosols, *Appl. Opt.* **9**, 2673 (1970).
14. G. W. KATTAWAR and G. N. PLASS, Radiance and polarization of light reflected from optically thick clouds, *Appl. Opt.* **10**, 74 (1971).
15. J. E. HANSEN, Multiple scattering of polarized light in planetary atmospheres, *J. Atmos. Sci.* **28**, 120 (1971).

Chapter 10

OPTICAL PROPERTIES
OF PLANETARY ATMOSPHERES

THE equations obtained in the preceding chapters (especially Chapter 9) may be used to determine the optical properties of planetary atmospheres. For this purpose we must compare the theoretical results with the characteristics of the observed radiation from the planet. Interpretation of the photometric observations allows us to obtain information about the phase function $x(\gamma)$, the single scattering albedo λ, and the optical thickness of the atmosphere τ_0. From this information we may draw conclusions about the nature of the aerosol particles in the atmosphere. Polarimetric observations may also be used to deduce these properties.

Study of planetary spectra provides another kind of information, including data on the abundance of different molecules in the atmosphere and the relative distribution of molecules and aerosol particles with altitude. Our knowledge in this regard is, of course, much more complete for those planets for which observational conditions are most favorable.

In the present chapter we present examples of the application of the theories developed previously to the determination of the optical parameters of planetary atmospheres. The first two sections give an interpretation of the photometric and polarimetric observations of Venus. In the two following sections we briefly examine the atmospheres of the Earth and of Mars, and in the last section we consider the atmospheres of the giant planets. More detailed information on the atmospheres of the planets may be found in other books [1–4].

10.1. Interpretation of the Photometric Observations of Venus

The planet for which observational conditions are the most favorable is Venus, since the phase angle varies between $0°$ and $180°$ as seen from Earth. Many astronomers have observed Venus photometrically, obtaining both the distribution of brightness across the disc at various phase angles α and the phase curve, which shows the change in brightness with changing α. We shall be concerned only with the interpretation of the phase curve.

The theoretical dependence of the stellar magnitude of the planet on phase angle is given by the following expression:

$$\frac{\pi}{2}\left(\frac{r_1 \Delta}{r_2 R}\right)^2 2.512^{m_\odot - m} = \int_{\alpha - (\pi/2)}^{\pi/2} \cos(\alpha - \omega)\cos\omega \, d\omega \int_0^{\pi/2} \varrho(\eta, \zeta, \varphi) \cos^3 \psi \, d\psi, \qquad (10.1)$$

which is obtained by substituting (9.8) into (9.11). The reflection coefficient ϱ entering this equation will in general depend on τ_0, $x(\gamma)$, λ, and a. For Venus, however, we may set $\tau_0 = \infty$. By comparing equation (10.1) with the observational phase curve for Venus, we may attempt to determine the quantities $x(\gamma)$ and λ.

We shall write the reflection coefficient in the form

$$\varrho(\eta, \zeta, \varphi) = \frac{\lambda}{4} \frac{x(\gamma)}{\eta + \zeta} + \Delta\varrho(\eta, \zeta, \varphi), \tag{10.2}$$

where the first term represents first-order scattering and the second corresponds to scattering of higher orders. The scattering angle γ is related to the phase angle α by

$$\gamma = \pi - \alpha. \tag{10.3}$$

We shall use an approximate method to determine $\Delta\varrho$, in which we replace the actual phase function $x(\gamma)$ by the two-term function $x_*(\gamma) = 1 + x_1 \cos \gamma$, where x_1 is the first-order coefficient of an expansion of the actual phase function in Legendre polynomials. That is,

$$x_1 = \tfrac{3}{2} \int\limits_0^\pi x(\gamma) \cos \gamma \sin \gamma \, d\gamma. \tag{10.4}$$

In this case it follows from equations (2.47)–(2.49) that the quantity $\Delta\varrho$ may be expressed in terms of the auxiliary functions $\varphi_0^0(\eta)$, $\varphi_1^0(\eta)$, and $\varphi_1^1(\eta)$ as

$$\Delta\varrho = \frac{\lambda}{4} \frac{\varphi_0^0(\eta) \, \varphi_0^0(\zeta) - x_1\varphi_1^0(\eta) \, \varphi_1^0(\zeta) - 1}{\eta + \zeta} + \frac{\lambda x_1}{4} \frac{\varphi_1^1(\eta) \, \varphi_1^1(\zeta) \cos \varphi + \cos \alpha}{\eta + \zeta}. \tag{10.5}$$

Since we shall verify below that the role of true absorption is very small in the Venus atmosphere $(1 - \lambda \ll 1)$, we may retain terms of order $\sqrt{1 - \lambda}$ in the expression for $\Delta\varrho$ and neglect terms of order $1 - \lambda$. Then, using equations (2.152) and (2.153) and substituting for $\cos \varphi$ according to equation (9.1), we obtain

$$\Delta\varrho = \frac{1}{4(\eta + \zeta)} \left\{ \varphi(\eta) \, \varphi(\zeta) \left[1 - 3 \sqrt{\frac{1 - \lambda}{3 - x_1}} \, (\eta + \zeta) \right] \right.$$

$$\left. + x_1 \frac{\varphi_1^1(\eta)\varphi_1^1(\zeta)}{\sqrt{(1 - \eta^2)(1 - \zeta^2)}} \, (\eta\zeta - \cos \alpha) - 1 + x_1 \cos \alpha \right\}, \tag{10.6}$$

where the function $\varphi(\eta)$ is determined by equation (2.44) and the function $\varphi_1^1(\eta)$ by equation (2.52) (in both cases for $\lambda = 1$).

Substitution of equation (10.2) into equation (10.1) gives

$$x(\pi - \alpha) \, F(\alpha) + G(\alpha) = H(\alpha), \tag{10.7}$$

where $F(\alpha)$ is determined by equation (9.77),

$$G(\alpha) = \int\limits_{\alpha - \pi/2}^{\pi/2} \cos \omega \cos(\alpha - \omega) \, d\omega \int\limits_0^{\pi/2} \Delta\varrho \cos^3 \psi \, d\psi, \tag{10.8}$$

and

$$H(\alpha) = \frac{\pi}{2}\left(\frac{r_1 \Delta}{r_2 R}\right)^2 2.512^{m_\odot - m}.$$ (10.9)

Introducing (10.6) into (10.8), we find

$$G(\alpha) = G_0(\alpha, x_1) - G_1(\alpha)\sqrt{\frac{1-\lambda}{3-x_1}},$$ (10.10)

where we have set

$$G_0(\alpha, x_1) = \frac{1}{4}\int_{\alpha-\pi/2}^{\pi/2} \frac{\cos\omega\cos(\alpha-\omega)}{\cos\omega+\cos(\alpha-\omega)}\, d\omega \int_0^{\pi/2}\left[\varphi(\eta)\,\varphi(\zeta)\right.$$

$$\left. + x_1 \frac{\varphi_1^1(\eta)\,\varphi_1^1(\zeta)}{\sqrt{(1-\eta^2)(1-\zeta^2)}}(\eta\zeta - \cos\alpha) - 1 + x_1\cos\alpha\right]\cos^2\psi\, d\psi,$$ (10.11)

$$G_1(\alpha) = \frac{3}{4}\int_{\alpha-\pi/2}^{\pi/2}\cos\omega\cos(\alpha-\omega)\, d\omega\int_0^{\pi/2}\varphi(\eta)\,\varphi(\zeta)\cos^3\psi\, d\psi.$$ (10.12)

We may then determine the phase function $x(\gamma)$ and the parameter λ from equation (10.7) (together with equation (10.4) and the normalization condition for the phase function). The function $H(\alpha)$ entering (10.7) may be found from the observational data, while the functions $F(\alpha)$ and $G(\alpha)$ may be computed according to equations (9.77) and (10.10)–(10.12). Table 9.7 presents values of the functions $G_0(\alpha, x_1)$, $G_1(\alpha)$, and $F(\alpha)$.

Evaluation of the quantities $x(\gamma)$ and λ for the Venus atmosphere with the aid of equation (10.7) was first carried out many years ago [5] (see also *TRT*, § 5 of Chapter 10). Computation of the function $G(\alpha)$ was undertaken using the approximate expression for the reflection coefficient given in Section 8.4. The relation between the stellar magnitude of Venus and phase angle was taken according to Müller in the form

$$m = -4.71 + 0.01322\alpha + 0.000000425\alpha^3$$ (10.13)

for $24° \leq \alpha \leq 156°$ and a Venus–Earth distance of one astronomical unit. It was shown that the phase function $x(\gamma)$ in the atmosphere of Venus is strongly elongated in the forward direction ($x_1 = 1.3$). From this it was deduced that the light scattering in the atmosphere of Venus is principally the result of particles which are large compared with the wavelength. Of course, molecules also play some role in the light scattering, but judging from their contribution to the phase function, this role is relatively small.

In addition to $x(\gamma)$, the albedo for single scattering λ was determined. This quantity was found to be very close to unity ($\lambda = 0.989$), which illustrates the small role of true absorption in the Venus atmosphere.

More recently N. P. Barabashov and V. I. Ezerskii [1] used the same method to determine the quantities $x(\gamma)$ and λ by using the phase curve for Venus obtained by Danjon. They obtained a phase function which was still more forward-directed ($x_1 = 1.71$) and a single scattering albedo which was still closer to unity ($\lambda = 0.995$).

Equation (10.7) with $G(\alpha)$ given by equations (10.10)–(10.12) has also been used to determine the optical properties of the Venus atmosphere [6]. In this case, a mean of the values from Müller and Danjon was used for the phase curve $m(\alpha)$. The resulting values of $x(\gamma)$ and λ differed only slightly from those which had been obtained earlier [5].

In determining the phase function by the above method we have made no assumption concerning its form. Nonetheless, this method leads to qualitatively correct results (a phase function which is strongly forward directed, small true absorption). We must not expect great precision in these results, however, since the phase function as determined from equation (10.7) is the difference between two quantities of different nature, one obtained from observation and the other from theory, with both quantities containing different sources of error.

It would seem that a more accurate procedure for solving the present problem would consist of the theoretical computation of the planetary phase curve for a given form of the phase function and a subsequent choice of the parameters characterizing this function (including λ) to provide the best agreement with observations. Such an interpretation of the Venus phase curve obtained by Knuckles et al. [7] in U, B, V and by Irvine [8] in ten spectral bands has led to phase functions which are still more elongated than those found previously, and also more similar to those found on the basis of the polarimetric observations (see the following section). The quantity $1 - \lambda$ was found to be of order 0.001–0.01.

Knowledge of the phase function $x(\gamma)$ and the single scattering albedo λ in different portions of the spectrum allows us in principle to determine the nature of the scattering particles. We shall not dwell on this point, however, since at the present time no firm conclusions have been reached. An example of one attempt in this direction is the work of Arking and Potter [9].

Since the entire phase curve is known for Venus, we may determine the value of the spherical albedo from the observational data. Comparison of this value of A_s with theoretical values also provides some knowledge of the optical properties of the atmosphere. We may use for this purpose equation (2.119), which gives A_s for the case of small true absorption. We have already pointed out that we may set $D = 8.5$ in this equation. As a result, we obtain

$$A_s = 1 - 4 \sqrt{\frac{1-\lambda}{3-x_1}} + 8.5 \cdot \frac{1-\lambda}{3-x_1}. \tag{10.14}$$

According to Harris [3], the spherical albedo of Venus in the visual portion of the spectrum is 0.76, so that we obtain from equation (10.14): $1 - \lambda = 0.005(3 - x_1)$. For the value $x_1 = 2.1$, which follows from an analysis of the polarimetric data, we find $1 - \lambda = 0.0045$.

Recently, E. G. Yanovitskii [10] has obtained an equation for the spherical albedo which is correct to order $(1 - \lambda)^{3/2}$, and thus adds an additional term to equation (10.14). Using this equation and the observational values of A_s contained in the previously mentioned article by Irvine, he determined values of $1 - \lambda$ as a function of wavelength.

10.2. Interpretation of the Polarimetric Observations of Venus

The first curve of the degree of polarization for Venus as a function of phase angle was contained in the data of Lyot (see [2]). He compared this curve with experimental results giving dependence of the degree of polarization on scattering angle for radiation singly scattered by water droplets. In this comparison the experimental data were reduced by a factor of 3.5 to take account of the fact that multiply-scattered radiation (assumed to be unpolarized) was included in the Venus observations. Since the theoretical and observational curves were in satisfactory agreement for a droplet diameter of approximately 2.5 μ, Lyot put forward the hypothesis that the clouds of Venus consisted of water droplets. Such an interpretation is not reliable, however, since the fraction of multiply-scattered light in the total planetary radiation may depend strongly on the phase angle.

We shall now attempt to interpret the results of Lyot (see [11]). As in Section 9.5, we shall assume that the observed polarization of the radiation diffusely reflected by the planet results principally from the scattering of first order. We shall use the observational data to specify the total energy reflected by the planet. Neglecting for simplicity the scattering by molecules, we then obtain from equation (9.78) or (9.83)

$$p(\alpha) = \frac{y_A(\pi - \alpha)}{H(\alpha)} \tag{10.15}$$

for the degree of polarization, where $y_A(\gamma)$ is given by equation (9.70) and $H(\alpha)$ is determined from the observational results by equation (10.9).

The quantity $y_A(\gamma)$ depends on the index of refraction m and the size distribution of the particles. We shall take an index of refraction $m = 1.5$ corresponding to quartz dust, which van de Hulst has pointed out as one possible component of the Venus atmosphere. Detailed tables [12] of the quantities $i_1(\gamma)$ and $i_2(\gamma)$ for values of $z = 2\pi a/\lambda = 0(0.2)$ 159 exist for such particles. The values of $i_1(\gamma)$ and $i_2(\gamma)$ from these tables were averaged according to equation (9.68), in which for simplicity we set $\Psi(a) = $ constant over a certain interval of radius a. In other words, the quantities

$$I_k(\gamma) = \int_{z_1}^{z_2} i_k(\gamma) \, dz \qquad (k = 1, 2) \tag{10.16}$$

were computed, following which the functions $x_A(\gamma)$ and $y_A(\gamma)$ were found from equations (9.69) and (9.70).

Knowledge of $y_A(\gamma)$ allows us to use equation (10.15) to construct a theoretical curve of the degree of polarization versus phase angle for the light from the planet (we label this

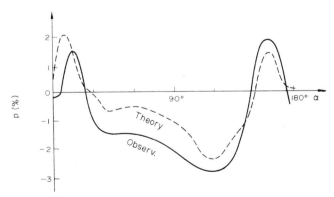

FIG. 10.1. Observational (Lyot) and theoretical curves of percentage polarization P versus phase angle α for Venus.

curve "theoretical", although it depends also on the observational data through $H(\alpha)$). These theoretical curves are different for different combinations of z_1 and z_2. Figure 10.1 illustrates such a curve for the case $z_1 = 7$ and $z_2 = 17$, where the function $H(\alpha)$ has been taken from the data of Müller and Danjon for the magnitude m_v. The results obtained by Lyot for the degree of polarization for Venus are shown on the same figure. We see that the theoretical curve reproduces the principal features of the observational curve. Consequently, the presence of particles with an index of refraction $m = 1.5$ in the Venus atmosphere provides a reasonable explanation for the observed polarization. For the given values of the parameters z_1 and z_2, the mean particle diameter is on the order of 2 μ.

Obviously, the particles producing the observed polarization also determine the phase function in the atmosphere of the planet. In particular, for particles with $z_1 = 7$ and $z_2 = 17$ the phase function has the form shown in Fig. 10.2.

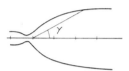

FIG. 10.2. Polar diagram showing the phase function $x(\gamma)$ for Venus which matches the polarimetric observations of Lyot.

This phase function is very similar to that which was obtained from the phase curve of Venus (see the preceding section), although it is more forward directed. The parameter x_1 for this case is found from equation (10.4) to have the value 2.1.

The preceding results refer to the visual portion of the spectrum. The observations show, however, that the polarization of Venus changes strongly with wavelength. The interpretation of these observations provides the opportunity for obtaining additional information about the atmosphere of the planet. We shall now use the equations derived in Section 9.5 to interpret the observational data of Gehrels and Samuelson [13]. The data, which is illustrated in Fig. 10.3, consists of values for the degree of polarization at different phase angles for three portions of the Venus spectrum: the infrared (effective wavelength of about 9900 Å), green (5600 Å), and ultraviolet (3590 Å). We see from Fig. 10.3 that for phase angles close to 90°, the degree of polarization (interpreted as an algebraic quantity) increases for decreasing wavelength $\tilde{\lambda}$. This may be understood by recalling that the role of Rayleigh scattering rapidly increases with decreasing $\tilde{\lambda}$, since the quantity δ which enters equations (9.78) and (9.83) obeys the law

$$\delta \sim \frac{1}{\tilde{\lambda}^4}. \qquad (10.17)$$

In the interpretation of the observational data in Fig. 10.3 we may not, however, simply use the results obtained above for the visual spectrum while only changing the quantity δ. It is necessary to keep in mind that in transforming from one part of the spectrum to another the quantity $y_A(\gamma)$ changes because of changing values of z, and $H(\alpha)$ changes as a result of different values for m_v and $m_\odot$ (by m_v we mean the stellar magnitude of Venus).

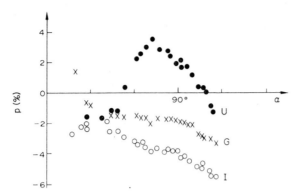

FIG. 10.3. Percentage polarization p versus phase angle α in the Ultraviolet, Green, and Infrared according to Gehrels and Samuelson [13].

Since Rayleigh scattering plays the predominent role at ultraviolet wavelengths, we shall determine the theoretical degree of polarization for this case. On the basis of the results obtained above for visual light, we concluded that the particles producing the polarization had an index of refraction $m = 1.5$ and size parameters $z_1 = 7$ and $z_2 = 17$. Such particles have values of z in the range from $z_1 = 10$ to $z_2 = 25$ for ultraviolet radiation. For these limits, values of $y_A(\gamma)$ may be found with the aid of equations (9.70) and (10.16). The quantity $H(\alpha)$ may then be determined from values of m_v reported for the ultraviolet part of the spectrum in [7]. Finally, the theoretical degree of polarization for various values of δ may be computed according to equation (9.78) (that is, for Model I). The results of such computations are given in Fig. 10.4.

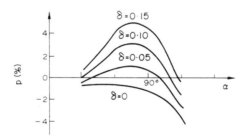

FIG. 10.4. Percentage polarization versus phase angle α for different fractions δ of Rayleigh scattering.

It is evident from a comparison of Figs. 10.3 and 10.4 that we must take $\delta \approx 0.12$ in order to obtain agreement between theory and observation. In other words, in ultraviolet light the fraction of radiation scattered by molecules is 12%. On the basis of equation (10.17), we may then say that for visual radiation $\delta \approx 0.03$, so that the role of molecular scattering is small. If we had utilized Model II, we would have obtained from Table 9.6 that the optical thickness of the gas layer τ_0 is about 0.03 in the ultraviolet and about 0.007 in the visual.

Recently, important investigations of the change in polarization for Venus as a function of phase angle and wavelength have been published [14–15]. Interpretation of these results has been partially carried out by Coffeen [16]. He computed values of $i_1(\gamma)$ and $i_2(\gamma)$ for various values of the index of refraction m and averaged them over particle size by using the distribution function $\Psi(a) \propto a^{-2}$. In his comparison of theory and observation Coffeen assumed that the polarization resulted principally from single scattering, and concentrated his attention only on the sign of the polarization. For ease of visualization he constructed, for every m, curves of zero polarization in a diagram whose abscissa was phase angle and whose ordinate was the mean value of the parameter $z = 2\pi a/\lambda$. Examples of such diagrams are shown in Fig. 10.5. By comparing the diagrams with observations of the change in sign

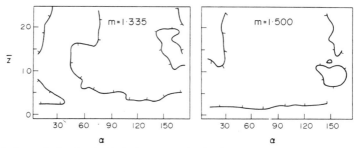

FIG. 10.5. Curves indicating the loci of zero polarization in the plane of mean particle size parameter $\bar{z} = 2\pi\bar{a}/\lambda$ and phase angle α for two choices of refractive index m.

of polarization with increasing α, we may decide whether agreement between theory and observation can be obtained for a given value of m. In particular, according to Lyot the polarization of Venus in visual light is initially positive ($\alpha \lesssim 25°$), then negative ($\alpha \lesssim 145°$), and finally again positive. It is clear from Fig. 10.5 that such behavior of the polarization cannot be produced by particles with $m = 1.335$, but may be caused by particles with $m = 1.5$ for sufficiently large values of $\bar{z}$. In this way, Coffeen concluded that the particles in the atmosphere of Venus have an index of refraction in the interval from 1.43 to 1.55 and a mean diameter on the order of 2.5 μ. This conclusion agrees with the results obtained in an earlier study [11]. Undoubtedly, continued interpretation of this data [14–15] will lead to new knowledge concerning the nature of the particles in the Venus atmosphere.

10.3. The Atmosphere of the Earth

Determination of the optical properties of the terrestrial atmosphere has great practical importance, so that these properties have been studied very thoroughly. The methods and results obtained have been set forth in a number of monographs (see, for example, [2] and [17]). Here we shall briefly examine only those characteristics of the atmosphere of the Earth which may also be determined for the atmospheres of the other planets.

The usual method for finding the optical thickness of the terrestrial atmosphere consists in measuring the brightness of the Sun at various elevations above the horizon. If the zenith angle of the Sun is ϑ_0, then the intensity of solar radiation penetrating through the atmosphere is given by

$$I = I_0 e^{-\tau_0 \sec \vartheta_0}, \qquad (10.18)$$

where I_0 is the intensity at the upper boundary of the atmosphere. If we have at least two different values of I corresponding to different solar zenith distances, we may use equation (10.18) to eliminate I_0 and determine τ_0. The optical thickness of the terrestrial atmosphere is of order 0.3 for a clear sky, and results from light scattering by both molecules and aerosol particles. Since the scattering coefficient for molecules decreases very rapidly with increasing wavelength (being inversely proportional to the fourth power of the wavelength), τ_0 in the red portion of the spectrum is normally significantly less than in the violet. From the dependence of τ_0 on wavelength we may estimate the ratio of the molecular to the aerosol component of the atmosphere.

Determination of the atmospheric optical thickness by equation (10.18) requires observations separated by significant intervals of time. During such a time interval, the state of the atmosphere may change significantly. Methods for finding τ_0 from observations taken at a single moment are therefore important. One such method is based on measuring the ratio of the illumination of a horizontal element of area by direct solar radiation to its illumination by light scattered in the atmosphere. Obviously, the smaller is this ratio, the larger is τ_0.

An approximate expression for this ratio may be obtained from the equations presented in Chapter 8. From equations (8.104) and (8.82) we find that the total illumination of the Earth's surface (caused by both direct solar radiation and diffuse atmospheric radiation) is

$$E = 2\,\frac{1+\frac{3}{2}\zeta+(1-\frac{3}{2}\zeta)e^{-\tau_0/\zeta}}{4+(3-x_1)(1-a)\tau_0}\,E_0, \tag{10.19}$$

where E_0 is the solar flux at the upper boundary of the atmosphere and $\zeta = \cos\vartheta_0$. The illumination at the Earth's surface by direct solar radiation is clearly

$$E_1 = E_0 e^{-\tau_0/\zeta}. \tag{10.20}$$

Dividing (10.20) by (10.19), we obtain an equation for the quantity E_1/E. Having obtained this quantity from observation, we may determine the optical thickness of the atmosphere τ_0. It is true that the quantity E_1/E also depends on a and x_1, but this dependence is rather weak for small values of τ_0. This method for obtaining τ_0 does not, however, give results of high precision, because of the approximate nature of equation (10.19).

When the sky is cloudy, the optical thickness may be very great (on the order of 100 or larger). In this case, instead of (10.19) we obtain for the total illumination of the surface

$$E = 2\,\frac{1+\frac{3}{2}\zeta}{4+(3-x_1)(1-a)\tau_0}\,E_0, \tag{10.21}$$

which allows us to evaluate the optical thickness of clouds. For this purpose we need to know the observational value of E/E_0, which, roughly speaking, equals the ratio of the surface illumination under cloudy conditions to that for a clear sky. In addition, one must know the albedo of the Earth's surface (which equals approximately 0.2 in the summer and 0.8 in the presence of snow) and the the value of the parameter x_1 which characterizes the forward elongation of the phase function for clouds (approximately 2–2.5).

The phase function for the cloudless terrestrial atmosphere may be determined by measuring the brightness of the sky at different angular separations from the Sun. The sky brightness is proportional to the transmission coefficient of the atmosphere, which may be written in the form

$$\sigma(\eta, \zeta, \varphi) = \frac{x(\gamma)}{4} \frac{e^{-\tau_0/\eta} - e^{-\tau_0/\zeta}}{\eta - \zeta} + \Delta\sigma,$$
(10.22)

where the first term on the right side represents single scattering and the second corresponds to scattering of higher orders (including reflection of light by the Earth's surface).

We may use the approximate expression

$$\Delta\sigma = \frac{[(1-a)R(\eta, \tau_0) + 2a]R(\zeta, \tau_0)}{4 + (3 - x_1)(1 - a)\tau_0} - \frac{1}{2}(e^{-\tau_0/\eta} + e^{-\tau_0/\zeta}) - (3 + x_1)\frac{\eta\zeta}{4}\frac{e^{-\tau_0/\eta} - e^{-\tau_0/\zeta}}{\eta - \zeta}$$
(10.23)

for $\Delta\sigma$, where $R(\zeta, \tau_0)$ is given by equation (8.82). Equation (10.23) corresponds to equation (8.101) for the quantity σ averaged over azimuth, without the term corresponding to single scattering.

In order to determine the phase function $x(\gamma)$ from equation (10.22), we must find the quantity σ from observation and compute $\Delta\sigma$ from equation (10.23). Usually σ is obtained in relative units. It is then necessary to use the normalization condition (1.4) to completely specify $x(\gamma)$.

Figure 10.6 illustrates the phase function for the terrestrial atmosphere found by this method. The presence of aerosol particles in the atmosphere causes the phase function to be very different from that for Rayleigh scattering. Of course the phase function may significantly differ from place to place, and at a given location from time to time, as a result of the weather. For the phase function in Fig. 10.6 the parameter $x_1 = 0.63$.

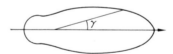

FIG. 10.6. Polar diagram showing the phase function for the terrestrial atmosphere.

Very often the phase function is computed from equation (10.22) without inclusion of the term $\Delta\sigma$. The resulting errors are particularly significant at those angles for which $x(\gamma)$ has small values.

For a completely overcast sky the transmission coefficient is given by

$$\sigma(\eta, \zeta) = \frac{[(1-a)(1 + \frac{3}{2}\eta) + 2a](1 + \frac{3}{2}\zeta)}{4 + (3 - x_1)(1 - a)\tau_0},$$
(10.24)

which may be deduced from (10.22) and (10.23) for $\tau_0 \gg 1$. According to equation (10.24), the brightness distribution across a cloudy sky does not depend on the phase function. This equation is an approximation; exact calculations, however, lead to approximately the same result. This may be seen from Table 7.5, which contains values of the function $u_0(\eta)$, which is proportional to $\sigma(\eta, \zeta)$. The phase function of clouds must therefore be found by a procedure other than observation of the distribution of brightness across a cloudy sky. The phase function turns out to be strongly elongated in the forward direction.

The equations presented in this section refer to the case of pure scattering ($\lambda = 1$). For a clear sky this is approximately the situation throughout the entire visible spectrum (apart from molecular bands). The scattering within clouds may also be taken as approximately pure scattering, as may be seen from the fact that the observed brightness distribution for a cloudy sky agrees with that given by equation (10.24). The same conclusion is indicated by the large albedo obtained for clouds from aircraft observations (approximately 0.6–0.8).

10.4. The Atmosphere of Mars

We know that the phase angle for Mars varies within the limits 0° to 47°. Observations of Mars thus provide less information about its atmosphere than do observations of Venus, for which the phase angle takes all possible values. Consequently, it is not possible to determine completely the phase function for the Martian atmosphere.

The optical thickness of the Martian atmosphere is very small, so that the surface of the planet is visible. Determination of τ_0 and the other characteristics of the atmosphere therefore requires simultaneous determination of the reflecting properties of the surface.

For small τ_0 it is sufficient to retain only primary scattering in the expression for the reflection coefficient $\varrho(\eta, \zeta, \varphi)$. Taking account of radiation scattered in the atmosphere itself and radiation reflected by the surface, we obtain for the case of pure scattering

$$\varrho(\eta, \zeta, \varphi) = \frac{x(\gamma)}{4} \frac{1 - e^{-\tau_0(1/\eta + 1/\zeta)}}{\eta + \zeta} + ae^{-\tau_0(1/\eta + 1/\zeta)}. \tag{10.25}$$

When $\tau_0 \ll 1$ equation (10.25) reduces to

$$\varrho(\eta, \zeta, \varphi) = \frac{x(\gamma)\tau_0}{4\eta\zeta} + a\left[1 - \tau_0\left(\frac{1}{\eta} + \frac{1}{\zeta}\right)\right]. \tag{10.26}$$

Mars is usually observed at opposition, so that the phase angle is zero. In this case $\eta = \zeta$, $\varphi = \pi$ and equation (10.26) takes the form

$$\varrho(\eta, \eta, \pi) = \frac{x(\pi)\tau_0}{4\eta^2} + a\left(1 - 2\frac{\tau_0}{\eta}\right). \tag{10.27}$$

The distribution of brightness across the Martian disc at different wavelengths may be obtained from observation. Comparison of the reflection coefficient so obtained with equation (10.27) allows us to determine the values of three parameters: the optical thickness of the atmosphere τ_0, the albedo of the surface a, and the quantity $x(\pi)$. Such determinations have been carried out many times (see the book by G. de Vaucouleurs [18] and the article by I. K. Koval' [19]) and lead to the conclusion that the optical thickness of the Martian atmosphere is approximately a factor of 10 less than that for the Earth (i.e. on the order of 0.03 for Mars), while the surface albedo is on the order of 0.1–0.3. The value of τ_0 decreases noticeably with increasing wavelength. This decrease is slower, however, than for the case of pure molecular scattering, which provides evidence for the presence of aerosol particles in the Martian atmosphere.

We should point out that it is very difficult to determine separately the three parameters by the above procedure. In addition, equation (10.26) which is normally used for this purpose is not exact, since it does not contain all terms of order τ_0. An expression for the reflection coefficient which does contain all these terms has been derived in Section 9.3 (equation (9.35)). For $\lambda = 1$ it can be rewritten in the form

$$\varrho(\eta, \zeta, \varphi) = \frac{x(\gamma)\tau_0}{4\eta\zeta} + a\left[1 - \tau_0\left(\frac{1}{\eta} + \frac{1}{\zeta}\right)\right]$$

$$+ \frac{a\tau_0}{2\zeta}\int_0^1 p(\eta', \zeta)\, d\eta' + \frac{a\tau_0}{2\eta}\int_0^1 p(\eta, \zeta')\, d\zeta' + a^2\tau_0\int_0^1 d\eta'\int_0^1 p(-\eta', \zeta')\, d\zeta'. \qquad (10.28)$$

The quantity $p(\eta, \zeta)$ entering (10.28) is expressed in terms of the phase function by equation (7.48).

In the violet portion of the spectrum, the optical thickness of the Martian atmosphere is rather large (on the order of unity). Interpretation of the observational data in this case requires utilization of an equation for the reflection coefficient which includes scattering of all orders. An approximate expression (8.100) for the reflection coefficient averaged over azimuth was obtained in Chapter 8 for $\lambda = 1$. Using this equation, we may obtain the following approximate expression for the desired function $\varrho(\eta, \zeta, \varphi)$:

$$\varrho(\eta, \zeta, \varphi) = \frac{x(\gamma)}{4}\frac{1 - e^{-\tau_0(1/\eta + 1/\zeta)}}{\eta + \zeta} + \Delta\varrho, \qquad (10.29)$$

where

$$\Delta\varrho = 1 - \frac{(1-a)\,R(\eta, \tau_0)\,R(\zeta, \tau_0)}{4 + (3 - x_1)(1-a)\tau_0} + [(3 + x_1)\eta\zeta - 2(\eta + \zeta)]\frac{1 - e^{-\tau_0(1/\eta + 1/\zeta)}}{4(\eta + \zeta)} \qquad (10.30)$$

and $R(\zeta, \tau_0)$ is given by equation (8.82). It is possible that future study of the Martian atmosphere will require consideration of true absorption ($\lambda < 1$) in the atmosphere and non-Lambertian reflection by the surface.

Analysis of the polarimetric as well as the photometric observations may provide knowledge of the Martian atmosphere. Polarization of the radiation from Mars results from both scattering of light in the atmosphere and reflection by the surface. If the latter effect is neglected, the degree of polarization for the radiation emerging from the entire disc of Mars may be determined in the same way as in Section 9.5 for Models I and II. In the present case, we shall assume that molecules and aerosol particles are mixed in a constant ratio in an atmosphere of optical thickness τ_0. We may then easily obtain the following expression for the degree of polarization:

$$p(\alpha) = \frac{F(\alpha) - F_1(\tau_0, \alpha)}{H(\alpha)}\left[\delta\frac{3}{4}\sin^2\alpha + (1 - \delta)y_A(\pi - \alpha)\right], \qquad (10.31)$$

which generalizes (9.78). The quantity δ is the ratio of the scattering coefficient for molecules to the total scattering coefficient, while the quantities $H(\alpha)$, $F(\alpha)$, and $F_1(\tau_0, \alpha)$ are given by equations (9.11), (9.77), and (9.82) respectively.

As we have stated, equation (10.31) relates to the case when polarization of light reflected by the surface is negligible. We may consider that this situation applies for Mars at phase angles close to 26°. This result follows from observations in the infrared portion of the spectrum, for which the optical thickness of the atmosphere is very small. It apparently applies also for other regions of the spectrum, a result which is supported by experiments on the polarization of light reflected by various terrestrial materials.

We therefore expect equation (10.31) to apply approximately to Mars when $\alpha = 26°$. Two unknown quantities enter this equation, the atmospheric optical thickness τ_0 and the parameter δ. The dependence of these quantities on wavelength, however, may be assumed to be known. Both quantities may therefore be determined with the aid of equation (10.31), if the degree of polarization for the planet is found from observations in different wavelengths.

A. V. Morozhenko has obtained detailed observations of the polarization of Mars which have been interpreted by I. N. Minin [20] on the assumption that the atmosphere is purely molecular. Minin found that the optical thickness for a wavelength 5600 Å is 0.03, and the atmospheric pressure at the surface is 13 millibars. Subsequently, A. V. Morozhenko [21] undertook interpretation of the same data with the assumption that the atmosphere consisted of molecules plus aerosol particles with an index of refraction $m = 1.5$. He found at a wavelength of 3550 Å an optical thickness of the gaseous component equal to 0.016 and of the aerosol component equal to 0.048. A value of 11 millibars was found for the surface pressure. Note that measurements of the atmospheric pressure by the Mariner spacecraft gave values of 4–7 millibars.

The most valuable information concerning the Martian atmosphere may, of course, be obtained from spectral observations. Infrared spectra reveal a large number of bands due to carbon dioxide gas, which were first discovered by Kuiper (see [2]). Bands of water vapor and other molecules have also been found. An important study of the infrared spectrum of Mars is contained in the work of V. I. Moroz (see [4]).

The theory set forth in Sections 9.3 and 9.4 may be used to interpret the Martian spectrum. The number of absorbing molecules may be determined from measurements of the equivalent width of absorption lines. This is particularly easy in the case of weak lines (for example, using equation (9.55)). These investigations provide information on the chemical composition of the atmosphere.

Such studies lead to the conclusion that the principle constituent of the Martian atmosphere is carbon dioxide gas. Various other molecules (such as H_2O) are present in much smaller quantities. It was earlier believed that considerable molecular nitrogen was present in the Martian atmosphere but was not observable because of the lack of absorption features in the region of the spectrum observable from Earth. Observations carried out by the Mariner spacecraft have shown, however, that the relative content of nitrogen is less than 1%.

10.5. Atmospheres of the Giant Planets

Relatively few studies have been made of the giant planets, with Jupiter investigated in somewhat more detail than the others. Our knowledge of this planet is discussed in the books referred to above [3, 4] and also in the monograph by V. G. Teifel' [22].

Because of the small range of phase angles observable for Jupiter (from 0 to 12°), we always see almost the entire disc of this planet. Observations give the brightness distribution across the disc at different wavelengths. By comparing this distribution with theoretical calculations for $\tau_0 = \infty$, we may attempt to find the quantities λ and $x(\gamma)$. Reliable estimates are hardly possible, however, because different combinations of the parameters lead to almost identical brightness distributions on the disc.

Absolute photometry of the Jovian disc was first carried out by V. V. Sharonov (see his book cited in Chapter 9). The change of brightness from the center of the disc to the limb which he found was compared with theoretical results for isotropic scattering. Satisfactory agreement was obtained for a single scatterng albedo $\lambda = 0.969$. This value must obviously be considered as a lower limit for λ, since for a given value of the reflection coefficient, λ must increase with increasing forward elongation of the phase function. Recently V. M. Loskutov has interpreted the brightness distribution for the Jovian disc with the aid of equations for the reflection coefficient for a three-term phase function (see Section 9.1). As an example, the best agreement between theory and observation was obtained for a dark zone on the disc and red wavelengths with $\lambda = 0.975$ and $x_1 = 1.3$.

The polarization at various locations on the Jovian disc has been measured by Lyot and Dollfus [3]. In the polar regions of the planet, the polarization is negative and rather large (on the order of 7%). Its absolute value quickly decreases with distance from the poles. This behavior of the polarization may be explained by multiple scattering of light in a Rayleigh atmosphere. According to van de Hulst (see [2]), it follows that we may assume the presence of a gaseous atmosphere in the polar regions of optical thickness $\tau_0 \approx 0.8$ lying above the planet's cloud deck.

The polarization at the center of the Jovian disc is also negative. The degree of polarization decreases from 0% at $\alpha = 0°$ to -0.6% at $\alpha = 12°$. To interpret this data we require a theory of multiple scattering by aerosol particles including polarization. Since such a theory has not yet been constructed, it is usual to assume that the polarization is a result of primary scattering only. The resulting equations for the degree of polarization were presented in Section 9.5. If we assume that a gaseous layer of optical thickness τ_0 overlies a cloudy layer, then the degree of polarization is given by equation (9.80). For $\tau_0 \ll 1$ this equation may be rewritten as

$$p = \frac{1}{4\varrho}\left[\frac{3\tau_0}{4\eta\zeta}\sin^2\alpha + \frac{y_A(\pi-\alpha)}{\eta+\zeta}\right]. \qquad (10.32)$$

V. M. Loskutov [23] has used equation (10.32) to compute the degree of polarization p on the assumption that the aerosol particles have an index of refraction $m = 1.38$ (appropriate for liquid ammonia). He obtained the quantity $y_A(\gamma)$ for the distribution of particle sizes given by equation (1.7), and took $\varrho = 0.59$ on the basis of observations. Comparison of the theoretical and observational values of p showed that they agreed for a mean particle radius on the order of 0.3 μ. The corresponding optical thickness of the upper gaseous layer was very small ($\tau_0 < 0.2$).

Broad molecular bands of ammonia (NH_3) and methane (CH_4) are present in the Jovian spectrum, especially in the red and infrared. Judging from the intensity of these bands, the abundance of these gases in the atmosphere is quite considerable. Since ammonia and methane contain hydrogen, it is natural to assume that a large quantity of molecular hydrogen is present in the atmosphere of Jupiter. The detection of molecular hydrogen is only possible through quadrupole transitions, whose probability is very small. Nonetheless, lines of H_2 arising from the quadrupole transitions have been found in the Jovian spectrum, thus verifying the expectations.

Evaluation of the abundance of different molecules in the Jovian atmosphere is usually based on the assumption that the molecular bands are formed in a purely absorbing atmosphere above the clouds. Such estimates do not merit much confidence, however, and we shall not present them here. In actuality, the molecular bands originate both in the atmosphere above the clouds and in the clouds themselves, so that the formation of the bands is dependent not only upon absorption processes by the molecules, but also upon scattering by the aerosol particles. The equivalent widths of the absorption lines must be computed with the aid of equations which take account of both of these processes. Such equations were given in Section 9.3 for two models of an atmosphere. The important question of the choice of a model may be decided from the nature of the observed variation in equivalent width between the center of the disc and the limb.

Most investigators have assumed that the predominant element in the Jovian atmosphere is hydrogen (present in H_2, NH_3, and CH_4). The amount of heavy gases in the atmosphere of this planet is relatively small. Evidence for this is provided by the determination of the mean molecular weight in the upper layers of the atmosphere. Such a determination may be made from observations of the occultation of a star by the planetary atmosphere. As the planet moves in front of the star, the brightness of the star gradually decreases because of differential refraction. From the curve of decreasing brightness we may determine the density distribution in the atmosphere, which depends on the temperature T and the mean molecular weight μ. Such observations for Venus have resulted in a value of $\mu = 38$ (for $T = 230°K$). For Jupiter the mean molecular weight turns out to be much smaller, $\mu = 3.3$ if we take $T = 86°K$. Consequently, it is possible that in addition to molecular hydrogen, the Jovian atmosphere may contain a considerable amount of helium.

In many respects, the more distant planets Saturn, Uranus, and Neptune are similar to Jupiter. As with Jupiter, intense bands of methane are present in the spectra of these planets. Quadrupole lines of molecular hydrogen have also been found. Observational data for Saturn, Uranus, and Neptune may be found both in monographs (see [3] and [4]) and in separate articles. However, the data are insufficient for reliable theoretical interpretation.

References

1. N. P. BARABASHOV, *Investigation of Physical Conditions on the Moon and Planets*, Izd. Khar'k. Univ., 1952.
2. *The Atmospheres of the Earth and Planets*, ed. G. P. KUIPER, Univ. of Chicago Press, 1947.
3. *Planets and Satellites*, ed. G. P. KUIPER and B. M. MIDDLEHURST, Univ. of Chicago Press, 1961.
4. V. I. MOROZ, *Fizika Planet*, Izd. "Nauka", Moscow, 1967 [NASA Technical Translation TTF 515, April 1968].
5. V. V. SOBOLEV, On the optical properties of the Venus atmosphere, *Astron. Zh.* 21, No. 5 (1944).
6. V. V. SOBOLEV, Investigation of the atmosphere of Venus I, *Astron. Zh.* 41, 97 (1964) [*Sov. Astron. A.J.* 8, 71 (1964)].
7. C. F. KNUCKLES, M. F. SINTON and W. M. SINTON, UBV photometry of Venus, *Lowell Obs. Bull.* 5, 153 (1961).

8. W. M. IRVINE, Monochromatic phase curves and albedos for Venus, *J. Atmos. Sci.* **25**, 610 (1968).
9. A. ARKING and J. POTTER, The phase curve of Venus and the nature of its clouds, *J. Atmos. Sci.* **25**, 617 (1968).
10. E. G. YANOVITSKII, On the spherical albedo of a planetary atmosphere, *Astron. Zh.* **49**, 844 (1972) [*Sov. Astron. A.J.* **16**, 687 (1973)].
11. V. V. SOBOLEV, Investigation of the atmosphere of Venus II, *Astron. Zh.* **45**, 169 (1968) [*Sov. Astron. A.J.* **12**, 135 (1968)].
12. R. H. GIESE, E. DE BARY, K. BULLRICH and C. D. VINNEMANN, *Tabellen der Streufunktionen und des Streuquerschnittes Homogener Kügelchen nach der Mie'schen Theorie*, Akademie Verlag, Berlin, 1962.
13. T. GEHRELS and R. E. SAMUELSON, Polarization-phase relations for Venus, *Astrophys. J.* **134**, 1022 (1961).
14. A. DOLLFUS and D. L. COFFEEN, Polarization of Venus. I. Disk observations, *Astron. and Astrophys.* **8**, 251 (1970).
15. D. L. COFFEEN and T. GEHRELS, Wavelength dependence of polarization. XV. Observations of Venus, *Astron. J.* **74**, 433 (1969).
16. D. L. COFFEEN, Wavelength dependence of polarization. XVI. Atmosphere of Venus, *Astron. J.* **74**, 446 (1969).
17. *Radiatsionnye Kharakteristiki Atmosfery i Zemnoi Poverkhnosti*, ed. by K. YA. KONDRAT'EV, Gidrometeoizdat, 1969.
18. G. DE VAUCOULEURS, *Physique de la Planete Mars*, Paris, 1951.
19. I. K. KOVAL', Results of photometric observations of Mars in 1954 at the Khar'kov Astronomical Observatory, *Astron. Zh.* **34**, 412 (1957) [*Sov. Astron. A. J.* **1**, 404 (1957)].
20. I. N. MININ, An optical model for the atmosphere of Mars, *Astron. Zh.* **44**, 1284 (1967) [*Sov. Astron. A.J.* **11**, 1024 (1968)].
21. A. V. MOROZHENKO, The atmosphere of Mars according to polarimetric observations, *Astron. Zh.* **46**, 1087 (1969) [*Sov. Astron. A.J.* **13**, 852 (1970)].
22. V. G. TEIFEL', *The Atmosphere of the Planet Jupiter*, Izd. "Nauka", Moscow, 1969.
23. V. M. LOSKUTOV, On the interpretation of the polarimetric observations of Jupiter, *Astron. Zh.* **48**, 1046 (1971) [*Sov. Astron. A.J.* **15**, 828 (1972)].

Addendum

Since 1972 when the Russian edition of this monograph was published, important progress has been made in both observations of planetary atmospheres and in the interpretation of observational data. Relevant recent research will now be briefly described. Where specific references are not cited below, the reader may refer to such reviews as [1] and [2].

As predicted by the author in Section 10.2, continued interpretation of the excellent polarization data for Venus as a function of phase angle and wavelength has led to new and exciting discoveries concerning the Venus clouds. Accurate, numerical multiple-scattering calculations, including rigorously the effect of polarization, have been made for homogeneous cloud models consisting of particles suspended in a molecular (Rayleigh scattering) atmosphere [3]. Comparison with the observations demonstrates convincingly that the resonances in the Venus polarization curves can only be understood as the result of scattering by *spherical* particles. Consequently, at least the uppermost cloud particles must be *liquid*. The refractive index m, effective radius a, and size dispersion are surprisingly well determined: $m = 1.46 \pm 0.015$ in the ultraviolet ($\lambda = 0.365\ \mu$), decreasing to $m = 1.43 \pm 0.015$ in the infrared ($\lambda = 0.99\ \mu$), and $a \approx 1\ \mu$ (see Fig. 10.7). These properties are quite uniform across the disc of Venus. The cloud droplets are thus neither water nor quartz. In fact the only substance thus far suggested which matches all the data is a concentrated solution of sulfuric acid (H_2SO_4–H_2O). Interestingly enough, the Junge aerosol layer in the terrestrial stratosphere also contains a large fraction of sulfuric acid.

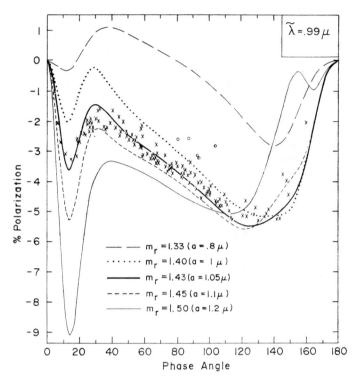

FIG. 10.7. Observations (refs. [14] and [15], Chapter 10) and theoretical computations of the polarization of sunlight reflected by Venus at wavelength $\tilde{\lambda} = 0.99$ micron. The x's and 0's refer to different time periods. Theoretical curves for polydisperse spherical particles with refractive index m and an effective particle radius a chosen to yield closest agreement with observations for the entire range $0.365 \leq \tilde{\lambda} \leq 0.99$ μ (from [3], courtesy of the *Journal of the Atmospheric Sciences*).

Spacecraft observations [4, 5, 6 and earlier references therein] have vastly increased our knowledge of the detailed structure of the Venus atmosphere below the visible clouds, and have clearly demonstrated time variable structure and circulation patterns in the uppermost clouds.

The most recent Mariner spacecraft studies of Mars indicate that clouds of H_2O ice, as well as very probably CO_2 ice haze, are sometimes present in the Martian atmosphere. The atmospheric composition, like that of Venus, is almost entirely carbon dioxide. The "yellow" clouds appear to have a significant content of silicates, appropriate to their classic identification as "dust" [7].

The important question of the relative helium abundance in the atmospheres of the giant planets is still unsolved. The recent detection of ethane (C_2H_6) and quite probably acetylene (C_2H_2) in the Jovian atmosphere emphasizes the potential role of organic chemical reactions to the coloration and structure of such atmospheres [8].

Problems which will undoubtedly receive increasing attention in the near future with the aid of light scattering theory include the nature of Saturn's rings, the atmosphere of Titan, and possible Raman scattering in the atmosphere of Uranus [1].

Bibliography for Addendum

1. *Exploration of the Planetary System (I.A.U. Symposium 65)*, ed. A. Woszczyk and W. Iwaniszewska, D. Reidel Pub. Co., Netherlands, 1974.
2. R. L. Newburn, Jr. and S. Gulkis, A survey of the outer planets Jupiter, Saturn, Uranus, Neptune, Pluto, and their satellites, *Space Sci. Revs.* **14,** 179 (1973).
3. J. E. Hansen and J. W. Hovenier, Interpretation of the polarization of Venus, *J. Atmos. Sci.*, **31,** 1137 (1974).
4. G. Fjeldbo, A. I. Kliore and V. R. Eshleman, The neutral atmosphere of Venus as studied with the Mariner V radio occultation experiments, *Astron. J.* **76,** 123 (1971).
5. B. C. Murray, M. I. S. Belton, C. E. Danielson, M. E. Davies, D. Gault, B. Hapke, B. O'Leary, R. G. Strom, V. Suomi and N. Trask, Venus: atmospheric motion and structure from Mariner 10 pictures, *Science* **183,** 1307 (1974).
6. M. Ya. Marov, V. S. Avduevsky, N. F. Borodin, A. P. Ekonomov, V. V. Kerzhanovich, V. P. Lysov, B. Ye. Moshkin, M. K. Rozhdestvensky and O. L. Ryabov, Preliminary results on the Venus atmosphere from the Venera 8 descent module, *Icarus* **20,** 407 (1973).
7. *J. Geophys. Research* **78,** No. 20, 1973 (special issue on Mariner 9).
8. S. T. Ridgway, Jupiter: identification of ethane and acetylene, *Astrophys. J.* **187,** L 41 (1974).

Chapter 11

SPHERICAL ATMOSPHERES

Two assumptions are usually made in the study of radiative transfer in planetary atmospheres: (1) that the atmosphere consists of plane-parallel layers, and (2) that these layers are illuminated by parallel solar radiation over their entire extent. We have made these assumptions in all of the preceding chapters. They are actually valid, however, only for those portions of the atmosphere for which the Sun is sufficiently high above the horizon. If the Sun is close to the horizon or below the horizon, it is necessary to take the spherical nature of the atmosphere into account.

The development of the theory of light scattering in a spherical atmosphere is of considerable importance. The theory is necessary for the study of the brightness of a planet close to the terminator and for twilight phenomena in the Earth's atmosphere. It is particularly important for the interpretation of photometric observations of the atmospheres of the Earth and planets carried out from spacecraft.

At the present time, the theory of light scattering in a spherical atmosphere is still in a preliminary stage of development. This is understandable in part because of the difficulty of the problem, and also because it has become clear only in the last few years that its solution is necessary for certain applications.

The problem of radiative transfer in a spherical atmosphere also arises in the study of stars. The problem in that case is much simpler, however, than in the case of planets because of the difference in the distribution of energy sources in the two cases. In the stellar case both the energy sources and the atmosphere itself have spherical symmetry. In the case of the planets the spherical atmosphere is illuminated from without by parallel radiation, and the overall spherical symmetry is lost.

11.1. The Integral Equation for the Source Function in the Case of Isotropic Scattering

For simplicity, we shall begin by assuming that the light scattering in the planetary atmosphere is isotropic. Although this is not actually true, examination of this case nonetheless allows us to clarify the essential nature of the problem. Moreover, the results for this case have practical significance, since with their aid we may obtain an approximate solution to the problem for anisotropic light scattering. This may be accomplished by computing the first-order scattering exactly (for the actual phase function of the atmosphere), and the higher-order scattering approximately (by assuming isotropic scattering).

We shall assume that a planet of radius R is illuminated by parallel solar radiation with flux πS through an element of area perpendicular to the direction towards the Sun. We shall designate the distance of an arbitrary point in the atmosphere from the center of the planet by r, and the angle between the radius vector of this point and the direction towards the Sun by ψ (Fig. 11.1).

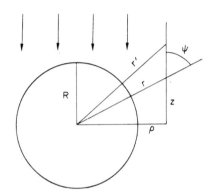

FIG. 11.1. Definition of coordinates for an isotropically scattering spherical atmosphere [surrounding a planet of radius R. Arrows indicate direction of incident solar radiation.

Let $\alpha(r)$ be the volume coefficient of absorption and $\lambda(r)$ be the single scattering albedo ("the particle albedo"). We shall, furthermore, set $\varepsilon(r, \psi)$ equal to the volume emission coefficient, so that in the usual manner

$$\varepsilon(r, \psi) = \alpha(r)B(r, \psi). \tag{11.1}$$

We may write the following integral equation for the determination of the source function $B(r, \psi)$ (see [1]):

$$B(r, \psi) = \frac{\lambda(r)}{2} \int_{R}^{\infty} \alpha(r') \, r'^2 \, dr' \int_{0}^{\pi} B(r', \psi') K(r, r', \psi, \psi') \sin \psi' \, d\psi' + B_1(r, \psi), \tag{11.2}$$

where $\alpha(r) B_1(r, \psi)$ is the volume emission coefficient corresponding to first-order scattering (that is, scattering of direct solar radiation). The quantity B_1 is clearly

$$B_1(r, \psi) = \frac{\lambda(r)}{4} \, S e^{-T(r, \psi)}, \tag{11.3}$$

where $T(r, \psi)$ is the optical distance along a ray from the Sun to the given point in the atmosphere. We shall set z equal to the distance along such a ray, taking $z = 0$ at $\psi = \pi/2$. We then have

$$T(r, \psi) = \int_{z}^{\infty} \alpha(r') \, dz', \tag{11.4}$$

where $z = r \cos \psi$ and $r'^2 = r^2 \sin^2 \psi + z'^2$ (see Fig. 11.1). Equation (11.4) may therefore be rewritten in the form

$$T(r, \psi) = \int_{r \cos \psi}^{\infty} \alpha(\sqrt{\{r^2 \sin^2 \psi + z'^2\}}) \, dz'. \tag{11.5}$$

This relation is valid both for $\psi \leq \pi/2$ and $\psi > \pi/2$.

The kernel of the integral equation (11.2) is given by the formula

$$K(r, r', \psi, \psi') = \frac{1}{2\pi} \int e^{-t(r, r')} \frac{d\Phi}{s^2(r, r')}, \tag{11.6}$$

where $s(r, r')$ is the distance between points with coordinates (r, ψ, Φ) and (r', ψ', Φ') and $t(r, r')$ is the optical distance between these points (Φ is the azimuthal angle). The limits of the integration in (11.6) must be determined by taking account of the presence of the planetary surface or by choosing $\alpha(r) = \infty$ for $r < R$.

Equation (11.2) may be significantly simplified if we make the usual assumption that the atmosphere consists of plane-parallel layers. Such an assumption is based only on the geometry of the atmosphere (in particular, that its thickness is much less than the radius of the planet), and does not depend on the conditions of its illumination. In what follows, we shall assume that the atmosphere consists of plane-parallel layers which are, however, illuminated by the Sun at every location in the same manner as they would be for a spherical atmosphere. In other words, we shall retain the first of the assumptions indicated in the opening paragraph of this chapter, but not the second.

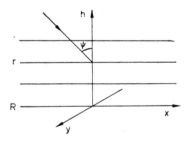

FIG. 11.2. Definition of Cartesian coordinates in a spherical atmosphere.

The position of a point in the atmosphere will now be characterized by its altitude $h = r - R$ and its coordinates x and y in a horizontal plane (Fig. 11.2). The coordinate x is chosen such that

$$x = R\psi. \tag{11.7}$$

We have pointed out that the quantity B_1 must be determined from equations (11.3) and (11.5). It is necessary in these equations, however, to transform from the angle ψ to the coordinate x with the aid of equation (11.7). As a result, B_1 will be a function of r and x, as will the source function B.

In the usual manner, we introduce the optical depth τ and the optical thickness of the atmosphere τ_0 according to the relations

$$\tau = \int_r^{\infty} \alpha(r') \, dr', \quad \tau_0 = \int_R^{\infty} \alpha(r) \, dr. \tag{11.8}$$

The single scattering albedo λ will now be a function of τ, and B and B_1 will be functions of τ and x.

We shall now transform equation (11.2) from spherical to Cartesian coordinates. We may integrate over the coordinate y, since B does not depend on y. The equation determining the function B is then

$$B(\tau, x) = \frac{\lambda(\tau)}{4\pi} \int_0^{\tau_0} d\tau' \int_{-\infty}^{\infty} B(\tau', x') \, K(\tau, \tau', x, x') \, dx' + B_1(\tau, x), \tag{11.9}$$

where

$$K(\tau, \tau', x, x') = \int_{-\infty}^{\infty} e^{-t} \frac{dy}{s^2}. \tag{11.10}$$

In the present case the distance between two points is given by

$$s^2 = (x-x')^2 + (y-y')^2 + (h-h')^2, \tag{11.11}$$

and the optical distance between the same points is

$$t = \frac{\tau - \tau'}{h' - h} s. \tag{11.12}$$

We may therefore write in place of (11.10)

$$K(\tau, \tau', x, x') = \int_{-\infty}^{\infty} e^{-(\tau-\tau')\sqrt{b^2+y^2}/(h'-h)} \frac{dy}{b^2+y^2}, \tag{11.13}$$

where we have defined

$$b^2 = (x-x')^2 + (h-h')^2. \tag{11.14}$$

Equation (11.13) may also be rewritten in the form

$$K(\tau, \tau', x, x') = \frac{2}{b} \int_1^{\infty} e^{-cw} \frac{dw}{w\sqrt{w^2-1}}, \tag{11.15}$$

where

$$c = b \frac{\tau - \tau'}{h' - h}. \tag{11.16}$$

The function

$$K_0(c) = \int_1^{\infty} e^{-cw} \frac{dw}{\sqrt{w^2-1}} \tag{11.17}$$

is the Bessel function of zeroth order and imaginary argument (sometimes referred to as the MacDonald function). Making use of (11.17), we find

$$K(\tau, \tau', x, x') = \frac{2}{b} \int_0^{\infty} K_0(c) \, dc. \tag{11.18}$$

In order to transform from r (or h) to the optical depth τ in the preceding equations, we must know the dependence of the absorption coefficient on altitude. Let us consider as an example

$$\alpha(r) = \alpha(R)\, e^{-h/H_*}, \tag{11.19}$$

where H_* is the geometric thickness of a homogeneous atmosphere with $\alpha = \alpha(R)$ and the same optical thickness as the given atmosphere (H_* is usually called the scale height). Using the definitions (11.8), we find

$$\tau = \tau_0 e^{-h/H_*}, \quad \tau_0 = \alpha(R)H_*, \tag{11.20}$$

which gives

$$h' - h = H_* \ln \frac{\tau}{\tau'}, \quad r = R + H_* \ln \frac{\tau_0}{\tau}. \tag{11.21}$$

For the determination of $B(\tau, x)$ we thus have the integral equation (11.9), in which the kernel is given by equation (11.18) and the inhomogeneous term by equations (11.3) and (11.5). To transform in these equations from the variables (r, ψ) to the variables (τ, x) we have equations (11.7) and (11.21) (the latter applies to the case of an absorption coefficient which exponentially decreases with altitude).

Once the source function $B(\tau, x)$ has been found from equation (11.9), the intensity of radiation may be determined with the aid of the transfer equation. In doing this we return to the spherical geometry; that is, we assume that the values of the source function found for the plane-parallel atmosphere coincide with the values at corresponding points in a spherical atmosphere. In this way we may determine the intensity of radiation emerging in directions only slightly inclined to the direction of the tangent to the planetary surface.

We have not included reflection of light by the planetary surface in equation (11.9), but it is not difficult to obtain an equation for the source function $B(\tau, x)$ which does include this effect (see [1]).

In the examination of light scattering in planetary atmospheres it is usual to make not only the assumption of plane-parallel layers, but also the assumption that such an atmosphere is illuminated over its entire extent by parallel radiation. In the preceding paragraphs we have not made the second of these assumptions and have obtained an equation for the source function which is more exact than the usual one. In particular, in contrast to the usual equation, the one derived above allows us to compute the radiation field in an atmosphere for values of the angle ψ close to $\pi/2$ (for positions of the Sun close to or below the horizon).

Clearly, if we also make the second assumption (which is possible only for angles ψ not close to $\pi/2$), we must obtain from equation (11.9) the usual equation for the source function. In that case the quantity B_1 is given by the expression

$$B_1(\tau, x) = \frac{\lambda(\tau)}{4}\, S e^{-\tau \sec \psi}, \tag{11.22}$$

and the angle ψ (or the coordinate x) does not change in the atmosphere. We may then take B outside the second integral sign in equation (11.9) and carry out the integration over x'.

It is not difficult to show that

$$\int_{-\infty}^{\infty} K(\tau, \tau', x, x')\, dx' = 2\pi E_1(|\tau - \tau'|). \tag{11.23}$$

Therefore, in place of equation (11.9) we find

$$B(\tau, x) = \frac{\lambda(\tau)}{2} \int_0^{\tau_0} B(\tau', x) E_1(|\tau - \tau'|)\, d\tau' + \frac{\lambda(\tau)}{4} S e^{-\tau \sec \psi}, \tag{11.24}$$

which is the usual integral equation for the source function in the theory of light scattering in planetary atmospheres (see Section 1.4).

11.2. The Basic Equations for Anisotropic Scattering

The integral equation for the source function obtained in the preceeding section is valid for isotropic scattering. We shall now consider radiative transfer in a spherical atmosphere for anisotropic scattering. Let us first write the basic equations determining the radiation field in this case (see [2] and [3]).

As above, we shall specify the position of a point in the atmosphere by the spherical coordinates r and ψ. The direction of radiation at the given point will be characterized by the angles ϑ and φ, where ϑ is the zenith distance (the angle between the direction of radiation and the radius vector) and φ is the corresponding azimuthal angle in a horizontal plane. Consequently, the intensity will be a function of r, ψ, ϑ, and φ (Fig. 11.3).

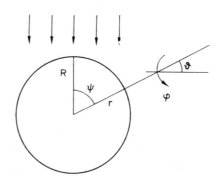

Fig. 11.3. Definition of coordinates for an anisotropically scattering spherical atmosphere surrounding a planet of radius R. Arrows indicate direction of incident solar radiation.

We shall begin from the equation of radiative transfer in the general form

$$\frac{dI}{ds} = \alpha(B - I). \tag{11.25}$$

Since I depends on the four variables we have mentioned, we must write

$$\frac{dI}{ds} = \frac{\partial I}{\partial r}\frac{dr}{ds} + \frac{\partial I}{\partial \psi}\frac{d\psi}{ds} + \frac{\partial I}{\partial \vartheta}\frac{d\vartheta}{ds} + \frac{\partial I}{\partial \varphi}\frac{d\varphi}{ds}. \tag{11.26}$$

We may easily obtain the relations

$$\frac{dr}{ds} = \cos \vartheta, \qquad \frac{d\psi}{ds} = \frac{\sin \vartheta \, \cos \varphi}{r},$$

$$\frac{d\vartheta}{ds} = -\frac{\sin \vartheta}{r}, \qquad \frac{d\varphi}{ds} = -\frac{\cot \psi \, \sin \vartheta \, \sin \varphi}{r}. \qquad (11.27)$$

We may then rewrite (11.25) as

$$\cos \vartheta \, \frac{\partial I}{\partial r} + \frac{\sin \vartheta \, \cos \varphi}{r} \frac{\partial I}{\partial \psi} - \frac{\sin \vartheta}{r} \frac{\partial I}{\partial \vartheta} - \frac{\cot \psi \, \sin \vartheta \, \sin \varphi}{r} \frac{\partial I}{\partial \varphi} = \alpha(B - I). \quad (11.28)$$

This is the desired equation of radiative transfer in a spherical planetary atmosphere.

The source function $B(r, \psi, \vartheta, \varphi)$ which enters equation (11.28) may be expressed in terms of the intensity by the usual relation

$$B(r, \psi, \vartheta, \varphi) = \frac{\lambda}{4\pi} \int_0^{2\pi} d\varphi' \int_0^{\pi} I(r, \psi, \vartheta', \varphi') \, x(\gamma') \sin \vartheta' \, d\vartheta' + B_1(r, \psi, \vartheta, \varphi), \quad (11.29)$$

where $x(\gamma)$ is the phase function and

$$\cos \gamma' = \cos \vartheta \, \cos \vartheta' + \sin \vartheta \, \sin \vartheta' \, \cos(\varphi - \varphi'). \qquad (11.30)$$

The function B_1 results from first-order scattering and is given by

$$B_1(r, \psi, \vartheta, \varphi) = \frac{\lambda}{4} \, Sx(\gamma)e^{-T}, \qquad (11.31)$$

where

$$\cos \gamma = -\cos \vartheta \, \cos \psi + \sin \vartheta \, \sin \psi \, \cos \varphi. \qquad (11.32)$$

The quantity T is the optical path from the Sun to the given point in the atmosphere and is determined by equation (11.5). The problem of determining the radiation field in the atmosphere thus requires the solution of the system of equations (11.28) and (11.29) for the functions I and B under appropriate boundary conditions.

From equations (11.28) and (11.29) we may obtain a single integral equation for the function B. This may be accomplished by finding an expression for I in terms of B from equation (11.28) and substituting this into (11.29). Since B depends on four variables, however, the resulting integral equation is rather complicated.

Because of the difficulty in obtaining an exact solution of equations (11.28) and (11.29), the determination of an approximate solution is of considerable interest. We may use for this purpose the approximate method set forth in Section 8.4. We shall now give the equations which follow from this procedure [3].

We reall that the above-mentioned method consists in computing the first-order scattering exactly and the higher-order scattering by an approximation in which only the first two terms of the Legendre expansion of the phase function are retained. More precisely in place of the actual phase function, we use the function

$$x(\gamma) = 1 + x_1 \cos \gamma, \tag{11.33}$$

and then replace the function (11.33) by the actual phase function in that term of the solution which corresponds to first-order scattering.

For the phase function (11.33) equation (11.29) becomes

$$B = \lambda J + \lambda x_1 H \cos \vartheta + \lambda x_1 G \sin \vartheta \cos \varphi$$
$$+ \frac{\lambda}{4} S(1 - x_1 \cos \vartheta \cos \psi + x_1 \sin \vartheta \sin \psi \cos \varphi) e^{-T}, \tag{11.34}$$

where we have introduced the notation

$$J = \int I \frac{d\omega}{4\pi}, \quad H = \int I \cos \vartheta \frac{d\omega}{4\pi}, \quad G = \int I \sin \vartheta \cos \varphi \frac{d\omega}{4\pi}. \tag{11.35}$$

The quantity B (and therefore also I) may thus be expressed in terms of three functions J, H, and G, each of which depends only on the two variables r and ψ.

To obtain equations which determine these functions we multiply equation (11.28) by $d\omega/4\pi$ and integrate over all directions. Using equation (11.11) we find

$$\frac{\partial H}{\partial r} + \frac{2}{r} H + \frac{1}{r} \frac{\partial G}{\partial \psi} + \frac{\cot \psi}{r} G = -\alpha(1 - \lambda) J + \alpha \frac{\lambda}{4} Se^{-T}. \tag{11.36}$$

Then, after multiplying (11.28) first by $\cos \theta (d\omega/4\pi)$ and then by $\sin \theta \cos \varphi (d\omega/4\pi)$, we obtain in a similar manner

$$(3 - \lambda x_1) H = -\frac{1}{\alpha} \frac{\partial J}{\partial r} - \frac{\lambda}{4} Sx_1 \cos \psi e^{-T}, \tag{11.37}$$

$$(3 - \lambda x_1) G = -\frac{1}{\alpha r} \frac{\partial J}{\partial \psi} + \frac{\lambda}{4} Sx_1 \sin \psi e^{-T}. \tag{11.38}$$

In the derivation of equations (11.37) and (11.38) we have used certain approximate relations similar to (8.57).

By substituting H and G from (11.37) and (11.38) into equation (11.36), we may obtain a single equation for the quantity J. Before doing this, however, we shall assume that the absorption coefficient α depends only on r and that $3 - \lambda x_1 = $ constant in the atmosphere (up to this point the quantities α, λ, and $x(\gamma)$ have been arbitrary functions of r and ψ). Proceeding in the indicated manner, we then find from the last three equations

$$\Delta J - \frac{\alpha'(r)}{\alpha(r)} \frac{\partial J}{\partial r} = \alpha^2 (k^2 J - f), \tag{11.39}$$

where Δ is the Laplacian operator,

$$k^2 = (3 - \lambda x_1)(1 - \lambda), \tag{11.40}$$

and

$$f = \frac{\lambda}{4} Se^{-T} \left[3 - \lambda x_1 + \frac{x_1 \sin \psi}{\alpha r} \frac{\partial T}{\partial \psi} - \frac{x_1 \cos \psi}{\alpha} \frac{\partial T}{\partial r} \right]. \tag{11.41}$$

With the aid of equation (11.5) the last expression may be rewritten in the form

$$f = \frac{\lambda}{4} Se^{-T}[3+(1-\lambda)\,x_1].\tag{11.42}$$

The boundary conditions for equation (11.39) must still be specified. In the plane-parallel case they took the form (8.60) and (8.98); analogous expressions may be written for the case of a spherical atmosphere. At the upper boundary we have

$$J = 2H,\tag{11.43}$$

while at the lower boundary

$$J+2H = a[J-2H+S\,\cos\psi\,e^{-T(R,\,\psi)}],\tag{11.44}$$

where a is the albedo of the planetary surface. When $\psi > \pi/2$ the last term on the right side of (11.44) vanishes, since $T(R, \psi) = \infty$.

Thus, for the determination of the quantity J we have equation (11.39) with the boundary conditions (11.43) and (11.44). Once J has been found, the quantities H and G may be obtained from equations (11.37) and (11.38).

To obtain the function B we must substitute J, H, and G into equation (11.34). In addition, in accordance with the method we have adopted, we must replace the function $1+x_1 \cos\gamma$ in the last term of this equation by the actual phase function $x(\gamma)$. The expression for B will thus have the form

$$B = \lambda J + \lambda x_1 H \cos\vartheta + \lambda x_1 G \sin\vartheta\,\cos\varphi + \frac{\lambda}{4} Sx(\gamma)\,e^{-T},\tag{11.45}$$

where x_1 is the first order coefficient of an expansion of the actual phase function in Legendre polynomials. After B has been determined, the intensity I is found from equation (11.28).

11.3. Solution of the Equations in Particular Cases

Before we may solve equation (11.39), we must specify the dependence of the absorption coefficient on altitude. It is instructive to consider the two following particular cases: (1) the absorption coefficient is constant in the atmosphere, and (2) the absorption coefficient decreases exponentially with altitude. The first of these conditions will hold approximately for completely overcast skies, and the second for clear-sky conditions. Equation (11.39) has been solved for these cases in a series of articles [4, 5, and others], and we shall now briefly examine the results obtained.

In the first case ($\alpha = $ constant), equation (11.39) for the determination of the basic quantity J takes the form

$$\varDelta J = \alpha^2 (k^2 J - f).\tag{11.46}$$

If we assume that the upper boundary of the atmosphere is a sphere of radius R_1, then the quantity f entering (11.46) is given by equation (11.42) in which

$$T = \alpha\,(\sqrt{R_1^2 - r^2 \sin^2\psi} - r\,\cos\psi).\tag{11.47}$$

Equation (11.47) applies to the portion of the atmosphere illuminated by the Sun. Within the unilluminated portion $T = \infty$, so that $f = 0$.

Equation (11.46) must be solved for the boundary conditions (11.43) (for $r = R_1$) and (11.44) (for $r = R$). A solution will not be obtained here in the general case, however (see [4]). We shall instead make the additional assumption that the optical thickness of the atmosphere $\alpha(R_1 - R)$ is very large. We may then assume that the atmosphere extends to the center of the planet. In other words, we consider the problem of the brightness of a homogeneous sphere illuminated by parallel radiation. The boundary condition (11.44) may now be replaced by the condition that the quantity J is bounded for $r = 0$.

Let us first find a particular solution of equation (11.46). Transforming from the spherical coordinates (r, ψ) to the cylindrical coordinates (ϱ, z) with the aid of $\varrho = r \sin \psi$, $z = r \cos \psi$, we have

$$f = \frac{\lambda}{4} S[3 + (1 - \lambda) x_1] e^{-\alpha(\sqrt{R_1^2 - \varrho^2} - z)} . \tag{11.48}$$

With the inhomogeneous term of equation (11.46) taking this form, it is convenient to seek a particular solution in the form

$$J^* = \frac{\lambda}{4} S[3 + (1 - \lambda) x_1] C(\varrho) e^{\alpha z}. \tag{11.49}$$

Substituting (11.49) into (11.46), we are led to the following equation for the determination of the function $C(\varrho)$:

$$C''(\varrho) + \frac{1}{\varrho} C'(\varrho) + \alpha^2 (1 - k^2) C(\varrho) = -\alpha^2 e^{-\alpha \sqrt{R_1^2 - \varrho^2}}. \tag{11.50}$$

The solution of equation (11.50) which is not singular at the origin has the form

$$C(\varrho) = \alpha^2 \frac{\pi}{2} \int_0^\varrho [J_0(\alpha \varrho \sqrt{1 - k^2}) N_0(\alpha \varrho' \sqrt{1 - k^2})$$

$$- N_0(\alpha \varrho \sqrt{1 - k^2}) J_0(\alpha \varrho' \sqrt{1 - k^2})] e^{-\alpha \sqrt{R_1^2 - \varrho'^2}} \varrho' d\varrho', \tag{11.51}$$

where J_0 and N_0 are Bessel functions of zeroth order.

In order to find the general solution of equation (11.46), we set

$$J = y + J^*. \tag{11.52}$$

We then obtain for the determination of y the equation

$$\Delta y = \alpha^2 k^2 y. \tag{11.53}$$

The solution of equation (11.53) may be sought as an expansion in Legendre polynomials:

$$y = \sum_0^\infty D_n(r) P_n (\cos \psi). \tag{11.54}$$

Substituting (11.54) into (11.53), we find the following equation for $D_n(r)$:

$$D_n''(r) + \frac{2}{r} D_n'(r) - \left[\alpha^2 k^2 + \frac{n(n+1)}{r^2} \right] D_n(r) = 0. \tag{11.55}$$

The solution of equation (11.55) is expressed in terms of Bessel functions of purely imaginary argument as

$$D_n(r) = \frac{a_n}{\sqrt{(r)}} I_{n+(1/2)}(\alpha kr) + \frac{b_n}{\sqrt{(r)}} K_{n+(1/2)}(\alpha kr), \tag{11.56}$$

where a_n and b_n are arbitrary constants. For the case of pure scattering ($\lambda = 1$, $k = 0$), we find instead of (11.56)

$$D_n(r) = a_n r^n + b_n r^{-n-1}. \tag{11.57}$$

The condition that J be finite at $r = 0$ requires $b_n = 0$. We therefore obtain

$$J = \frac{\lambda}{4} S[3 + (1-\lambda) x_1] C(r \sin \psi) e^{\alpha r \cos \psi}$$

$$+ \frac{1}{\sqrt{(r)}} \sum_0^\infty a_n I_{n+(1/2)}(\alpha kr) P_n(\cos \psi) \qquad \text{(for } \lambda < 1) \tag{11.58}$$

and

$$J = \tfrac{3}{4} SC(r \sin \psi) e^{\alpha r \cos \psi} + \sum_0^\infty a_n r^n P_n(\cos \psi) \qquad \text{(for } \lambda = 1). \tag{11.59}$$

The constants a_n entering (11.58) and (11.59) may be found from the boundary condition (11.43), which may be rewritten with the aid of (11.37) in the form

$$\frac{3 - \lambda x_1}{2} J = -\frac{1}{\alpha} \frac{\partial J}{\partial r} - \frac{\lambda}{4} Sx_1 \cos \psi e^{-T} \qquad \text{(for } r = R_1). \tag{11.60}$$

We should point out that the problem of light scattering in a homogeneous sphere was also considered by Giovanelli and Jefferies [6]. They obtained equation (11.46) for the case of isotropic scattering and solved it for arbitrary $f(r, \psi)$.

For the second of the two special cases mentioned above we have

$$\alpha(r) = \alpha(R) e^{-(r-R)/H_*}, \tag{11.61}$$

and equation (11.39) takes the form

$$\frac{\partial^2 J}{\partial r^2} + \left(\frac{2}{r} + \frac{1}{H_*}\right) \frac{\partial J}{\partial r} + \frac{1}{r^2 \sin^2 \psi} \frac{\partial}{\partial \psi}\left(\sin \psi \frac{\partial J}{\partial \psi}\right) = \alpha^2(k^2 J - f), \tag{11.62}$$

where f is determined by equation (11.42) with

$$T = \alpha(r) \int_{r \cos \psi}^\infty e^{-(\sqrt{r^2 \sin^2 \psi + z^2} - r)/H_*} dz. \tag{11.63}$$

Equation (11.62) simplifies considerably if we note that the scale height H_* is much less than the radius R of the planet. Then, in place of (11.62) we obtain

$$\frac{\partial^2 J}{\partial r^2} + \frac{1}{H_*} \frac{\partial J}{\partial r} + \frac{1}{R^2} \frac{\partial^2 J}{\partial \psi^2} = \alpha^2(k^2 J - f). \tag{11.64}$$

Introducing the optical depth τ and the optical distance t in a horizontal plane with the aid of the relations

$$\tau = \alpha(r)\, H_*, \quad t = \alpha(R)\, R\psi, \tag{11.65}$$

we may rewrite (11.64) as

$$\frac{\partial^2 J}{\partial \tau^2} + \left(\frac{\tau_0}{\tau}\right)^2 \frac{\partial^2 J}{\partial t^2} = k^2 J - f, \tag{11.66}$$

where τ_0 is the optical thickness of the atmosphere.

The transformation from equation (11.62) to (11.64) or (11.66) corresponds in actuality to the assumption that the atmosphere consists of plane-parallel layers. As with the integral equation (11.9), however, we assume that this atmosphere is illuminated by solar radiation in the same way as the original spherical atmosphere. Equation (11.64) has been solved by O. I. Smoktii [7], who obtained numerical values for the quantity J.

Recently, the problem of light scattering in a spherical atmosphere has been solved by means other than transforming to the approximate equation (11.39). G. I. Marchuk and G. A. Mikhailov [8] used a Monte Carlo method in which data on the phase function and the absorption coefficient as a function of altitude were taken from observation. T. A. Germogenova, L. I. Koprova, and T. A. Sushkevich [9] solved the system of equations (11.28) and (11.29) by numerical methods.

The "principles of invariance" may be applied to the solution of the problem of diffuse light reflection by a spherical atmosphere. Bellman et al. [10] obtained in this way a functional equation for the determination of the reflection coefficient.

11.4. The Case of an Absorption Coefficient Exponentially Decreasing with Altitude

Let us assume that the dependence of the absorption coefficient on altitude in the atmosphere is given by equation (11.61). In this case equation (11.66) may be used to determine the function J. We shall now find an approximate solution to this equation, following the procedure in [5].

We first consider the quantity $T(\tau, \psi)$, which is the optical path of the direct solar radiation between the upper boundary of the atmosphere and the point with coordinates (τ, ψ). This quantity appears in the exponent of the inhomogeneous term in equation (11.66); it is determined in the present case by equation (11.63). In order to simplify this equation, we shall make use of the fact that $H_*/R \ll 1$.

We shall initially assume that $\psi \le \pi/2$. Then equation (11.63) may be rewritten in the form

$$T = \alpha(r) \int_r^\infty e^{-(r'-r)/H_*} \frac{r'\, dr'}{\sqrt{(r'^2 - r^2 \sin^2 \psi)}}. \tag{11.67}$$

Setting $r' = r + H_* x$, we obtain

$$T = \alpha(r)\, H_* \int_0^\infty e^{-x} \frac{(r + H_* x)\, dx}{\sqrt{(r^2 \cos^2 \psi + 2rH_* x + H_*^2 x^2)}}. \tag{11.68}$$

Using the first of equations (11.65) and setting $\tau/\tau_0 = u$, we find

$$T(\tau, \psi) = \tau b(u, \psi), \tag{11.69}$$

where

$$b(u, \psi) = \int_0^\infty e^{-x} \frac{\left(1+\dfrac{H^*}{r} x\right) dx}{\sqrt{\left(1-\sin\psi+\dfrac{H_*}{r} x\right)\left(1+\sin\psi+\dfrac{H_*}{r} x\right)}}, \tag{11.70}$$

$$\frac{r}{H_*} = \frac{R}{H_*} - \ln u. \tag{11.71}$$

For $H_* \ll r$ the function $b(u, \psi)$ may be expanded in the series

$$b(u, \psi) = b_0(u, \psi) + b_1(u, \psi) + \cdots, \tag{11.72}$$

where

$$b_0(u, \psi) = \frac{1}{\sqrt{(1+\sin\psi)}} \int_0^\infty e^{-x} \frac{dx}{\sqrt{1-\sin\psi+\dfrac{H_*}{r} x}} \tag{11.73}$$

and

$$b_1(u, \psi) = \frac{1+2\sin\psi}{2(1+\sin\psi)^{3/2}} \frac{H_*}{r} \int_0^\infty e^{-x} \frac{x \, dx}{\sqrt{1-\sin\psi+\dfrac{H_*}{r} x}}. \tag{11.74}$$

Introducing the notation

$$p = \frac{r}{H_*} (1-\sin\psi), \tag{11.75}$$

we find in place of (11.73) and (11.74)

$$b_0(u, \psi) = g(p) \sec\psi, \tag{11.76}$$

$$b_1(u, \psi) = \left[1+\left(\frac{1}{2p}-1\right) g(p)\right] \frac{1+2\sin\psi}{2(1+\sin\psi)^2} \cos\psi, \tag{11.77}$$

where

$$g(p) = 2\sqrt{p}\, e^p \int_{\sqrt{p}}^\infty e^{-z^2} \, dz. \tag{11.78}$$

We note that for $p \gg 1$, we may use the expansion

$$g(p) = 1 - \frac{1}{2p} + \frac{3}{(2p)^2} - \cdots \tag{11.79}$$

for the calculation of $g(p)$. Therefore, for angles ψ not close to $\pi/2$, we obtain approximately $T = \tau \sec\psi$.

16*

It follows from (11.76) and (11.77) that

$$b_0\left(u, \frac{\pi}{2}\right) = \sqrt{\frac{\pi r}{2H_*}}, \qquad b_1\left(u, \frac{\pi}{2}\right) = \frac{3H^*}{8r} b_0\left(u, \frac{\pi}{2}\right). \tag{11.80}$$

With the aid of equations (11.72) and (11.80) we may find the quantity $2T(\tau,\pi/2) = 2\tau b(u,\pi/2)$ which represents the total optical path in the atmosphere along a ray which penetrates to optical depth τ at its closest approach to the planetary surface.

If $\psi > \pi/2$, we obtain for $T(\tau, \psi)$

$$T(\tau, \psi) = 2T\left(\tau_1, \frac{\pi}{2}\right) - T(\tau, \pi - \psi), \tag{11.81}$$

where τ_1 is the greatest optical depth penetrated by a particular ray being considered. Using equation (11.69) for $T(\tau, \psi)$ and setting $u_1 = \tau_1/\tau_0$, we obtain in place of (11.81)

$$b(u, \psi) = 2\frac{u_1}{u} b\left(u_1, \frac{\pi}{2}\right) - b(u, \pi - \psi). \tag{11.82}$$

For u_1 we find from (11.71) and (11.75)

$$\ln \frac{u_1}{u} = p = (1 - \sin \psi)\left(\frac{R}{H_*} - \ln u\right). \tag{11.83}$$

Table 11.1 presents values of the function $b(u, \psi)$ for various values of u and ψ. The calculations were carried out for $R/H_* = 800$, which approximately applies to the Earth. Since H_*/R is very small, only the first term was taken in the expansion (11.72), so that the computation was in fact for $b_0(u, \psi)$. Observe that for $\psi \leq \pi/2$ at the upper boundary of the atmosphere, $b_0(u, \psi) = b_0(0, \psi) = \sec \psi$.

TABLE 11.1. VALUES OF THE FUNCTION $b(u, \psi)$

ψ \ u	0	10^{-5}	10^{-4}	10^{-3}	10^{-2}	10^{-1}	1
80°	5.76	5.59	5.59	5.59	5.58	5.58	5.58
82	7.18	6.81	6.81	6.81	6.80	6.80	6.80
84	9.57	8.76	8.75	8.75	8.73	8.73	8.73
86	14.3	12.0	12.0	12.0	12.0	12.0	12.0
87	19.1	14.7	14.7	14.7	14.7	14.7	14.7
88	28.6	18.7	18.7	18.7	18.7	18.7	18.7
89	57.3	25.0	24.9	24.9	24.9	24.9	24.9
90		35.7	35.7	35.6	35.6	35.5	35.4
91		55.8	55.7	55.7	55.5	55.3	
92		98.4	98.1	97.8	97.4	97.0	
93		203	202	201	200	198	
94		504	500	496	493	489	
96		6 080	5 990	5910	5830		
98		192 000	188 000				

It follows from Table 11.1 that for a given value of ψ, $b(u, \psi)$ is almost constant. Only in the very outer layers of the atmosphere with very small values of u does b increase with altitude. The same conclusion could be reached on the basis of (11.83), from which it may be seen that the quantities p and u_1/u for large R/H_* are almost independent of u if u is not too small.

The function $T(\tau, \psi)$ may thus be represented with an accuracy sufficient for many applications as the product of optical depth τ and the quantity b, where the latter depends only on ψ and plays the role of a "generalized secant". Analogously, the function f, determined by equation (11.42), may be approximately represented as

$$f = \frac{\lambda}{4} S[3 + (1-\lambda) x_1] e^{-\tau b(\psi)}. \tag{11.84}$$

We return now to the solution of equation (11.66). We recall that it was obtained on the assumption that the atmosphere consists of plane-parallel layers. In the usual treatment of the problem, it is also assumed that the inhomogeneous term in this equation is given by (11.84), with $b = \sec \psi$. This corresponds to the assumption that the quantity $b = \sec \psi$ has the same value throughout the atmosphere as at the point under consideration; consequently, we may neglect the derivative with respect to t in equation (11.66). The latter approximation is very accurate for angles ψ which are sufficiently far from $\pi/2$. We shall, however, apply it to the entire atmosphere, so that equation (11.66) will be approximately replaced by

$$\frac{d^2 J}{d\tau^2} = k^2 J - f, \tag{11.85}$$

with the inhomogeneous term given by equation (11.84) and values of $b(\psi)$ obtained for the given location from equation (11.70) or from Table 11.1. The error resulting from this procedure increases as ψ approaches $\pi/2$, but equation (11.85) may still be used as a first approximation even for angles ψ close to $\pi/2$.

The solution of equation (11.85) takes the form

$$J = Ce^{k\tau} + De^{-k\tau} - \frac{1}{2k} \int_0^\tau f(\tau') \left[e^{k(\tau-\tau')} - e^{-k(\tau-\tau')} \right] d\tau' \quad \text{(for } \lambda < 1\text{)}, \tag{11.86}$$

$$J = C + D\tau - \int_0^\tau f(\tau')(\tau - \tau') \, d\tau' \quad \text{(for } \lambda = 1\text{)}, \tag{11.87}$$

where C and D are constants to be determined from the boundary conditions (11.43) and (11.44).

Once the function $J(\tau, \psi)$ is known, the radiation intensity at any point in the atmosphere may be found with the aid of the equations derived in Section 11.2. In what follows we shall determine the intensity for two cases of particular interest in practical applications.

1. *The brightness of a planet close to the terminator*

We shall assume that the planet is surrounded by an atmosphere of infinitely great optical thickness. From the fact that J does not increase as $\tau \rightarrow \infty$, we find for the constant C which enters equation (11.86)

$$C = \frac{1}{2k} \int_0^\infty f(\tau) \, e^{-k\tau} \, d\tau. \tag{11.88}$$

For simplicity, we shall assume that the scattering in the atmosphere is isotropic. In this case $k = \sqrt{3(1-\lambda)}$ and we obtain from the boundary condition (11.43) for $x_1 = 0$

$$D = -\frac{3-2k}{3+2k} C. \tag{11.89}$$

Substitution of (11.88) and (11.89) into (11.86) leads to the following expression for J:

$$J = \frac{1}{2k} \int_0^\infty f(\tau') \left[e^{-k|\tau - \tau'|} - \frac{3-2k}{3+2k} e^{-k(\tau+\tau')} \right] d\tau'. \tag{11.90}$$

Let us find the brightness distribution along the intensity equator of the planet. The intensity emerging from the atmosphere at an angular distance ϑ from the center of the disc is

$$I(\vartheta, \psi) = \lambda \int_0^\infty \left[J(\tau, \psi) + \frac{S}{4} e^{-T(\tau, \psi)} \right] e^{-\tau \sec \vartheta} \sec \vartheta \, d\tau. \tag{11.91}$$

Substituting (11.90) into (11.91) and using the approximate expression (11.84) for $x_1 = 0$, we obtain after some small transformations

$$I(\vartheta, \psi) = I_1(\vartheta, \psi) + \frac{3}{4} S \frac{\lambda^2}{1-k^2 \cos^2 \vartheta} \left(\frac{2+3 \cos \vartheta}{2k+3} \frac{1}{b+k} - \frac{\cos^2 \vartheta}{1+b \cos \vartheta} \right), \tag{11.92}$$

where

$$I_1(\vartheta, \psi) = \frac{\lambda}{4} S \frac{1}{1+b \cos \vartheta} \tag{11.93}$$

is the intensity resulting from first-order scattering. Note that the sum (or the difference) of the angles ψ and ϑ is just the phase angle (the angle at the planet between the directions to the Sun and the Earth).

Table 11.2 presents values of I and I_1 determined from equations (11.92) and (11.93) for phase angle $90°$ (in units of $10^{-3} S$). The calculations were carried out for the case of pure scattering ($\lambda = 1$). For comparison, values of the quantity I_* determined by equation (11.92) for $b = \sec \psi$ are given; I_* is the intensity obtained from the usual theory neglecting atmospheric curvature.

TABLE 11.2. BRIGHTNESS OF A PLANET CLOSE TO THE TERMINATOR FOR ISOTROPIC
SCATTERING

ψ	I_1	I	I_*	ψ	I_1	I	I_*
80°	38.5	148	145	90°	6.85	21.7	0
82	32.2	120	113	91	4.42	13.6	0
84	25.8	92.6	82.9	92	2.52	7.71	0
86	19.2	65.8	54.6	93	1.24	3.76	0
87	15.9	52.7	40.7	94	0.505	1.52	0
88	12.7	41.6	26.8	96	0.0425	0.127	0
89	9.65	31.0	13.3	98	0.0133	0.0396	0

The computed values of I may be compared with observational results for Venus. Such a comparison shows that the computed brightness of the planet beyond the terminator decreases considerably more rapidly than the observations. The difference probably reflects inaccuracies in the theory. At the same time, it is necessary to keep in mind inaccuracies in the observational data, since the observed brightness of the planet beyond the terminator is strongly affected by photographic irradiation.

2. Brightness of the zenith

We shall assume that the atmosphere has a finite optical thickness τ_0 and overlies a surface with albedo a. In addition, we shall restrict ourselves to the case of pure scattering ($\lambda = 1$). The quantity J must then be obtained from equation (11.87). As in the preceding example, we shall take $x_1 = 0$, which corresponds to isotropic scattering or Rayleigh scattering.

The constants C and D entering equation (11.87) may be found with the aid of the boundary conditions (11.43) and (11.44), in which $3H = \partial J/\partial \tau$, as follows from (11.37). These conditions show that $C = 2D/3$ and D is given by

$$[\tfrac{4}{3}+(1-a)\,\tau_0]\,D = (1-a)\int_0^{\tau_0} f(\tau)(\tau_0-\tau)\,d\tau$$

$$+\tfrac{2}{3}(1+a)\int_0^{\tau_0} f(\tau)\,d\tau + aSe^{-T}\cos\psi. \qquad (11.94)$$

The last term in equation (11.94) vanishes for $\psi > \pi/2$.

Knowledge of J allows us to determine the radiation intensity in the atmosphere. In particular, we may find the brightness at the zenith as observed from the surface of the planet. Writing this brightness as $I(\psi)$, where ψ is the zenith distance of the Sun, we have

$$I(\psi) = \int_0^{\tau_0} J(\tau, \psi)\, e^{-\tau_0+\tau}\,d\tau + I_1(\psi), \qquad (11.95)$$

where $I_1(\psi)$ is the zenith brightness produced by first-order scattering. Substituting (11.87) into (11.95), we find

$$I(\psi) = D(\tau_0 - \tfrac{1}{3} + \tfrac{1}{3}e^{-\tau_0}) + \int_0^{\tau_0} f(\tau)(1 - \tau_0 + \tau - e^{-\tau_0 + \tau})\, d\tau + I_1(\psi). \tag{11.96}$$

For the function $f(t)$ we shall take the approximate expression (11.84), which in the present case ($\lambda = 1$) takes the form

$$f(\tau) = \tfrac{3}{4} S e^{-\tau b(\psi)}. \tag{11.97}$$

Of course, for $\psi > \pi/2$ this expression is valid only in that part of the atmosphere illuminated by the Sun. Nonetheless, we shall now apply it to the entire atmosphere. The resulting error will not be large, since for an optical thickness of the atmosphere that is not very small, the quantity f as given by equation (11.97) will be very small in the unilluminated region. As a result, the approximate value will be close to the actual value $f = 0$.

Substituting equation (11.97) into (11.96), we obtain

$$I(\psi) = D(\tau_0 - \tfrac{1}{3} + \tfrac{1}{3}\, e^{-\tau_0})$$
$$+ \frac{3S}{4b}\left[\left(1 + \frac{1}{b}\right)(1 - e^{-b\tau_0}) - \tau_0 - \frac{b}{b-1}(e^{-\tau_0} - e^{-b\tau_0})\right] + I_1(\psi), \tag{11.98}$$

where

$$I_1(\psi) = x(\psi)\,\frac{S}{4}\,\frac{e^{-\tau_0} - e^{-b\tau_0}}{b-1}. \tag{11.99}$$

For the constant D we now have the expression

$$[\tfrac{4}{3} + (1-a)\tau_0]\,D = \frac{3S}{4b}\left[(1-a)\left(\tau_0 - \frac{1}{b} + \frac{1}{b}\,e^{-b\tau_0}\right)\right.$$
$$\left. + \tfrac{2}{3}(1+a)(1 - e^{-b\tau_0}) + \tfrac{4}{3}ae^{-b\tau_0}b\,\cos\psi\right], \tag{11.100}$$

which follows from substitution of (11.97) into (11.94).

Table 11.3 presents values of the zenith brightness computed from equation (11.98) for a Rayleigh phase function (in units of $10^{-3}\,S$). Values of $I(\psi)$ are given for two values of the surface albedo of the planet ($a = 0.2$ and $a = 0.8$), corresponding approximately to summer and winter conditions.

The brightness at the zenith for various positions of the Sun close to the horizon has been determined observationally in many studies. A comparison of the computed and observed values shows general agreement between them.

When the Sun is below the horizon, the zenith brightness results principally from light scattering in the upper layers of the atmosphere. As the zenith distance of the Sun increases further, the zenith brightness decreases in a manner which depends on the changing optical properties of the atmosphere with altitude. Consequently, from measurements of the brightness of the twilight sky, we may determine the structure of the upper layers of the atmosphere. More details of this important method for studying these layers are contained in the book by G. V. Rosenberg [11].

TABLE 11.3. BRIGHTNESS AT THE ZENITH FOR SCATTERING ACCORDING
TO THE RAYLEIGH PHASE FUNCTION

ψ	$\tau_0 = 0.1$		$\tau_0 = 0.3$	
	$a = 0.2$	$a = 0.8$	$a = 0.2$	$a = 0.8$
80°	17.3	21.4	36.2	45.8
82	16.0	19.2	31.0	38.6
84	14.4	16.7	25.3	30.6
86	12.2	13.7	18.9	22.4
87	11.0	12.1	15.6	18.4
88	9.33	10.2	12.1	14.3
89	7.54	8.14	9.00	10.6
90	5.56	5.95	6.26	7.39
91	3.62	3.87	3.94	4.67
92	2.04	2.19	2.20	2.61
93	1.00	1.07	1.08	1.41
94	0.402	0.432	0.435	0.515
96	0.0338	0.0362	0.0365	0.0432
98	0.0107	0.0113	0.0114	0.0134

11.5. Spacecraft Observations of Planets

The theory of light scattering in a spherical atmosphere may be used to interpret the results of observations of the planets from spacecraft. Such observations allow us to determine the intensity emerging from various parts of the planet at different wavelengths. Comparison of theory and observation allows us to make certain conclusions concerning the structure of the atmosphere (such as the variation with altitude of different atmospheric components).

If an observer is located at a certain altitude above the surface of a planet, the intensity reaching him is determined by the equation

$$I = \int_0^{T_0} Be^{-T_1} dT_1 + I_R e^{-T_0}, \tag{11.101}$$

where B is the source function, T_1 is the optical distance from the radiating volume element to the observer, T_0 is the optical distance from the planetary surface to the observer, and I_R is the intensity of radiation reflected by the surface. For directions which do not intersect the planetary surface, the second term in equation (11.101) is absent, and T_0 is the total optical path through the atmosphere to the observer.

We have already obtained the approximate equation (11.45) for the source function. Substituting it into (11.101), we find

$$I = I_1 + \Delta I + I_R e^{-T_0}, \tag{11.102}$$

where

$$I_1 = \frac{\lambda}{4} Sx(\gamma) \int_0^{T_0} e^{-T - T_1} dT_1 \tag{11.103}$$

and

$$\Delta I = \lambda \int_0^{T_0} (J + x_1 H \cos \vartheta + x_1 G \sin \vartheta \cos \varphi) \, e^{-T_1} \, dT_1. \tag{11.104}$$

The quantity I_1 represents first-order scattering, and ΔI corresponds to scattering of higher orders.

In order to actually carry out the integration in equations (11.103) and (11.104), we must express the coordinates r and ψ and the angles ϑ and φ in terms of quantities characterizing the position of the observer and his direction of view for each point along the path through the atmosphere defined by the direction of observation.

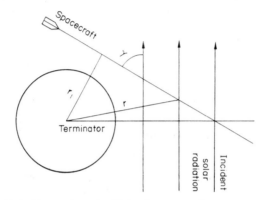

FIG. 11.4. Observation of twilight phenomena from a spacecraft.

The case when the spacecraft is located in the portion of the atmosphere which is not illuminated by the Sun and is undertaking observations of the twilight sky is of particular interest (Fig. 11.4). Such observations may provide valuable knowledge of the structure of the upper atmospheric layers. As we have already mentioned in the preceding section, these layers have been studied by twilight observations from the Earth's surface. Spacecraft observations have a different character, however, because of their altitude above the surface, and new results may be obtained from them. For such observations, the angle γ between the direction of solar radiation and the direction of radiation reaching the observer is generally small. Due to the forward elongation of the phase function, the quantity I_1 in the direction of observation is therefore relatively large. In contrast, the quantity ΔI is more or less equally distributed with respect to direction. Consequently, we may assume that the intensity I results principally from first-order scattering. More detailed calculations lead to the same conclusion [8].

Thus, in the present case we need retain only the first term in equation (11.102). Assuming in the interest of generality that the phase function depends on altitude, we obtain for the intensity reaching the observer

$$I = \frac{S}{4} \int_{T_*}^{T_0} x(\gamma, r) e^{-T - T_1} \, dT_1, \tag{11.105}$$

where T_* designates the optical distance along the path of observation from the observer to the boundary of the illuminated region. For simplicity, we have assumed in equation (11.105) that $\lambda = 1$.

The optical distances entering equation (11.105) may be found with the aid of expressions analogous to (11.4). For example, T_1 is given by the equation

$$T_1 = T_0 - \int_r^\infty \alpha(r') \frac{r' \, dr'}{\sqrt{(r'^2 - r_1^2)}}, \tag{11.106}$$

where r_1 is the minimum distance from the path of observation to the center of the planet (see Fig. 11.4). In order to compute T_1 from equation (11.106) (and also the quantities T, T_* and T_0 from analogous expressions), the law expressing the absorption coefficient as a function of altitude must be given. Let us assume that the atmosphere consists of two components, molecules and aerosols. Designating the corresponding absorption coefficients by $\alpha_M(r)$ and $\alpha_A(r)$, we have

$$\alpha(r) = \alpha_M(r) + \alpha_A(r). \tag{11.107}$$

We might, for example, take $\alpha_M(r)$ and $\alpha_A(r)$ as decreasing exponential functions of altitude with differing values of the scale height.

The phase function which enters equation (11.105) may be written in the form

$$x(\gamma, r) = \frac{\alpha_M(r)x_M(\gamma) + \alpha_A(r)x_A(\gamma, r)}{\alpha_M(r) + \alpha_A(r)}, \tag{11.108}$$

where $x_M(\gamma)$ and $x_A(\gamma, r)$ are the phase functions for scattering of light by molecules and by aerosols, respectively. The quantity $x_M(\gamma)$ is given by

$$x_M(\gamma) = \tfrac{3}{4}(1 + \cos^2 \gamma), \tag{11.109}$$

while $x_A(\gamma, r)$ must be defined by experimental data.

As an example of the determination of the optical characteristics of the terrestrial atmosphere from spacecraft observations, we may cite the work of G. V. Rosenberg and V. V. Nikolaeva-Tereshkova [12]. Those authors used photographs of the twilight aureole of the Earth obtained from the spacecraft "Vostok-6". By comparing the measured and theoretical brightness of the aureole, they determined the structure of aerosol layers in the stratosphere.

References

1. I. N. MININ and V. V. SOBOLEV, On the theory of light scattering in planetary atmospheres, *Astron. Zh.* **40**, 496 (1963) [*Sov. Astron. A.J.* **7**, 379 (1963)].
2. J. LENOBLE and Z. SEKERA, Equation of radiative transfer in a planetary spherical atmosphere, *Proc. Nat. Acad. Sci.U.S.A.* **47**, 372 (1961).
3. V. V. SOBOLEV and I. N. MININ, Light scattering in a spherical atmosphere. I, *Iskusstven. Sputniki Zemli* No.14 (1962).
4. I. N. MININ and V. V. SOBOLEV, Scattering of light in a spherical atmosphere. II, *Kosmichesk. Issled.* **1**, 287 (1963).
5. I. N. MININ and V. V. SOBOLEV, Scattering of light in a spherical atmosphere. III, *Kosmichesk. Issled.* **2**, 610 (1964).

6. R. G. GIOVANELLI and J. T. JEFFERIES, Radiative transfer with distributed sources, *Proc. Phys. Soc.* B **69**, 1077 (1956).

7. O. I. SMOKTII, Multiple scattering of light in an inhomogeneous spherically symmetric planetary atmosphere, *Izv. Akad. Nauk SSSR Fiz. Atmosfer. Okeana* **3**, 496 (1967) [*Atmos. and Oceanic Phys.* **3**, 281 (1967)].

8. G. I. MARCHUK and G. A. MIKHAILOV, Results of the solution of some problems in atmospheric optics by a Monte Carlo method, *Izv. Akad. Nauk SSSR Fiz. Atmosfer. Okeana* **3**, 394 (1967) [*Atmos. and Oceanic Phys.* **3**, 227 (1967)].

9. T. A. GERMOGENOVA, L. I. KOPROVA and T. A. SUSHKEVICH, Investigation of the angular distribution and spectral structure of the terrestrial radiation field for characteristic model spherical atmospheres, *Izv. Akad. Nauk SSSR Fiz. Atmosfer. Okeana* **5**, 1266 (1969) [*Atmos. and Oceanic Phys.* **5**, 731 (1969)].

10. R. E. BELLMAN, H. H. KAGIWADA, R. E. KALABA and S. UENO, Diffuse reflection of solar rays by a spherical shell atmosphere, *Icarus* **11**, 417 (1969).

11. G. V. ROSENBERG, *Sumerki*, Fizmatgiz, Moscow, 1963.

12. G. V. ROSENBERG and V. V. NIKOLAEVA-TERESHKOVA, Stratospheric aerosol according to spacecraft measurements, *Izv. Akad. Nauk SSSR Fiz. Atmosfer. Okeana* **1**, No. 4 (1965) [*Atmos. and Oceanic Phys.* **1**, 228 (1965)].

CONCLUDING REMARKS

DEVELOPMENT of the theory of light scattering in planetary atmospheres began in the 1940s. Subsequently, however, interest in the subject decreased sharply and only recently have efforts been renewed, following the general increase of interest in the planets. The main content of the present book consists of results obtained in this field during the last few years. We shall now summarize the most important of these results and mention some of the problems which remain unsolved.

Of paramount interest for various applications of the theory is the intensity of radiation diffusely reflected and diffusely transmitted by the atmosphere. Chapters 2–4 are devoted to the determination of these quantities. Results are presented which are well established theoretically. Particular attention, however, is given to asymptotic expressions which have been obtained recently for the reflection and transmission coefficients. Some of these equations refer to the case of small true absorption $(1 - \lambda \ll 1)$, others to the case of an atmosphere of large optical thickness $(\tau_0 \gg 1)$. The unusual simplicity of the expressions in these two cases make them quite useful in practical applications. All of these results apply to homogeneous atmospheres (with the exception of Section 3.5). Since in real atmospheres the phase function $x(\gamma)$ and the single scattering albedo λ depend upon altitude, it is necessary to generalize the theory to inhomogeneous atmospheres.

Chapters 5 and 6 are central to the theory. In them we introduce the fundamental functions $\Phi^m(\tau)$ and $\Phi^m(\tau, \tau_0)$ which are determined by equations (5.80) and (6.15), respectively. Knowledge of $\Phi^m(\tau)$ completely determines the radiation field in a semi-infinite atmosphere, while knowledge of $\Phi^m(\tau, \tau_0)$ achieves the same for an atmosphere of finite optical thickness τ_0. Specifically, we may find the intensity emerging from the atmosphere for an arbitrary distribution of energy sources if the auxiliary functions are known. The auxiliary functions $\varphi_i^m(\eta)$ for a semi-infinite atmosphere may be easily expressed in terms of the function $H^m(\eta)$, and the auxiliary functions $\varphi_i^m(\eta, \tau_0)$ and $\psi_i^m(\eta, \tau_0)$ for $\tau_0 < \infty$ in terms of the functions $X^m(\eta, \tau_0)$ and $Y^m(\eta, \tau_0)$. The function $H^m(\eta)$ may be expressed in terms of $\Phi^m(\tau)$, and the functions $X^m(\eta, \tau_0)$ and $Y^m(\eta, \tau_0)$ in terms of the function $\Phi^m(\tau, \tau_0)$.

It is shown in Chapter 6 that, after $\Phi^m(\tau, \tau_0)$ has been determined, it is relatively easy to find the quantities $I^m(\tau, \eta, \zeta, \tau_0)$, the coefficients of an expansion of the intensity in a cosine series in azimuth. The most important of these is $I^0(\tau, \eta, \zeta, \tau_0)$, for which we need only know $\Phi^0(\tau, \tau_0)$. This importance is due to the fact that the density and flux of radiation in the atmosphere may be expressed in terms of I^0. In addition, I^0 represents the total intensity in an

235

atmosphere illuminated at normal incidence, and also the total intensity in the deep layers of an atmosphere of large optical thickness. In general, however, it is necessary to determine all the quantities I^m in order to specify the total intensity. The relative magnitude of the coefficients I^m generally decreases with increasing m, the speed of this decrease depending to a large degree upon τ_0. The behavior of the I^m has been insufficiently studied up to the present, but definitely should be thoroughly investigated because of the extremely great significance of these coefficients.

In Chapter 7 we obtain linear integral equations for the coefficients of reflection and transmission and also for the auxiliary functions in terms of which those coefficients may be expressed. In certain respects these equations have significant advantages over the non-linear integral equations introduced in Chapters 2 and 3.

The equations contained in Chapter 5–7 have been solved for various particular cases in a number of papers. As a result, tables of $\Phi^m(\tau)$, $H^m(\eta)$, and related functions, as well as the reflection and transmission coefficients, have been obtained. A portion of these tables is presented in the present book. The numerical solution of the indicated equations does not present great difficulty; nonetheless, the finding of more efficient methods of solution would be quite desirable.

This book is basically devoted to the exact solution of the problem of light scattering in planetary atmospheres. In the interpretation of the observational data, however, it is often sufficient to use approximate expressions for the intensity of scattered radiation. Such expressions are given in Chapter 8. Some of them have already been broadly applied. In efforts to obtain new approximate formulas, particular attention should be paid to the case of greatly elongated phase functions.

The solution to the problem of diffuse reflection by a planetary atmosphere allows us to determine the characteristics of the radiation emerging from a planet. Chapter 9 presents formulas giving the brightness distribution across the disc of a planet and its total brightness as a function of phase angle. Equations are also derived for the profile and equivalent width of lines in planetary spectra. With the aid of these equations the values of various quantities may be computed and compared with corresponding values obtained from observation (for example, the value of the equivalent width of a molecular band in the spectrum of the given planet). In the same chapter the problem of the polarization of planetary radiation is also briefly examined. Further progress in this area requires additional work on the solution of the equation of transfer for polarized radiation in a medium containing aerosol particles.

By comparing the theoretical and observational values of the quantities characterizing the radiation from a planet, we may deduce the optical properties of the planetary atmosphere. Examples of such determinations are given in Chapter 10. More recent observations of the planets have already yielded additional valuable results (especially spectroscopic and polarimetric data). In the near future, further studies will undoubtedly appear which interpret the results of planetary observations on the basis of light scattering theory. The reliability of such interpretations must be carefully considered, since the characteristics of the radiation scattered by media with different optical properties may be quite similar. It is thus important to solve the inverse problem of the theory of light scattering, which consists of determining the optical properties of a medium when the intensity of scattered radiation is given.

In the last chapter we examine the problem of light scattering in a spherical atmosphere, which has assumed particular practical importance following the launching of spacecraft. The analytical solution of this problem is quite cumbersome. Consequently, it will probably be solved in the future by numerical methods (in particular, by Monte Carlo methods).

The present book is written primarily for research workers in planetary science. The first eight chapters, however, have a much wider value. These chapters consider general problems of light-scattering theory, which is necessary not only for the study of planetary atmospheres but also in other branches of science.

The theory of light scattering is used extensively in the physics of the terrestrial atmosphere; in fact, many problems of atmospheric optics could not be solved without it (e.g. the determination of the optical parameters of the atmosphere; see Section 10.3). In addition, use of this theory allows us to solve a number of problems which have great practical significance. One of these is the "problem of visibility", which consists in the calculation of the distance at which objects are visible to an observer on the surface of the Earth or at an arbitrary altitude above the surface.

The theory of light scattering is also used in studying optical phenomena in the sea. One of the main problems in the optics of the sea is in determining the radiation field at large depths. The equations obtained in Sections 2.1 and 2.4 solve this problem. Certain other problems of marine optics are solved with the aid of the theory in the book *TRT*.

The theory of light scattering is used quite extensively in physics and chemistry. Important properties of many substances may be determined by their ability to scatter radiation. Such substances include powders, colloidal solutions, paints, tree leaves, and others. Comparison of the theoretical and measured values of the intensity of the scattered radiation allows us to determine the optical properties of the given substance; that is, the coefficients of scattering and true absorption and the phase function. From these properties we may estimate the concentration of particles, their sizes, and their physical nature. Spectrophotometric investigations give the most valuable results. In some cases such studies are carried out for optically thin layers of a substance, and in other cases (particularly when the role of true absorption is small, but must be determined) for layers of large optical thickness. These methods are utilized not only in scientific laboratories, but also in industry.

We have already mentioned that the theory of light scattering is in many respect close to the theory of particle transport (in particular, for neutrons). Consequently, certain results contained in the first eight chapters of this book may be used in studying such problems.

APPENDIX

Light Scattering in an Inhomogeneous Atmosphere[†]

V. V. SOBOLEV

Abstract

The problem of the diffuse reflection and transmission of radiation by an isotropically scattering atmosphere is considered. The albedo for single scattering is first assumed to be an arbitrary function of optical depth, and is then specialized to be the step function. The resulting two-layer atmosphere is considered in detail. Three methods are given for the solution of the problem when such an atmosphere is composed of a layer of finite optical thickness overlying a semi-infinite medium.

It is usual in the theory of stellar and planetary atmospheres to assume that the atmosphere consists of plane-parallel layers with the same optical properties (see [1, 2]). In reality, however, such atmospheric properties as the phase function and albedo for single scattering will change with altitude. The scattering of light in such atmospheres has received relatively little attention [3–8].

In the present article we consider the diffuse reflection and transmission of light by an atmosphere within which the single scattering albedo λ (the "particle albedo") is a function of optical depth τ. We shall assume that the phase function is isotropic.

Let the atmosphere be illuminated by parallel radiation, incident at an angle $\arccos \zeta$ to the normal and having incident flux πS through an area perpendicular to the direction of incidence. The source function $B(\tau, \zeta)$ is then determined by the equation

$$B(\tau, \zeta) = \frac{\lambda(\tau)}{2} \int_0^{\tau_0} E_1(|\tau - t|) B(t, \zeta) \, dt + \frac{\lambda(\tau)}{4} S e^{-\tau/\zeta}, \qquad (1)$$

where τ_0 is the optical thickness of the atmosphere.

---

† This article appeared in the *Astronomicheskii Zhurnal*, **51**, 50 (1974). The reader interested in the topic discussed may also wish to refer to the article "Spectrum of a planet with a two-layer atmosphere", *Doklady Akad. Nauk SSSR*, **211**, 63 (1973), reprinted on p. 245.

We shall designate the reflection and transmission coefficients of the atmosphere by $\varrho(\eta, \zeta)$ and $\sigma(\eta, \zeta)$, respectively, where η is the cosine of the angle of reflection or transmission. The physical significance of these coefficients lies in the fact that the quantities $S\varrho(\eta, \zeta)\zeta$ and $S\sigma(\eta, \zeta)\zeta$ are the intensities of radiation diffusely reflected and diffusely transmitted by the atmosphere. The quantities $\varrho(\eta, \zeta)$ and $\sigma(\eta, \zeta)$ may be expressed in terms of the source function $B(\tau, \zeta)$ through the equations

$$S\varrho(\eta, \zeta)\zeta = \int_0^{\tau_0} B(\tau, \zeta)\, e^{-\tau/\eta}\, \frac{d\tau}{\eta}, \tag{2}$$

and

$$S\sigma(\eta, \zeta)\zeta = \int_0^{\tau_0} B(\tau, \zeta)\, e^{-(\tau_0-\tau)}\, \frac{d\tau}{\eta}. \tag{3}$$

In order to find ϱ and σ, we first differentiate equation (1) with respect to τ, obtaining the following equation for the function $B'(\tau, \zeta)$:

$$B'(\tau, \zeta) = \frac{\lambda(\tau)}{2} \int_0^{\tau_0} E_1(|\tau - t|)\, B'(\tau, \zeta)\, dt + \frac{\lambda(\tau)}{2} E_1(\tau)\, B(0, \zeta)$$

$$- \frac{\lambda(\tau)}{2} E_1(\tau_0 - \tau)\, B(\tau_0, \zeta) - \frac{\lambda(\tau)}{4\zeta} Se^{-\tau/\zeta} + \frac{\lambda'(\tau)}{\lambda(\tau)} B(\tau, \zeta). \tag{4}$$

After multiplying (4) by $B(\tau, \eta)/\lambda(\tau)$, integrating over τ from 0 to τ_0, and using (1) and (2), we find

$$4(\eta+\zeta)\, \varrho(\eta, \zeta) = \lambda(0)\, C(0, \eta)\, C(0, \zeta) - \lambda(\tau_0)\, C(\tau_0, \eta)\, C(\tau_0, \zeta) + \int_0^{\tau_0} \lambda'(\tau)C(\tau, \eta)C(\tau, \zeta)\, d\tau, \tag{5}$$

where we have set

$$B(\tau, \zeta) = \frac{\lambda(\tau)}{4} SC(\tau, \zeta). \tag{6}$$

Equation (5) may also be written in the form

$$4(\eta+\zeta)\varrho(\eta, \zeta) = -\int_0^{\tau_0} \lambda(\tau)\frac{d}{d\tau}\, [C(\tau, \eta)\, C(\tau, \zeta)]\, d\tau. \tag{7}$$

The quantities $C(0, \zeta)$ and $C(\tau_0, \zeta)$ entering equation (5) may clearly be expressed in terms of the coefficients of reflection and transmission by

$$C(0, \zeta) = 1 + 2\zeta \int_0^1 \varrho(\eta, \zeta)\, d\eta, \tag{8}$$

and

$$C(\tau_0, \zeta) = e^{-\tau_0/\zeta} + 2\zeta \int_0^1 \sigma(\eta, \zeta)d\eta. \tag{9}$$

In order to find an equation analogous to (5) for the quantity $\sigma(\eta, \zeta)$, we must introduce the equation

$$B_*(\tau, \zeta) = \frac{\lambda(\tau)}{2} \int_0^{\tau_0} E_1(|\tau - t|) B_*(t, \zeta) dt + \frac{\lambda(\tau)}{4} S e^{-(\tau_0 - \tau)/\zeta}, \qquad (10)$$

which determines the source function $B(\tau, \zeta)$ for an atmosphere illuminated from below. In this case the coefficients of reflection and transmission take the form

$$\varrho_*(\eta, \zeta) = \frac{1}{S} \int_0^{\tau_0} B_*(\tau, \zeta) e^{-(\tau_0 - \tau)/\eta} \frac{d\tau}{\eta\zeta}, \qquad (11)$$

and

$$\sigma_*(\eta, \zeta) = \frac{1}{S} \int_0^{\tau_0} B_*(\tau, \zeta) e^{-\tau/\eta} \frac{d\tau}{\eta\zeta}. \qquad (12)$$

Multiplying equation (4) by $B_*(\tau, \eta)/\lambda(\tau)$, integrating over τ from 0 to τ_0, and using (10), (3) and (12), we find

$$4(\eta - \zeta) \sigma(\eta, \zeta) = \lambda(0) C(0, \zeta) C_*(0, \eta) - \lambda(\tau_0) C(\tau_0, \zeta) C_*(\tau_0, \eta)$$

$$+ \int_0^{\tau_0} \lambda'(\tau) C(\tau, \zeta) C_*(\tau, \eta) d\tau = - \int_0^{\tau_0} \lambda(\tau) \frac{d}{d\tau} [C(\tau, \zeta) C_*(\tau, \eta)] d\tau. \qquad (13)$$

In the derivation we have used the definition of $C_*(\tau, \zeta)$ in terms of $B_*(\tau, \zeta)$ analogous to (6), and also the relation

$$\sigma_*(\eta, \zeta) = \sigma(\zeta, \eta), \qquad (14)$$

which follows from (1) and (10). From (10), (11), and (12) we may show that

$$C_*(0, \zeta) = e^{-\tau_0/\zeta} + 2\zeta \int_0^1 \sigma_*(\eta, \zeta) d\eta, \qquad (15)$$

and

$$C_*(\tau_0, \zeta) = 1 + 2\zeta \int_0^1 \varrho_*(\eta, \zeta) d\eta. \qquad (16)$$

An equation analogous to (5) may also be obtained for the reflection coefficient $\varrho_*(\eta, \zeta)$. It takes the form

$$4(\eta + \zeta) \varrho_*(\eta, \zeta) = - \lambda(0) C_*(0, \eta) C_*(0, \zeta) + \lambda(\tau_0) C_*(\tau_0, \eta) C_*(\tau_0, \zeta)$$

$$- \int_0^{\tau_0} \lambda'(\tau) C_*(\tau, \eta) C_*(\tau, \zeta) d\tau = \int_0^{\tau_0} \lambda(\tau) \frac{d}{d\tau} [C_*(\tau, \eta) C_*(\tau, \zeta)] d\tau. \qquad (17)$$

An equation for the transmission coefficient $\sigma_*(\eta, \zeta)$ may be derived from (13) and (14).

17*

We have thus obtained equations for the reflection and transmission coefficients of the atmosphere when illuminated either from above (equations (5) and (13)) or from below (equations (14) and (17)). The quantities $C(0, \zeta)$, $C(\tau_0, \zeta)$, $C_*(0, \zeta)$, and $C_*(\tau_0, \zeta)$ entering these equations are given by the relations (8), (9), (15), and (16).

When $\lambda = $ constant, the equations which we have obtained reduce to those found by V. A. Ambartsumyan [9]. If, on the other hand, λ is a function of optical depth, the above equations are insufficient to determine the reflection and transmission coefficients. Nonetheless, these equations may be used in certain cases. We shall now apply them to the case of a multi-layered atmosphere.

Let $\lambda = \lambda_1$ for optical depths between 0 and τ_1, $\lambda = \lambda_2$ in the interval from τ_1 to $\tau_2, \ldots,$ and $\lambda = \lambda_n$ in the interval from τ_{n-1} to $\tau_n = \tau_0$. Equations (5), (13), (14), and (17) then take the form

$$4(\eta + \zeta) \varrho(\eta, \zeta) = \sum_{k=0}^{n} (\lambda_{k+1} - \lambda_k) C(\tau_k, \eta) C(\tau_k, \zeta), \tag{18}$$

$$4(\eta - \zeta) \sigma(\eta, \zeta) = \sum_{k=0}^{n} (\lambda_{k+1} - \lambda_k) C_*(\tau_k, \eta) C(\tau_k, \zeta), \tag{19}$$

$$4(\eta + \zeta) \varrho_*(\eta, \zeta) = - \sum_{k=0}^{n} (\lambda_{k+1} - \lambda_k) C_*(\tau_k, \eta) C_*(\tau_k, \zeta), \tag{20}$$

$$4(\zeta - \eta) \sigma_*(\eta, \zeta) = \sum_{k=0}^{n} (\lambda_{k+1} - \lambda_k) C(\tau_k, \eta) C_*(\tau_k, \zeta), \tag{21}$$

where we have defined $\lambda_0 = \lambda_{n+1} = 0$.

We see that the reflection and transmission coefficients of a multi-layered atmosphere have a completely determined structure. The four functions of two arguments are expressed in terms of $2n+2$ auxiliary functions of a single argument. Upon substitution of equation (18)–(21) into the relations (8), (9), (15) and (16) we obtain four equations for the determination of the auxiliary functions. Additional equations may be found by examining the boundary conditions between the layers. If the reflection and transmission coefficients of each layer are considered as known, then these equations may be written quite simply.

As an example we shall consider a two-layer atmosphere. According to equations (18)–(21) we then have

$$4(\eta + \zeta) \varrho(\eta, \zeta) = \lambda_1 C(0, \eta) C(0, \zeta) + (\lambda_2 - \lambda_1) C(\tau_1, \eta) C(\tau_1, \zeta) - \tag{22}$$
$$- \lambda_2 C(\tau_0, \eta) C(\tau_0, \zeta),$$

$$4(\eta - \zeta) \sigma(\eta, \zeta) = \lambda_1 C(0, \zeta) C_*(0, \eta) + (\lambda_2 - \lambda_1) C(\tau_1, \zeta) C_*(\tau_1, \eta) - \tag{23}$$
$$- \lambda_2 C(\tau_0, \zeta) C_*(\tau_0, \eta),$$

$$4(\eta + \zeta) \varrho_*(\eta, \zeta) = - \lambda_1 C_*(0, \eta) C_*(0, \zeta) - (\lambda_2 - \lambda_1) C_*(\tau_1, \zeta) C_*(\tau_1, \zeta) + \tag{24}$$
$$+ \lambda_2 C_*(\tau_0, \eta) C_*(\tau_0, \zeta),$$

$$4(\zeta - \eta) \sigma_*(\eta, \zeta) = \lambda_1 C(0, \eta) C_*(0, \zeta) + (\lambda_2 - \lambda_1) C(\tau_1, \eta) C_*(\tau_1, \zeta) - \tag{25}$$
$$- \lambda_2 C(\tau_0, \eta) C_*(\tau_0, \zeta).$$

The four coefficients of reflection and transmission for a two-layer atmosphere are thus expressed in terms of six auxiliary functons. The equations expressing the conditions at the atmospheric boundaries and at the boundary between the layers may be used to determine these functions. When the atmosphere is illuminated from above, these equations have the form

$$\varrho(\eta, \zeta) = \varrho_1(\eta, \zeta) + 2 \int_0^1 \beta(\eta', \zeta)\sigma_1(\eta, \eta',)\eta' d\eta' + \beta(\eta, \zeta)e^{-\tau_1/\eta}, \tag{26}$$

$$\alpha(\eta, \zeta) = \sigma_1(\eta, \zeta) + 2 \int_0^1 \beta(\eta', \zeta)\varrho_1(\eta, \eta')\eta' d\eta', \tag{27}$$

$$\beta(\eta, \zeta) = 2 \int_0^1 \alpha(\eta', \zeta)\varrho_2(\eta, \eta')\eta' d\eta' + \varrho_2(\eta, \zeta)e^{-\tau_1/\zeta}, \tag{28}$$

$$\sigma(\eta, \zeta) = 2 \int_0^1 \alpha(\eta', \zeta)\sigma_2(\eta, \eta')\eta' d\eta' + \sigma_2(\eta, \zeta)e^{-\tau_1/\zeta}. \tag{29}$$

In these last equations $S\alpha(\eta, \zeta)\zeta$ and $S\beta(\eta, \zeta)\zeta$ are the radiation intensities at the boundary between the layers traveling upwards and traveling downwards, respectively; $\varrho_1(\eta, \zeta)$ and $\sigma_1(\eta, \zeta)$ are the reflection and transmission coefficients of the upper layer; and $\varrho_2(\eta, \zeta)$ and $\sigma_2(\eta, \zeta)$ are the corresponding coefficients for the lower layer.

It is clear that equations (26)–(29) completely determine the quantities $\varrho(\eta, \zeta)$ and $\sigma(\eta, \zeta)$ which we are seeking. This may be accomplished by finding the functions $\alpha(\eta, \zeta)$ and $\beta(\eta, \zeta)$ from equations (27) and (28) and substituting the results into equations (26) and (29). Note that equations (27) and (28) have been studied in detail by S. D. Gutshabash [10].

Equations (26)–(29) may also be used to find the auxiliary functions which enter the relations (22)–(25). Setting $\zeta = 0$ in the former equations we may obtain the quantities $\varrho(\eta, 0)$ and $\sigma(\eta, 0)$. This allows us to find in turn the functions $C(0, \eta)$ and $C_*(0, \eta)$ from the expressions

$$4\eta\varrho(\eta, 0) = \lambda_1 C(0, \eta), \quad 4\eta\sigma(\eta, 0) = \lambda_1 C_*(0, \eta), \tag{30}$$

which follow from (22) and (23). We may determine the functions $C(\tau_0, \eta)$ and $C_*(\tau_0, \eta)$ in a similar manner, with the aid of the equations analogous to (26)–(29) for an atmosphere illuminated from below. The functions $C(\tau_1, \eta)$ and $C_*(\tau_1, \eta)$ may be found by using equation (23) for $\eta = \zeta$ and the equation which results from substitution of (22) into (8).

The particular case of a two-layer atmosphere consisting of a layer with finite optical thickness τ_1 and single scattering albedo $\lambda = \lambda_1$ overlying a semi-infinite medium with $\lambda = \lambda_2$ merits special attention. For such an atmosphere we need determine only the reflection coefficient, which we shall denote by $\varrho(\eta, \zeta, \tau_1)$. On the basis of equation (22) it may be written in the form

$$\varrho(\eta, \zeta, \tau_1) = \frac{\lambda_1\varphi(\eta, \tau_1)\varphi(\zeta, \tau_1) + (\lambda_2 - \lambda_1)\psi(\eta, \tau_1)\psi(\zeta, \tau_1)}{4(\eta + \zeta)}, \tag{31}$$

where we have set $\varphi(\eta, \tau_1) = C(0, \eta)$ and $\psi(\eta, \tau_1) = C(\tau_1, \eta)$. The auxiliary functions $\varphi(\eta, \tau_1)$ and $\psi(\eta, \tau_1)$ which enter (31) may be found by the method described above.

APPENDIX

The problem of diffuse reflection of light by an inhomogeneous atmosphere has previously been considered both by Sobolev [3] and by Bellman and Kalaba [4]. It was found that in the present particular case of a two-layer atmosphere the reflection coefficient is determined by the equation

$$\varrho(\eta, \zeta, \tau) = \frac{\lambda_1}{4} \int_0^\tau \varphi(\eta, \tau')\varphi(\zeta, \tau')\,e^{-(\tau-\tau')(\zeta^{-1}+\eta^{-1})}\frac{d\tau'}{\eta\zeta} + \varrho_2(\eta, \zeta)\,e^{-\tau(\eta^{-1}+\zeta^{-1})}, \qquad (32)$$

where

$$\varphi(\eta, \tau) = 1 + 2\eta \int_0^1 \varrho(\eta, \zeta, \tau)\,d\eta, \qquad (33)$$

$$\varrho_2(\eta, \zeta) = \frac{\lambda_2}{4}\frac{\varphi_2(\eta)\,\varphi_2(\zeta)}{\eta+\zeta}, \qquad (34)$$

and $\varphi_2(\eta)$ is the Ambartsumyan function for $\lambda = \lambda_2$. Thus, with the aid of equation (32) the reflection coefficient may be expressed in terms of the auxiliary function $\varphi(\eta, \tau)$, which may itself be found from the equation which results from the substitution of (32) into (33).

This latter method for determining the quantity $\varrho(\eta, \zeta, \tau_1)$ requires knowledge of only one auxiliary function, which, however, must be determined for all values of τ between 0 and τ_1. The method described earlier for the determination of $\varrho(\eta, \zeta, \tau_1)$ requires two auxiliary functions, but it is sufficient to know their values for the given optical thichness τ_1 of the upper layer.

We may, moreover, present still a third method for solving the problem of diffuse light reflection by the two-layer atmosphere under consideration. This procedure consists of using equation (31) together with equations for the auxiliary functions $\varphi(\eta, \tau)$ and $\psi(\eta, \tau)$ which include an integration over τ. Such equations are easily obtained from (31) and (32) and take the form

$$\varphi(\eta, \tau) = \varphi_2(\eta) - \frac{\lambda_2-\lambda_1}{2}\int_0^\tau \psi(\eta, \tau')d\tau'\int_0^1 \psi(\zeta, \tau')\frac{d\zeta}{\zeta}, \qquad (35)$$

and

$$\psi(\eta, \tau) = \varphi_2(\eta)e^{-\tau/\eta} + \frac{\lambda_1}{2}\int_0^\tau e^{-(\tau-\tau')/\eta}\varphi(\eta, \tau')d\tau'\int_0^1 \psi(\zeta, \tau')\frac{d\zeta}{\zeta}. \qquad (36)$$

In the absence of the underlying semi-infinite medium (i.e., when $\lambda_2 = 0$ so that $\varphi_2(\eta) = 1$), equations (35) and (36) reduce to the well-known equations for the auxiliary functions through which the reflection and transmission coefficients of a homogeneous layer may be expressed (see [2], Chap. III).

References

1. S. Chandrasekhar, *Radiative Transfer*, Oxford, 1950.
2. V. V. Sobolev, *Perenos Luchistoi Energii v Atmosferakh Zvezd i Planet*, Gostekhizdat, Moscow, 1956 [*A Treatise on Radiative Transfer*, Van Nostrand, Princeton, N.J., 1963].

3. V. V. Sobolev, *Doklad. Akad. Nauk SSSR* **111**, 1000, 1956.
4. R. E. Bellman and R. E. Kalaba, *Proc. Nat. Acad. Sci. USA* **42**, 629, 1956.
5. S. Ueno, *Astrophys. J.* **132**, 729, 1960.
6. I. W. Busbridge, *Astrophys. J.* **133**, 198, 1961.
7. E. G. Yanovitskii, *Astron. Zh.* **38**, 912, 1961 [*Sov. Astron.—AJ* **5**, 697, 1962].
8. K. D. Abhyankar and A. L. Fymat, *Astrophys. J.* **158**, 315, 1969; **158**, 325, 1969; **159**, 1009, 1969; **158**, 1019, 1970.
9. V. A. Ambartsumyan, *Doklad. Akad. Nauk SSSR* **38**, 257, 1943.
10. S. D. Gutshabash, *Vest. Leningrad. Univ.*, No. 1, 158, 1957.

Spectrum of a Planet with a Two-layer Atmosphere[†]

V. V. Sobolev

In the theory of planetary spectra it is usual to assume (see, for examples, refs. 1–6) that the optical properties of an atmosphere do not change with altitude, that is, the atmosphere is assumed to be homogeneous. Actually, planetary atmospheres are not, and the variation in their optical properties often takes place in steps. Examples are transitions from gaseous composition to clouds or from one type of cloud to another. It is therefore desirable to examine the problem of spectrum formation in an atmosphere consisting of several different layers. In the present short article the problem is solved for the special model of a two-layer atmosphere.

Let there be, above a layer of infinitely large optical thickness, a layer of small optical thickness τ having other properties. The optical properties of the lower layer shall be characterized by a phase function $x^*(\gamma)$ and a single scattering albedo λ^*, while the optical properties of the upper layer shall be characterized by a phase function $x(\gamma)$ and a single scattering albedo λ. Both of these phase functions are assumed to be arbitrary.

The quantities λ^*, λ, and τ apply to the continuous spectrum. The respective quantities for a frequency ν within a spectral line shall be designated by λ_ν^*, λ_ν, and τ_ν.

The quantities λ^* and λ_ν^* are determined by the equations

$$\lambda^* = \frac{\sigma^*}{\sigma^* + \varkappa^*}, \quad \lambda_\nu^* = \frac{\sigma^*}{\sigma^* + \varkappa^* + \varkappa_\nu^*}, \tag{1}$$

where σ^* is the scattering coefficient, $\varkappa^*$ is the coefficient of true absorption, and $\varkappa_\nu^*$ is the true-absorption coefficient of the molecules in the lower layer. For the quantities λ and λ_ν we analogously have

$$\lambda = \frac{\sigma}{\sigma + \varkappa}, \quad \lambda_\nu = \frac{\sigma}{\sigma + \varkappa + \varkappa_\nu}, \tag{2}$$

where σ, $\varkappa$, and $\varkappa_\nu$ are the respective coefficients in the upper layer.

The optical thicknesses of the upper layer, in the continuous spectrum and in the line, are respectively

† Reprinted from *Sov. Phys. Dokl.*, Vol. 18, No. 7, January 1974, by kind permission of the American Institute of Physics.

$$\tau = \int\limits_0^\infty (\sigma + \varkappa)dh, \qquad \tau_\nu = \int\limits_0^\infty (\sigma + \varkappa + \varkappa_\nu)dh, \tag{3}$$

where the altitude h is counted from the boundary between the two layers.

From Eqs. (2) and (3) we get

$$\lambda_\nu \tau_\nu = \lambda \tau. \tag{4}$$

We shall assume that the atmosphere is illuminated by parallel solar radiation incident at an angle arccos ζ to the normal and producing an illumination πS through an area perpendicular to the solar direction. We shall designate the intensity of the radiation diffusely reflected by the atmosphere in the continuous spectrum and in a line, respectively, by $I(\eta, \zeta, \varphi)$ and $I_\nu(\eta, \zeta, \varphi)$, where η is the cosine of the angle of reflection, and φ is the difference between the azimuths of the reflected and the incident rays. These intensities are usually represented in the form

$$I(\eta, \zeta, \varphi) = S\varrho(\eta, \zeta, \varphi)\zeta, \qquad I_\nu(\eta, \zeta, \varphi) = S\varrho_\nu(\eta, \zeta, \varphi)\zeta, \tag{5}$$

and the quantites ϱ and ϱ_ν are called coefficients of reflection.

The spectral line profile that interests us in a given location on the planetary disk is, as is well known, characterized by the quantity

$$r_\nu(\eta, \zeta, \varphi) = \frac{I_\nu(\eta, \zeta, \varphi)}{I(\eta, \zeta, \varphi)} = 1 - \frac{\delta_\nu(\eta, \zeta, \varphi)}{\varrho(\eta, \zeta, \varphi)}, \tag{6}$$

$$\delta_\nu(\eta, \zeta, \varphi) = \varrho(\eta, \zeta, \varphi) - \varrho_\nu(\eta, \zeta, \varphi). \tag{7}$$

In order to write the equation for the quantites ϱ and ϱ_ν, we shall take advantage of the smallness of the optical thicknesses τ and τ_ν. Disregarding quantities of the order of τ^2, we find

$$\varrho(\eta, \zeta, \varphi) = \varrho^*(\eta, \zeta, \varphi)\,[1 - \tau(1/\eta + 1/\zeta)]$$
$$+ \lambda\tau f\,[\varrho^*(\eta, \zeta, \varphi), x(\gamma)], \tag{8}$$

where $\varrho^*(\eta, \zeta, \varphi)$ is the coefficient of reflection of the lower layer in the continuous spectrum.

The first term on the right side of Eq. (8) takes into account the passage of the solar radiation through the upper layer, its reflection from the lower layer, and its second passage through the upper layer, while the second term takes into account all the scattering processes that occur in the upper layer. These processes are: (1) the scattering of radiation by the upper layer in the direction of the observer; (2) the scattering of radiation by this layer in the direction of the lower layer and subsequent reflection from it; (3) the reflection of radiation by the lower layer and subsequent scattering by the upper layer; (4) the reflection of radiation by the lower layer, scattering by the upper layer, and a second reflection by the lower layer. The equation for the function f can be given in an explicit form (see ref. 7); however, it will not be necessary for us in this paper.

Having written an equation for $\varrho_\nu(\eta, \zeta, \varphi)$ analogous to Eq. (8), and subtracting one from the other, we have

$$\delta_\nu(\eta, \zeta, \varphi) = \varrho^*(\eta, \zeta, \varphi)[1 - \tau(1/\eta + 1/\zeta)]$$
$$- \varrho_\nu^*(\eta, \zeta, \varphi)[1 - \tau_\nu(1/\eta + 1/\zeta)]$$
$$+ \lambda\tau f\,[\varrho^*(\eta, \zeta, \varphi), x(\gamma)] - \lambda_\nu\tau_\nu f\,[\varrho_\nu^*(\eta, \zeta, \varphi), x(\gamma)], \tag{9}$$

where $\varrho_\nu^*(\eta, \zeta, \varphi)$ is the coefficient of reflection of the lower layer in a spectral line. Substitution of Eq. (9) in Eq. (6) also gives us the desired quantity $r_\nu(\eta, \zeta, \varphi)$. We shall find the quantity $r_\nu(\eta, \zeta, \varphi)$ in three particular cases of the assumed model of an atmosphere.

Case I. Let us assume that there are no molecules in the lower layer that absorb radiation in the line under consideration, that is, $\varkappa_\nu^* = 0$. Then $\varrho_\nu^* = \varrho^*$ and the quantity f does not depend on the frequency ν. In this case making use of Eq. (4), from Eq. (9) we derive

$$\delta_\nu(\eta, \zeta, \varphi) = \varrho^*(\eta, \zeta, \varphi)(\tau_\nu - \tau)(1/\eta + 1/\zeta). \tag{10}$$

By substituting Eq. (10) in Eq. (6) and replacing ϱ with ϱ^* (since these quantities differ from each other by terms of the order of τ), we have

$$1 - r_\nu(\eta, \zeta, \varphi) = (1/\eta + 1/\zeta)(\tau_\nu - \tau). \tag{11}$$

This equation can be rewritten also in the form

$$1 - r_\nu(\eta, \zeta, \varphi) = (1/\eta + 1/\zeta)k_\nu N, \tag{12}$$

where k_ν is the coefficient of absorption calculated for one molecule, and N is the number of molecules in a column with a cross section of 1 cm².

We see that, in this case, the formation of a line is caused mainly by the absorption by molecules of light when it passes through the upper layer, is reflected from the lower layer, and for a second time passes through the upper layer. Not all the above-enumerated light-scattering processes play an appreciable part in the formation of a line. This conclusion is explained by the fact that, with the small optical thickness of the layer, the intensity of the scattered radiation is proportional to the quantity $\lambda_\nu \tau_\nu$, which (on the basis of Eq. (4)), does not depend on the frequency.

As is well known, Eq. (12) is widely used with the assumption that the upper layer consists only of absorbing molecules. As a matter of fact, it may be used even when scattering particles are present in this layer. It is obvious that the last assertion is correct also when the layer of small optical thickness is not above clouds, but directly over the planet's surface with irregular light reflection from the surface (provided the surface albedo is an order of magnitude greater than τ).

Case II. Let the role of true absorption in the lower layer be small both in the continuum and in the line, that is, $1 - \lambda^* \ll 1$ and $1 - \lambda_\nu^* \ll 1$. Then the coefficient of reflection of the lower layer in the continuous spectrum is represented by the asymptotic equation

$$\varrho^*(\eta, \zeta, \varphi) = \varrho_0^*(\eta, \zeta, \varphi) - 4\sqrt{\frac{1 - \lambda^*}{3 - x_1^*}}\, u_0^*(\eta)u_0^*(\zeta), \tag{13}$$

where $\varrho_0^*(\eta, \zeta, \varphi)$ and $u_0^*(\eta)$ are the coefficients of reflection and transmission of this layer in the case of pure scattering (that is, with $\lambda^* = 1$), and x_1^* is the first coefficient of the expansion for the phase function $x^*(\gamma)$ in Legendre polynomials (see chap. 2 of ref. 6). Equation (13) becomes more exact as $1 - \lambda^*$ becomes smaller. An asymptotic equation for the quantity $\varrho_\nu^*(\eta, \zeta, \varphi)$ is derived from Eq. (13) by replacing λ^* with λ_ν^*.

By substituting Eq. (13) and the analogous equation for ϱ_ν^* into Eq. (9), we see that in the expansions of the quantity f in powers of $\sqrt{1 - \lambda^*}$ and $\sqrt{1 - \lambda_\nu^*}$ it is possible to keep only the zero-order terms, that is, to simply replace ϱ^* and ϱ_ν^* with ϱ_0^* (since we are disregarding the terms of the order of $\tau \sqrt{1-\lambda^*}$ and $\tau_\nu \sqrt{1-\lambda_\nu^*}$). Then, applying Eq. (4), we find

$$\delta_\nu(\eta, \zeta, \varphi) = \varrho_*^0(\eta, \zeta, \varphi)(1/\eta+1/\zeta)(\tau_\nu-\tau)$$
$$+ \frac{4}{\sqrt{3-x_1^*}} u_0^*(\eta) u_0^*(\zeta)(\sqrt{1-\lambda_\nu^*} - \sqrt{1-\lambda^*}). \tag{14}$$

Substitution of Eq. (14) into Eq. (6) (with replacement of ϱ by ϱ_0^*) gives

$$1-r_\nu(\eta, \zeta, \varphi) = (1/\eta+1/\zeta)k_\nu N+C^*(\eta, \zeta, \varphi)(\sqrt{1-\lambda_\nu^*} - \sqrt{1-\lambda^*}), \tag{15}$$

where

$$C^*(\eta, \zeta, \varphi) = \frac{4}{\sqrt{3-x_1}} \frac{u_0^*(\eta) u_0^*(\zeta)}{\varrho_0^*(\eta, \zeta, \varphi)}. \tag{16}$$

Equation (15) also determines the profile of the absorption line in the case under consideration. The equivalent width of this line is

$$W(\eta, \zeta, \varphi) = \left(\frac{1}{\eta}+\frac{1}{\zeta}\right) N \int_0^\infty k_\nu d\nu + C^*(\eta, \zeta, \varphi)Q^*, \tag{17}$$

where

$$Q^* = \int_0^\infty (\sqrt{1-\lambda_\nu^*} - \sqrt{1-\lambda^*}) \, d\nu. \tag{18}$$

Equations (15) and (17) are a generalization of the corresponding equations derived previously (ref. 5) for the case of one semi-infinite layer with small true absorption (that is, with $N = 0$). This generalization appears obvious when there is located over the just-mentioned semi-infinite layer a purely molecular absorbing layer of small optical thickness. However, the equations remain valid when, in addition to the molecules, there are aerosol particles which scatter radiation.

In ref. 5 are given tables of the function $C^*(\eta, \eta, \pi)$ (pertaining to zero phase angle) for the simplest phase functions, and also a table of the quantity Q^* for a Lorentz profile of the coefficient of absorption in the line. For the utilization of Eqs. (15) and (17), it is desirable to compile a more detailed table of the quantities $C^*(\eta, \zeta, \varphi)$ and Q^*.

Case III. We shall assume that the phase functions in the two layers are equal, that is, $x^*(\gamma) = x(\gamma)$. In this case the upper layer differs from the lower layer only in single scattering albedo: in the upper layer it is λ and λ_ν, while in the lower layer it is λ^* and λ_ν^*. Hence Eq. (8) for the coefficient of reflection of a two-layer atmosphere in the continuous spectrum can be expediently written in the form

$$\varrho(\eta, \zeta, \varphi, \lambda, \lambda^*) = \varrho(\eta, \zeta, \varphi, \lambda^*)[1-\tau(1/\eta+1/\zeta)]+\lambda\tau f(\eta, \zeta, \varphi, \lambda^*), \tag{19}$$

where $\varrho(\eta, \zeta, \varphi, \lambda^*)$ is the coefficient of reflection of the lower layer.

In order to determine the function $f(\eta, \zeta, \varphi, \lambda^*)$, we assume $\lambda = \lambda^*$ in this relationship. Then the two-layer atmosphere becomes a uniform semi-infinite atmosphere and, consequently, $\varrho(\eta, \zeta, \varphi, \lambda^*, \lambda^*) = \varrho(\eta, \zeta, \varphi, \lambda^*)$. Thus, from Eq. (19) we derive

$$f(\eta, \zeta, \varphi, \lambda^*) = \frac{1}{\lambda^*} \varrho(\eta, \zeta, \varphi, \lambda^*) \left(\frac{1}{\eta} + \frac{1}{\zeta}\right). \tag{20}$$

Substitution of Eq. (20) into Eq. (19) results in the following equation for the coefficient of reflection of a two-layer atmosphere in the continuous spectrum:

$$\varrho(\eta, \zeta, \varphi, \lambda, \lambda^*) = \varrho(\eta, \zeta, \varphi, \lambda^*)[1 - \tau(1/\eta + 1/\xi)(1 - \lambda/\lambda^*)]. \tag{21}$$

An analogous equation can be written also for the coefficient of reflection in a spectral line. By dividing one equation by the other, we obtain

$$r_v(\eta, \zeta, \varphi) = r_v^*(\eta, \zeta, \varphi) \frac{1 - \tau_v(1/\eta + 1/\zeta)(1 - \lambda_v/\lambda_v^*)}{1 - \tau(1/\eta + 1/\zeta)(1 - \lambda/\lambda^*)}, \tag{22}$$

where $r_v^*(\eta, \zeta, \varphi)$ designates the quantity $r_v(\eta, \zeta, \varphi)$ for the lower layer.

It should be noted that the absorption line in case III can be strong (due to absorption in the lower layer), unlike the weak lines in cases I and II.

References

1. M. J. S. BELTON, D. M. HUNTEN, and R. M. GOODY, The Atmospheres of Venus and Mars (1968).
2. J. E. HANSEN, Astrophys. J. 158, 337 (1969).
3. A. UESUGI and W. M. IRVINE, Astrophys. J. 159, 127 (1970); 161, 243 (1970).
4. J. W. CHAMBERLAIN, Astrophys. J. 159, 137 (1970).
5. V. V. SOBOLEV, Astron. Zh. 49, 397 (1972) [Sov. Astron.—AJ 16, 324 (1972)].
6. V. V. SOBOLEV, The present monograph.
7. V. A. AMBARTSUMYAN, Nauchn. Tr. 1, Erevan (1960).

AUTHOR INDEX

SUBJECT INDEX

Absorption coefficient *see* Coefficient
Absorption lines 174, 180 et seq.
  equivalent width of 181, 183–6, 188, 207
  profile of 181, 183, 185–6
Albedo
  atmospheric (plane) 18, 34, 43–46, 56, 80–83, 139, 141, 151, 163–4, 166, 168
  geometric 178–80
  spherical 18, 34, 45–46, 82, 83–86, 139, 178–80
  surface 74, 79
  see also Isotropic scattering; Phase function; Single-scattering albedo
Ambartsumyan–Chandrasekhar equation 95
Ambartsumyan functions 32–35, 94, 104–6, 131, 141–2, 176, 184
Approximate methods 153 et seq.
  Eddington 22, 161 et seq., 217 et seq.
  Schwarzschild–Schuster 161
Asymptotic relations *see* Atmospheres of large optical thickness; Small true absorption; Deep layers
Atmospheric albedo *see* Albedo
Atmospheres
  of finite optical thickness 52–73, 107 et seq., 143–7
  of large optical thickness 60 et seq., 81–83, 84, 121 et seq., 158, 235
  semi-infinite 24 et seq., 89 et seq., 126 et seq.
  spherical 213 et seq.
Auxiliary functions 21–22
  see also Ambartsumyan functions

Basic problem 8 et seq.
Brightness distribution across planetary disk 175–7
Boundary conditions
  for approximate method of Eddington 162, 221
  for basic problem 11
  for surface reflection 75

Characteristic function 93
Coefficient
  of absorption 1–2
  of absorption, decreasing exponentially with altitude 224 et seq.
  of emission 1, 6–8

of reflection 16–17, 29–33, 47, 53 et seq., 79, 126 et seq.
of relative transmission 35 et seq., 48–50, 133, 135, 140, 142–3, 153 et seq.
of scattering 1, 180
of transmission 16–17, 53 et seq., 79, 126 et seq.
of true absorption 1, 180

Deep layers, radiation field in 24 et seq., 41 et seq., 101–2, 103, 160

Earth 74, 177, 202–5, 226, 232–3
Emission coefficient *see* Coefficient
Equation
  of radiative equilibrium 1, 8
  of radiative transfer 1, 5–8, 159, 218–19
Exponential integral 14

Flux integral 19, 146, 157
Fundamental function 99 et seq., 109 et seq., 120 et seq.

Geometric albedo *see* Albedo

H-functions 94 et seq., 133 et seq.
  for isotropic scattering 95

Illumination 16, 56–57, 80–83, 151, 166, 168
Inhomogeneous atmospheres 66 et seq., 224 et seq.
Intensity of radiation 5–6, 131–3, 161–4
Invariance principles 21, 22–23, 30, 224
Isotropic reflection by planetary surface 78–80, 83–84
Isotropic scattering 4, 13, 26 et seq., 88, 95, 113, 125, 155, 160, 184–5
  albedo for 35
  Ambartsumyan functions for 32–33, 50, 55, 59–60, 66, 117
  brightness of a planet near the terminator for 228–9
  coefficient of reflection for 32, 56, 63–65
  fundamental function for 115 et seq.
  spherical atmospheres with 213 et seq.

255

256 OTHER TITLES IN THE SERIES